# Statistics for Biology and Health

*Series Editors*
M. Gail, K. Krickeberg, J. Samet, A. Tsiatis, W. Wong

# Statistics for Biology and Health

David Siegmund
Benjamin Yakir

# The Statistics of
# Gene Mapping

 Springer

David Siegmund
Department of Statistics
Stanford University
Stanford, CA 94305
USA
dos@stat.stanford.edu

Benjamin Yakir
Department of Statistics
The Hebrew University of Jerusalem
Jerusalem, Israel 91905
msby@pluto.mscc.huji.ac.il

*Series Editors*

M. Gail
National Cancer Institute
Rockville, MD 20892
USA

K. Krickeberg
Le Chatelet
F-63270 Manglieu
France

J. Sarnet
Department of Epidemiology
School of Public Health
Johns Hopkins University
615 Wolfe Street
Baltimore, MD 21205-2103
USA

A. Tsiatis
Department of Statistics
North Carolina State
   University
Raleigh, NC 27695
USA

W. Wong
Department of Statistics
Stanford University
Stanford, CA 94305-4065
USA

Library of Congress Control Number: 2006938272

ISBN-10: 0-387-49684-X
ISBN-13: 978-0-387-49684-9

e-ISBN-10: 0-387-49686-6
e-ISBN-13: 978-0-387-49686-3

Printed on acid-free paper.

Printed in the United States of America

9 8 7 6 5 4 3 2 1

springer.com

Dedicated to Lily and Sandra for their love and support

# Preface

The ultimate goal of gene mapping is to identify the genes that play important roles in the inheritance of particular traits (phenotypes) and to explain the role of those genes in relation to one another and in relation to the environment. In practice this ambition is often reduced to identifying the genomic neighborhoods where one or a small number of the important genetic contributors to the phenotype are located.

Gene mapping takes place in many different organisms for many different reasons. In humans one is particularly interested in inherited or partly inherited diseases, and hopes that an identification of the responsible genes can lead in the relatively short run to better diagnostics and in the long run to strategies to alleviate the disease. In these cases the phenotype can be qualitative, whether an individual is affected with a particular disease, or it can be quantitative, say the level of a biomarker like cholesterol level, blood pressure, or body mass index, which is known or thought to be related to the disease. In plants or animals gene mapping can be of interest in its own right, or to produce more vigorous hybrid plants of agricultural value or farm animals that are more productive or more disease resistant. It can also involve model organisms, e.g., the plant arabidopsis, inbred strains of mice, or baker's yeast (*S. cerevisiae*), where one hopes to gain basic knowledge yielding insights that are broadly applicable.

The traits, or phenotypes, can be essentially any reproducible quality of an organism, and the goal of the mapping need not even be an actual gene. An example of considerable recent interest is a phenotype measuring the level of expression of some gene or genes under given experimental conditions, with the goal of discovering whether the expression of the gene is controlled by the immediate "upstream" region of the gene (*cis* control) or by some other genomic region, perhaps a master control region for a number of genes that must work in coordination (*trans* control).

There is variability in the expression of essentially any phenotype. There is also variability in the inheritance of genotypes, first because of Mendel's laws, but equally important for gene mapping because of recombination. The level of variability is such that gene mapping necessarily has a statistical component, which is the subject of this book. At its simplest, gene mapping involves the correlation of the phenotype with the genotype of genetic markers, which are themselves hoped to be located close to, hence correlated with, the genes (or genomic regions) of interest.

Although gene mapping was practiced for a good part of the twentieth century, the subject has changed and grown substantially since the late 1980s. In the mid twentieth century the number of suitable markers was small, on the order of a few per genome (and the genomic location of these markers was often imprecise). As a consequence the principal impediment to gene mapping was the usually large genomic distance between gene and marker, which leads to small correlations between phenotype and marker genotype. The statistical model developed in human genetics to deal with this situation assumed that the mode of inheritance of a trait could be adequately modeled. The model almost invariably involved a single gene with the mode of inheritance, usually dominant or recessive, assumed to be known, and the penetrance, i.e., the conditional probability of expressing the trait given the genotype, also known. The unknown parameter of interest was the genetic distance from gene to marker, as measured by the recombination fraction, which then allowed one to test whether the trait was unlinked (recombination fraction $= 1/2$) or linked (recombination fraction $< 1/2$) and to estimate the recombination fraction.

Since the explosion in the experimental techniques of molecular genetics in the late twentieth century, it has become possible to cover the genome with informative markers. As a consequence genes associated with many simple traits, which generate a large correlation between phenotype and markers that are close to the gene, have been mapped successfully. For complex traits, which may involve multiple genes, it is reasonable to assume that there is some marker close to the gene or genes influencing the trait, but the contribution of any particular gene may be small. This leads to small correlations between marker and phenotype, even if the marker is close to the gene; and this small correlation has now become the primary impediment to successful gene mapping.

The principal goal of this book is to explain the statistical principles of and problems arising in gene mapping. Particular emphasis is placed on the ideas that have arisen with the recent experimental developments leading to the availability of large numbers of molecular markers having known genomic locations and the desire to map progressively more complex traits. Indeed, the so-called "parametric" or "LOD" score method of human genetics, which was the established paradigm in human genetics from the time of the classical paper of N. Morton in 1955 [56] until quite recently and is still frequently used, is mentioned only briefly. (See Ott [57] for a thorough discussion.)

We have attempted to keep the formal statistical and computational requirements for reading the book as few as seems reasonable, with the hope that the book can be understood by a diverse audience. It is not, however, a handbook of methods, but a discussion of concepts, where we assume the reader wants to play an active role, particularly in performing computational experiments and in thinking about the meaning of the results. Mathematical details are omitted when we did not think they gave added insight into the scientific issues. Since they can range from routine to difficult, we do not recommend that the reader expend effort to fill in the omitted details, unless he or she finds that intellectual activity rewarding for its own sake.

The book is organized globally as follows. The first three chapters deal with basic statistical, computational, and genetic concepts that are used later in the book. The next five are concerned with mapping quantitative traits using data from crosses of inbred strains. Chapters 9–13 involve primarily human genetics. Of those, Chaps. 9 and 11 discuss gene mapping based on data from pedigrees. This is conceptually similar to the first half of the book although necessarily substantially more complicated. Chapters 12 and 13 discuss association analysis, where relations between individuals in pedigrees are replaced by relations in populations. The discussion here is substantially less complete, and is to some extent limited to pointing out certain difficulties arising from complicated and uncertain population history. Chapter 10 involves admixture mapping, which has some features in common with the earlier chapters on gene mapping based on meiotic recombination and others related to population based association analysis.

A more detailed road map is as follows.

Chapter 1 reviews basic statistical concepts that are used throughout the book and explores these concepts computationally. It can be skipped or read quickly by someone familiar with basic statistics and computation in the language R.

In Chap. 2, we introduce our basic model relating the phenotype as dependent variable to genotype(s) as independent variable(s). It is a simple linear regression model having its origins in the classical paper of Fisher [30]. A principal attraction of the model is that straightforward variations can be developed to deal with quantitative or qualitative traits, which can be dominant or recessive, and which can involve multiple genes that interact with one another and/or with the environment. These variations are discussed at different places in the book.

Chapter 3 deals with some fundamental concepts of population genetics, and provides an opportunity to introduce some new programming techniques. It contains some difficult material, and except for recombination, which plays a central role throughout the book, most of the chapter is referenced only occasionally. The reader may wish to read Chap. 3 selectively, to get a rough idea of its contents and be prepared to refer to it occasionally.

For the experimental genetics of crosses of inbred lines, one can use the regression model and standard statistical methods to develop direct robust tests for linkage between phenotype and marker genotype. In Chap. 4 we discuss the simplest case of testing a single marker, first when that marker is itself a gene affecting the trait, and then when the marker is linked to a gene affecting the trait; and we see quantitatively the (deleterious) effect of recombination between gene and marker on the power to determine if the marker is linked.

Since we will usually have no good idea where a gene affecting the trait is likely to be found, in Chap. 5 we introduce the notion of a genome scan, where we test a large number of markers distributed throughout the genome for linkage. This leads naturally to a problem of multiple comparisons that is solved both by computational methods and by theoretical results involving the maximum of a stochastic process. A systematic discussion of the power of a genome scan and of the related idea of confidence intervals for locating a linked gene as precisely as possible is given in Chap. 6.

Chapter 7 introduces in the simple context of experimental genetics the problem of missing information and one simple statistical idea to recapture that information. It turns out that in the context of that chapter, the solution is often more complicated than the problem warrants, but the problem appears again in a more complex form in Chaps. 9, 10, and 13, where missing information poses unavoidable difficulties that require progressively more complex algorithms to address.

Chapter 8 is concerned with more advanced problems in experimental genetics. Our goal here is to introduce the reader to these problems and point out which ones can in principle be dealt with by simple adaptations of methods developed up to that point and which ones pose more serious challenges.

Starting in Chap. 9, we discuss gene mapping in human genetics, where it is intrinsically more complicated, especially when there may be more than one gene and uncontrolled environmental conditions. Our discussion here is less complete than in the first eight chapters. It is essentially limited to pointing out how the theoretical framework of earlier chapters can be adapted to the more complex problems of human genetics and how the problem of missing information, which here moves to center stage, can be addressed. Chapters 9 and 11 are concerned with gene mapping based on inheritance within families. The concepts developed in Chaps. 4–8 are used, somewhat indirectly. Our presentation is designed to bring out the similarities while highlighting important differences. One important conclusion is that because of one's inability to perform breeding experiments in humans, family based methods for gene mapping are intrinsically less powerful than mapping in experimental genetics based on crosses of inbred lines.

Chapters 12 and 13 contain a brief introduction to gene mapping in populations, which is often called association analysis. It has the potential advantage over family based methods of substantially more power, providing one can successfully overcome some potential difficulties arising from the unknown population history. Our discussion is limited to describing a few simple models, the reasons they are attractive, and the potential pitfalls.

Chapter 10 is something of a bridge between the earlier chapters and the last two. While the methods discussed there are very similar to the methods of earlier chapters, and the issue of missing information is closely related to the same issue in Chap. 9, there are also complications of population history.

As indicated above, the classical parametric method of linkage analysis in human pedigrees is discussed only briefly. Also, in choosing to emphasize regression based methods, we have limited our discussion of likelihood methods, which can be very powerful when basic modeling assumptions are satisfied, to cases where they seemed to offer distinct advantages. The modeling assumptions, often in the form that phenotypes are normally distributed, can fail to hold, even approximately. When that happens, likelihood methods can be less robust than regression methods, although no statistical method should be regarded as so automatic that computer output is thought to speak for itself.

Finally, we would like to emphasize that the primary purpose of this book is didactic. The concepts and many related details have been published elsewhere by a large number of authors. We have provided some references to the scientific literature, largely for the purpose of introducing the reader to the substantial primary literature; but we have not tried to provide a complete scholarly bibliography.

For feedback in classes where preliminary versions of this book were used, we would like to thank students at The Hebrew University of Jerusalem, the Weizmann Institute, Stanford University, and the National University of Singapore. We also thank those universities, along with the Free University of Amsterdam, the Israel–U.S. Binational Science Foundation, the NIH, and the U.S. National Science Foundation for their support.

Stanford and Jerusalem,                              *David O. Siegmund*
October 2006                                          *Benjamin Yakir*

# Contents

---

**Part I Background and Preparations**

---

# List of Notations and Terminology

QTL = Quantitative trait locus.

$y$ = Phenotype.

$x$ = Copy number of a given allele.

$\alpha$ = Additive effect.

$\delta$ = Dominance effect.

$\tilde{\alpha}$ = Additive effect in orthogonalized model.

$\tilde{\delta}$ = Dominance effect in orthogonalized model.

$\sigma_A^2$ = Additive variance.

$\sigma_D^2$ = Dominance variance.

$\sigma_e^2$ = Residual variance.

$\sigma_y^2$ = Variance of the phenotype.

IBD = Identical by descent.

$J$ = Number of alleles shared IBD.

$\chi$ = Indicator of population source.

$\tau$ = Position of QTL.

$p$ = Frequency of QTL.

$g$ = Penetrance or the number of generations, depending on the context.

$t, s$ = Positions of genetic markers.

$f$ = Frequency of a genetic marker.

$\theta$ = Recombination fraction for a single meiosis.

$\Delta$ = Distance between (equally spaced) markers.

$\varphi = 2\theta(1-\theta)$ = Probability that IBD state of two half siblings changes between markers separated by a recombination fraction $\theta$.

$\phi$ = Probability density function for a standard normal distribution.

$\Phi$ = Cumulative distribution function for a standard normal distribution.

$\mu$ = Mean value.

$\xi$ = Noncentrality parameter.

$I$ = Indicator of an event.

$\nu$ = Function associated with overshoot in approximations.

$\mathrm{T}$ = Transition probability matrix.

$\pi$ = Stationary distribution, written as a column vector.

# Part I

# Background and Preparations

# 1

# Background in Statistics

Statistics is the science that formalizes the process of making inferences from observations. Basic to this process of formalization is the concept of a statistical model. In general, a statistical model is an attempt to provide a mathematical simplification of the mechanism that produced the observations. Statistical models are useful since they allow investigation and optimization of the process of analyzing the observations in a context that is wider than the context of the outcome of the specific trial that is being analyzed. For example, it opens the door to considerations such as: "Had we had the opportunity to try a specific inferential procedure on other datasets, all generated by the same statistical mechanism as the one we observe, how would our procedure perform on the average? What would be the probability of drawing an incorrect conclusion?" Such questions are impossible to address unless we adopt a wider point of view.

Exploration of the properties of statistical procedures may be conducted using mathematical and/or computational tools. In this book we will use mathematical approximations and computerized Monte-Carlo simulations to explore problems of statistical genetics. Monte-Carlo simulation can help one explore scenarios where variability plays a role. The basic idea behind such simulations is to generate a sequence of random datasets and use them to mimic the actual distribution of repeated sampling of the data in the investigation of the statistical procedure. The simulation in this book will be conducted in the R programming environment. We start the chapter with a small introduction to R and then proceed with a discussion of statistical models and statistical inference in a general framework.

Once a statistical model is set, quantities that connect the observations and the model can be computed. A central one is the likelihood function. The likelihood function is the probability density function of the observations given the statistical model. It is the key for the derivation of efficient inferential tools. In most cases, the statistical model is not confined to a unique distribution function but can be any member of a given family of such distributions. In such a case it is useful to think of the likelihood function as a function of the parameters that determine the distribution within the family. Varying the values of the parameters will change the value of the likelihood function according to the probability of the observations under the new distribution.

Throughout this book we will introduce statistical models that may fit different scenarios in statistical genetics. We start this chapter by introducing three basic models, which apply not only in genetics: the normal model, binomial model, and Poisson model. In order to illustrate the basic concepts in statistical inference we will consider statistical testing of hypotheses as our primary example. In this context another distribution, the chi-square distribution, will also be introduced. The properties of a statistical test, e.g., its significance level and power, will be discussed. A brief introduction to regression and the concept of a stochastic process, both given a central role in the book, is provided. Later in the chapter we will examine general approaches for constructing statistical tests.

## 1.1 Introduction to R

R is a freely distributed software for data analysis. In order to introduce R we quote the first paragraphs from the manual *Introduction to R*, written by W. N. Venables, D. M. Smith, and the R Development Core Team. (The full text, as well as access to the installation of the software itself, are available online at http://cran.r-project.org/):

> "R is an integrated suite of software facilities for data manipulation, calculation and graphical display. Among other things it has
> - an effective data handling and storage facility,
> - a suite of operators for calculations on arrays, in particular matrices,
> - a large, coherent, integrated collection of intermediate tools for data analysis,
> - graphical facilities for data analysis and display either directly at the computer or on hardcopy, and
> - a well developed, simple and effective programming language (called S) which includes conditionals, loops, user defined recursive functions and input and output facilities. (Indeed most of the system supplied functions are themselves written in the S language.)
>
> The term environment is intended to characterize it as a fully planned and coherent system, rather than an incremental accretion of very specific and inflexible tools, as is frequently the case with other data analysis software.
>
> R is very much a vehicle for newly developing methods of interactive data analysis. It has developed rapidly, and has been extended by a large collection of packages. However, most programs written in R are essentially ephemeral, written for a single piece of data analysis."

The R system may be obtained as a source code or installed using a pre-compiled code on the Linux, Macintosh, or Windows operating system. Programming in R for this book was carried out under Windows. (A more-detailed explanation regarding the installation of R under Windows may be found at the URL http://www.biostat.jhsph.edu/~kbroman/Rintro/Rwin.html.)

After the installation of R under Windows an icon will be added to the desktop. It is convenient to have a separate working directory for each project. For that, one may copy the R icon into the new directory and set the working directory to be the new directory. (Right-click on the icon and choose Properties. Copy the path of the directory to the "start in:" box of the Shortcut slip. Consequently, the given directory will become the default location where R expects to find input and to where it saves its output.) Double clicking on the icon will set the R system going.

The R language is an interactive expression-oriented programming language. The elementary commands may consist of expressions, which are immediately evaluated, printed to the standard output, and lost. Alternatively, expressions can be assigned to objects, which store the evaluation of the expression. In the latter case the result is not printed out to the screen. These objects are accessible for the duration of the session, and are lost at the end of the session, unless they are actively stored. At the end of the session the user is prompted to store the entire workspace image, including all objects that were created during the session. If "Yes" is selected, then the objects used in the current session will be available in the next. If "No" is selected, then only objects from the last saved image will remain.

Commands are separated either by a semi-colon (;) or by a new line. Consider the following example, which you should try typing into the window of the R Console after the ">" prompt:

```
> x <- c(1,2,3,4,5,6)
> x
[1] 1 2 3 4 5 6
```

Note that the first line created an object named x (a vector of length 6, which stores the value 1, ..., 6). In the second line we evaluated the expression x, which printed out the actual values stored in x. In the formation of the object x we have applied the concatenation function "c". This function takes inputs and combines them together to form a vector.

Once created, an object can be manipulated in order to create new objects. Different operations and functions can be applied to the object. The resulting objects, in turn, can be stored with a new name or with the previous name. In the latter case, the content of the object is replaced by the new content. Continue the example:

```
> x*2
[1]  2  4  6  8 10 12
```

```
> x
[1] 1 2 3 4 5 6
> x <- x*2
> x
[1]  2  4  6  8 10 12
```

Observe that the original content of x was not changed due to the multiplication by two. The change took place only when we deliberately assigned new values to the object x using the assignment operator "<-".

Say we want to compute the average of the vector x. The function "mean" can be applied to produce:

```
> mean(x)
[1] 7
```

A more complex issue is to compute the average of a subset of x, say the values larger than 6. Selection of a sub-vector can be conducted via the vector index, which is accessible by the use of square brackets next to the object. Indexing can be implemented in several ways, including the standard indexing of a sequence using integers. An alternative method of indexing, which is natural in many applications, is via a vector with logical TRUE/FALSE components. Consider the following example:

```
> x > 6
[1] FALSE FALSE FALSE  TRUE   TRUE   TRUE
> x[x > 6]
[1]  8 10 12
> mean(x[x > 6])
[1] 10
```

The vector created by the expression "x > 6", and which is used in the example for indexing, is a logical vector of the same length as the vector x. Only the components of x having a "TRUE" value in the logical indexing vector are selected. In the last line of the example above the resulting object is used as the input to the function "mean", which produces the expected value of 10.

For comparison consider a different example:

```
> x*(x > 6)
[1]  0  0  0  8 10 12
> mean(x*(x > 6))
[1] 5
```

In this example we multiplied a vector of integers x with a vector of logical values "x > 6". The result is a vector of length 6 with zero components where the logical vector takes the value "FALSE" and the original values of x where the logical value takes the value "TRUE". Two points should be noted. Observe that R can interpret a product of a vector with integer components and a vector with logical components in a reasonable way. Standard programming languages may produce error messages in such circumstances. In this case, R translates the logical vector into a vector with integer values – one for "TRUE" and zero for "FALSE". The outcome, a product of two vectors with integer components, is a vector of the same type. A second point to make is that multiplication of two vectors using "*" is conducted term by term. It is not the inner product between vectors. A different operator is used in R in order to compute inner products.

As in any programming langauge, R requires experience – something that can be obtained only though practice. We will base simulations in this book on R. Starting with very simple examples, we will gradually present more sophisticated code. Our hope is that during that process any reader who did not have a previous exposure to R will learn to use the system and will share our appreciation of the beauty of the langauge and its usefulness for conducting simple simulations in statistics. Indeed, for understanding our exposition, an ability to *read* R is more or less necessary. To solve exercises, programs may be written in other programming languages. In the first chapters of the book we do not assume familiarity with R. Thus, detailed explanations will accompany code lines. These explanations will become less detailed as we progress through the chapters. A reader who is interested in a more systematic introduction to the system is encouraged to use any of the many introductory resources to R that can be found in the form of books or online documents. (Consult, for example, the contributed documentation in `http://cran.r-project.org/`.)

## 1.2 The Binomial, Poisson, and Normal Models

We now return to the main subject matter of this chapter by considering three popular statistical models: the *binomial*, the *normal*, and the *Poisson* random variables.

### The Binomial Model

Assume that the observations can be represented as a sequence of $n$ binary outcomes. In such a case, the possible outcomes may be classified as *success* or *failure*, or numerically coded as 1 or 0. Such a sequence is termed *Bernoulli trials* if the trials are statistically independent of each other (i.e., the probability of success in one trial is not affected by the outcomes in the other trials).

Suppose the probability of success is the same for all trials. Denote this probability by $p$ and let the random variable $X$ denote the total number of successes among the $n$ trials. Then $X$ is said to have a binomial distribution. For future reference, it will be helpful to observe that $X$ can be regarded as the sum of $n$ independent Bernoulli random variables, which are themselves the special case of binomial random variables with $n = 1$. The probability density function of $X$ is given by:

$$f(x) = \Pr(X = x) = \binom{n}{x} p^x (1-p)^{n-x}, \quad x = 0, 1, 2, \ldots, n \ .$$

The short notation $X \sim B(n, p)$ is used to refer to this distribution. The expectation of $X$ (i.e., the average value of $X$, denoted "$E(X)$") is equal to $np$, and its variance (denoted "$\text{var}(X)$") is equal to $np(1-p)$ (with $[np(1-p)]^{1/2}$ the standard deviation of $X$).

(A brief summary of some properties of expectations, variances, and co-variances can be found at the end of the chapter. While not strictly necessary for what follows, it will facilitate understanding of some calculations that otherwise must be accepted "on faith.")

## The Normal Distribution

The normal distribution – also known as the *Gaussian* distribution – is a very popular statistical model. The formula for the density of the normal distribution is given by:

$$f(x) = \frac{e^{-(x-\mu)^2/(2\sigma^2)}}{\sqrt{2\pi\sigma^2}}, \quad -\infty < x < \infty \ ,$$

which forms the famous bell shape. The parameter $\mu$ is the mean, or the location, of the distribution and $\sigma^2$ is its variance ($\sigma$ is the standard deviation or the scale parameter). In particular, when $\mu = 0$ and $\sigma^2 = 1$ the distribution is called the *standard* normal distribution. The density of the standard normal distribution is symbolized by "$\phi(x)$" and the cumulative distribution function (cdf) is symbolized by

$$\Phi(x) = \int_{-\infty}^{x} \phi(z) \mathrm{d}z \ .$$

We denote the fact that $X$ has a normal distribution with mean $\mu$ and variance $\sigma^2$ by the notation $X \sim N(\mu, \sigma^2)$.

An important property of the normal distribution is that if we add or subtract independent, normally distributed variables, the result is again normally distributed. Symbolically, if $X_1$ and $X_2$ are independent with $X_i \sim N(\mu_i, \sigma_i^2)$, then $X_1 \pm X_2 \sim N(\mu_1 \pm \mu_2, \sigma_1^2 + \sigma_2^2)$.

Statistics are quantities computed as functions of the observations. The distribution of a statistic can be quite complex. Surprisingly, it is frequently the case that the distribution of a statistic resembles the bell-shaped distribution of the normal random variable, provided that the sample size is large enough and the statistic is computed as an average or a function of averages. One form of this result will be stated more formally later in this chapter when we discuss the *central limit theorem* (CLT).

### The Poisson Distribution

The Poisson distribution is useful in the context of counting the occurrences of rare events. Like the binomial distribution, it takes integer values. As we will see later in this chapter, it can arise as an approximation to the binomial distribution when $p$ is small and $n$ is large.

We say that a random variable $X$ has a Poisson distribution with mean value $\lambda$ (written $X \sim \text{Poisson}(\lambda)$) if the probability function of $X$ has the form

$$f(x) = e^{-\lambda} \frac{\lambda^x}{x!} , \quad x = 0, 1, 2, \dots .$$

The expectation and the variance of $X$ are both equal to $\lambda$.

If $X_1$ and $X_2$ are independent and Poisson distributed with parameters $\lambda_1$ and $\lambda_2$, the sum $X_1 + X_2$ is Poisson distributed with parameter $\lambda_1 + \lambda_2$.

## 1.3 Testing Hypothesis

Statistical inference is used in order to detect and characterize meaningful signals that may be hidden in a environment contaminated by random noise. Hypothesis testing is a typical step in the process of making inferences. In this step one often tries to answer the simple question: "Is there any signal at all?" In other words, can the observed data be reasonably explained by a model for which there is no signal – only noise?

### 1.3.1 The Structure of a Statistical Test of Hypotheses

Assuming the statistical model has been set, we describe the process of testing a statistical hypothesis in three steps: (i) formulation of the hypotheses, (ii) specification of the test, and (iii) reaching the final conclusion. The first two steps are carried out on the basis of the statistical model, and in principal can be conducted prior to the collection of the observations. Only the third step involves the actual data.

**(i) Formulating the hypotheses**: A model corresponds to a family of possible distributions. Some of the distributions describe signals in the data, while others describe only noise (for example, when treatment and control data vary but are on average the same). The set of the distributions where there is no signal is called the *null hypothesis* or "$H_0$". The collection of distributions containing signals is denoted the *alternative hypothesis* or "$H_1$".

In many cases the classification into the two possible hypotheses can be based on some parameters. In such a case, we specify the hypotheses by a partition of the space of parameters that determine the distribution into two parts – one representing the null hypothesis and the other representing the alternative. For example, the hypotheses can often be formulated in terms of the mean value $\mu$ of some observations. The null hypothesis may correspond to $\mu = 0$ (denoted "$H_0 : \mu = 0$"). The alternative hypothesis may then correspond to the case where the expectation is not equal to zero. (The alternative in this case is called *two-sided* and it is denoted by "$H_1 : \mu \neq 0$".) In other cases there are scientific reasons why negative expectations cannot occur and the alternative corresponds to positive values of the expectation ("$H_1 : \mu > 0$"). Such an alternative hypothesis is termed *one-sided*.

**(ii) Specifying the test:** In this step one decides which statistic to use for the test and which values of the statistic should correspond to rejection of the null hypothesis. The selected statistic is called the *test statistic* and the set of values for which the null hypothesis is rejected is called the *rejection region*.

For example, for testing whether the population mean $\mu$ is equal to zero the average of the observations can be used as a test statistic. Large values of this statistic indicate that the null hypothesis is not correct when a one-sided alternative is tested. Similarly, large absolute values are an indication for rejection if a two-sided alternative is considered. The error of rejecting the null hypothesis when it is true is called a "Type I" error and is usually considered more serious than the other type of error (failure to reject the null hypothesis when it is false). Consequently, the selection of a threshold for the rejection region is determined from the distribution of the test statistic under the null hypothesis. The probability of a Type I error is called the *significance level* of the test. The threshold is set to meet a required significance level criteria, traditionally taken to be 5%.

**(iii) Reaching a conclusion:** After the stage is set, all that is left is to carry out the test. The value of the test statistic for the observed data set is computed. This is the *observed value* of the statistic. A verdict is determined based on whether the observed value of the statistic falls inside or outside the rejection region. Failing to reject the null hypothesis often means dropping the line of investigation and looking for new directions. Rejection of the null hypothesis is a trigger for the initiation of more statistical analysis aimed at a characterization of the signal.

### 1.3.2 Testing Genetic Identity by Descent of Affected Siblings

This book summarizes an array of strategies for learning relations between expressed heritable traits and genes – the carrier of the genetic information for the formation of proteins. One of these strategies, called the *affected sibpairs* (ASP) approach, calls for the collection of a large number of nuclear families, each with a pair of affected siblings that share the condition under investigation. Chapter 9 goes into details in describing statistical issues related to this design. Here we consider an artificial, but somewhat simpler, scenario where all the sibling pairs are actually half-siblings, who share only one parent in common, and we concentrate on a single gene, which may or may not contribute to the disease. The aim is to test the null hypothesis of no contribution.

The gene may be embodied in any one of several variant forms, called *alleles*. On autosomal chromosomes an individual carries two homologous copies of the gene, one inherited from the mother and the other from the father. Therefore, each offspring carries two versions of the given gene, which may not be identical in form. Still, one of the genes is an identical copy of one of the two homologous genes in the common parent while the other is a copy of one of the homologous genes in the other parent. Concentrate on the copies in the half-siblings that originated from the common parent. There are two possibilities: both half-siblings' copies emerge from a common ancestral source or else each was inherited from a different source. In the former case we say that the two copies are *identical by descent* (IBD), and in the latter case we say that they are not IBD. It is natural to model the IBD status of a given pair as a Bernoulli trial, with an IBD event standing for success. Counting the number of half-sibling pairs for which a gene is inherited IBD would produce a binomial random variable.

At a locus unrelated to the trait, Mendel's laws governing segregation of genetic material from parent to offspring will produce IBD or not with equal probabilities, since whichever gene the first child inherited, the second child has a 50% chance of inheriting the same gene. This probability of IBD is the probability of success when the gene does not contribute to the trait. Suppose, however, that the gene does contribute to the trait. Since both siblings share the trait one may reasonably expect an elevated level of sharing of genetic material within the pair, thus an elevated probability of IBD. Denote by $J$ the IBD count for a given pair, with $J = 0$ or $J = 1$, and let $\pi$ be the probability that $J = 1$. A natural formulation of the statistical hypothesis is given by $H_0 : \pi = 1/2$ versus $H_1 : \pi > 1/2$. Given a sample of $n$ pairs of half-siblings who share the trait, one may use as a test statistic the number of pairs that share an allele IBD. Since each pair can be regarded as a Bernoulli trial, if we also assume the parents are unrelated to one another, the trials are independent, so the sum is binomially distributed with paramters $n$ and $\pi$. One may standardize this binomially distributed statistic by subtracting out the expectation and dividing by the standard deviation, both computed under the null distribution: $B(n, 1/2)$. The standardized statistic is:

$$Z_n = \frac{\sum_{i=1}^n J_i - n/2}{(n/4)^{1/2}} .$$

A common recommendation is to reject the null hypothesis if $Z_n$ exceeds a threshold of $z = 1.645$, since according to the central limit theorem (discussed below) this will produce a significance level of about 0.05. Values of the test statistic $Z_n$ above that threshold lead to the rejection of the null hypothesis and to the conclusion that the gene contributes to the trait.

Let us investigate the significance level of the proposed test. Assume that a total of $n = 100$ pairs were collected and that in fact $\pi = 1/2$. Then the results of the test may look like this:

```
> n <- 100
> J <- rbinom(1,n,0.5)
> J
[1] 44
> Z <- (J-n/2)/sqrt(n/4)
> Z
[1] -1.2
> Z > 1.645
[1] FALSE
```

The number of pairs that share an IBD copy of the gene was 44 (which is actually less than the expected value of 50). This result was generated using the function "rbinom", which simulates the binomial distribution. The first argument of the function is the number of independent copies to produce; a single copy in our case. The second argument is the number of Bernoulli trials. In this example, the number is n, which was assigned a value of 100. The third argument is the probability of success, $\pi = 1/2$. The statistic Z was computed by standardizing the statistic J, which in this case equals the negative value -1.2. Obviously, the null hypothesis is not rejected. Note that the function "rbinom" simulates random occurrences of a binomial random variable. Running the same code again may produce different outcomes.

In order to evaluate the significance level of the test it is not enough to simulate a single trial. Consider the following code:

```
> J <- rbinom(10^6,n,0.5)
> Z <- (J-n/2)/sqrt(n/4)
> mean(Z > 1.645)
[1] 0.044226
```

In this case the function "rbinom" produces one million independent copies of the binomial distribution, all stored in a vector J of that length. Each of the components of the vector J is then standardized in the same way as the single number was in the previous example. The result is the vector Z, which contains the standardized values. The last line of code involves an application of the function "mean", which computes, as we have previously seen, the average value of its input. Note that the input here is a vector with logical TRUE/FALSE components. A component takes the value "TRUE" if the null hypothesis is rejected and "FALSE" when it is not. When introduced to the function "mean", the logical values are translated into numerical values: one for "TRUE" and zero for "FALSE". As a result, the function "mean" produces the relative frequency of rejecting the null hypothesis, which is an approximation of the significance level. Observe, that the resulting number is 0.044226. This is close, but not identical, to the nominal significance level of 0.05, which was based on the central limit theorem, discussed next.

## 1.4 Limit Theorems

The rationale behind the selection of 1.645 as a threshold in the test discussed above lies in the similarity between the standardized binomial distribution and the standard normal distribution. The given threshold is the appropriate threshold in the normal case. This similarity is justified by the central limit theorem (CLT). In this section we will formulate (without proof) the CLT in the context of sums of independent and identically distributed (i.i.d.) random variables. Actually, the scope of central limit theorems is much wider. It includes multivariate distributions as well as sums of non-identical and weakly dependent random variables. When rare events are considered, the Poisson distribution may provide a better approximation than the normal. A Poisson limit theorem will be presented here in the context of binomial random variables. Again, generalizations of the basic theorem in various directions exist.

The central limit theorem states that the distribution function of a standardized sum of independent and identically distributed random variables converges to the standard normal distribution function. More precisely (recall that $\Phi$ denotes the distribution function of the standard normal distribution):

**Central Limit Theorem:** Let $X_1, X_2, \ldots$, be a sequence of independent and identically distributed random variables. Denote the expectation of these random variables by $\mu$ and the variance by $\sigma^2$. Let $\bar{X} = n^{-1} \sum_{i=1}^{n} X_i$. Consider, for each $n$, the random variable:

$$Z_n = \frac{\sum_{i=1}^{n} X_i - n\mu}{(n\sigma^2)^{1/2}} = \frac{(\bar{X} - \mu)}{(\sigma/n^{1/2})} = \frac{1}{n^{1/2}} \sum_{i=1}^{n} \frac{X_i - \mu}{\sigma} \, .$$

Then, for any $-\infty < x < \infty$,

$$\lim_{n \to \infty} \Pr(Z_n \leq x) = \Phi(x) \, .$$

As an example of the application of the central limit theorem consider the binomial distribution. Recall that if $X \sim B(n, p)$, then $X$ can be represented as a sum of $n$ Bernoulli variables. Moreover, it is easy to see that the expectation of each of these Bernoulli variables is $p$ and the variance is $p(1 - p)$. Hence the distribution of $Z_n = (X - np)/[np(1 - p)]^{1/2}$ can be approximated by the standard normal distribution. In particular, when $n$ is large,

$$\Pr(Z_n > 1.645) \approx 1 - \Phi(1.645) = 0.05 \, .$$

For the example of the preceding section, where $n = 100$ and $\pi = 1/2$, we can actually calculate the exact probability by using the binomial distribution.

The inequality $Z_n > 1.645$ is equivalent to $J > 50 + 1.645 \times (100/4)^{1/2} = 58.225$. The R command "$1$ - $\texttt{pbinom(58.225,100,.5)}$" shows that the exact probability of this event is 0.044313, which is very close to the simulated value found earlier. In more complicated cases it may not be possible to evaluate a probability of interest exactly, and then simulation and/or approximations like those based on the CLT are especially useful.

The normal approximation to the binomial distribution works best when $p$ is not too close to 0 or to 1. When this is not the case the Poisson approximation will tend to produce better results. We can state the theorem that establishes the Poisson approximation as follows:

**Poisson Approximation:** Let $X_n \sim B(n, p_n)$ be a sequence of binomial random variables. Assume that the sequence of $p_n$ of success probabilities obeys the relation $np_n \to \lambda$, as $n \to \infty$, where $0 < \lambda < \infty$. Then, for any $x = 0, 1, 2, \ldots$,

$$\lim_{n\to\infty} \Pr(X_n = x) = \mathrm{e}^{-\lambda} \frac{\lambda^x}{x!} \ .$$

Note that the requirement $np_n \to \lambda$ is equivalent to stating that the probability of success, $p_n$, converges to zero at a rate that is inversely proportional to the number $n$ of Bernoulli trials. In practice, we have only a single value of $n$ and of $p$. The Poisson approximation is appropriate when $n$ is large, $p$ is small, and the product $\lambda = np$ is neither very large nor very small, say about 0.5 or 1 to about 5 or 6.

Let us demonstrate both the normal and the Poisson approximation in the binomial setting. The following lines of code will produce the plot in Fig. 1.1:

```
> n <- 100; p <- 0.5
> X <- rbinom(10^6,n,p)
> Z <- (X-n*p)/sqrt(n*p*(1-p))
> z <- seq(-4,4,by=0.01)
> plot(z,pnorm(z),type="l")
> lines(ecdf(Z))
> x <- z*sqrt(n*p*(1-p)) + n*p
> lines(z,ppois(x,n*p),type="s")
```

The first three lines of code require no explanation. They are essentially identical, with J replaced by X, to the code that was used in order to generate the distribution of the test statistic.

In the forth line we generate a sequence of numbers, ranging between -4 and 4, in jumps of size 0.01. This sequence is generated with the aid of the function "$\texttt{seq}$". The first argument to the function is the starting point of the sequence and the second argument is the ending point. The third argument is the jump size, and it is introduced using the name of the argument "$\texttt{by}$". The rule in the introduction of arguments to functions is that arguments may be set either by placing them in the same order in which they appear in

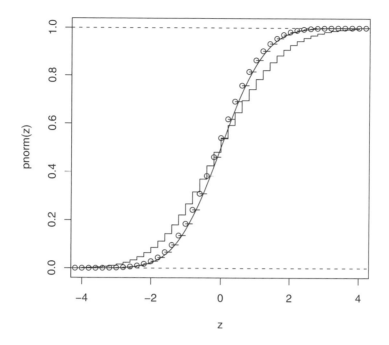

**Fig. 1.1.** Normal and Poisson approximations: $p - 0.5$, $n = 100$

the definition of the function, or by using the argument assignment format "par_name = par_value". If not all preceding arguments are assigned, then the argument must be assigned using the argument assignment format.

The subsequent line produces a plot. The function "plot" is a generic function for making plots. In its simplest application it requires as input a sequence of $x$ values and a sequence of $y$ values, both of the same length. It produces the appropriate scatter plot of the points. This basic behavior may be modified by setting arguments. For example, the argument "type" determines the plotting style. Setting its value to "l" will result in sequentially connecting the points by segments, which will produce a curve. The $y$ values are produced here by the function "pnorm". This function takes as input real values and produces as output the normal cumulative distribution function at these values. Execution of the code will result in opening a graphical window within R and the generation of the plot of the normal cumulative distribution function over the range of z.

The function "plot" is classified as a high-level plotting function, since it independently produces a plot. Low-level plotting functions add features to existing plots. The function "lines" is a low-level function. In its generic usage it adds lines to a plot. In its first appearance in this example it takes as input the output of the function "ecdf". This function calculates the empirical cumulative distribution function from a set of observation – the vector Z in this case. The empirical distribution is then added to the plot. Note that the empirical distribution is a step function with jumps, indicated as small circles, at the points of the data. (As a matter of fact, the application of the function "lines" at this line of the program is non-generic but is specific to output of the function "ecdf".)

For comparison, we would like to add the cumulative distribution function of the Poisson distribution to the plot. The vector x is the image of the vector z in the original scale of the binomial random variable (the integers between zero and $n$). The cumulative distribution function of the Poisson distribution in the original scale is computed with the aid of the function "ppois". The second argument to the function is the mean parameter of the Poisson distribution, which we equate with the mean of the binomial distribution. The function "lines" adds this cumulative distribution function to the plot. As for the function "plot", the default behavior of the function "lines" can be modified. Here we used the option "type="s"" in order to produce a step function.

Examine the plot in Fig. 1.1. The *circles and broken lines* correspond to the standardized binomial distribution, the *solid line* corresponds to the normal approximation, and the *step function* to the Poisson approximation. Observe that in this case the normal distribution is a good approximation to the binomial distribution. The Poisson distribution is not appropriate. In order to examine a scenario where the Poisson approximation is more appropriate one can re-run the same code as given here, but for values of p closer to zero. In Fig. 1.2 one may find the outcome of replacing "p = 0.5" by "p = 0.01". The Poisson distribution gives an excellent approximation to the binomial distribution but the normal approximation is poor.

## 1.5 Testing Equality of Two Binomial Parameters

An illustrative application of the central limit theorem that we will encounter again in Chap. 12 is the following. Suppose that $X_1$ and $X_2$ are independent and binomially distributed with parameters $n_1$, $p_1$ and $n_2$, $p_2$, respectively, and that we want to test the hypothesis $p_1 = p_2$. Let $\hat{p}_i = X_i/n_i$. Since $\hat{p}_i$ can be regarded as an average of $n_i$ independent Bernoulli random variables, which have mean value $p_i$ and variance $p_i(1 - p_i)$, it follows from the central limit theorem that if $n_1$ and $n_2$ are large, then $\hat{p}_i \sim N(p_i, p_i(1 - p_i)/n_i)$ for $i = 1$ and $i = 2$. Hence by the addition property of independent and (approximately) normal random variables, $[\hat{p}_1 - \hat{p}_2 - (p_1 - p_2)]/[p_1(1 - p_1)/n_1 + p_2(1 - p_2)/n_2]^{1/2}$ is approximately $N(0, 1)$. Under the hypothesis $p_1 = p_2 = p$, say, this simplifies to $(\hat{p}_1 - \hat{p}_2)/[p(1 - p)(1/n_1 + 1/n_2)]^{1/2}$. We cannot use this directly as a test statistic because it depends on the unknown parameter $p$, but under the null hypothesis that the two binomial distributions have the

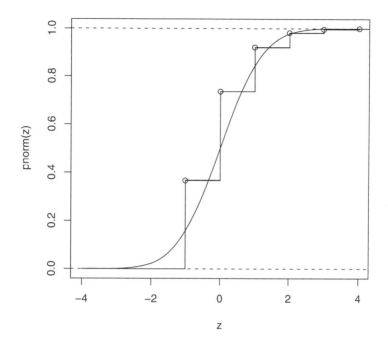

**Fig. 1.2.** Normal and Poisson approximations: $p = 0.01$, $n = 100$

same value of $p$, we can estimate $p$ by $\hat{p} = (X_1 + X_2)/(n_1 + n_2)$. In large samples, when the null hypothesis is true, the estimator of $p$ will be very close to the true value, so one can show that $Z = (\hat{p}_1 - \hat{p}_2)/[\hat{p}(1-\hat{p})(1/n_1 + 1/n_2)]^{1/2}$ is also approximately $N(0,1)$. We can use $Z$ to test the hypothesis $p_1 = p_2$ by rejecting that hypothesis if $|Z| > z$, where for a two-sided test the threshold for a 0.05 level of significance is $z = 1.96$.

## 1.6 Statistical Power

The formulation of the hypotheses in hypothesis testing is typically based on the scientific characteristics of the phenomena under investigation. In the specification of the test, however, the statistician has the freedom to choose which test statistic to use and how to set the rejection region. In some cases, if statistical considerations are introduced in the design of the trial, the statistician may also have input into the selection of design parameters such as sample size and measurement methods.

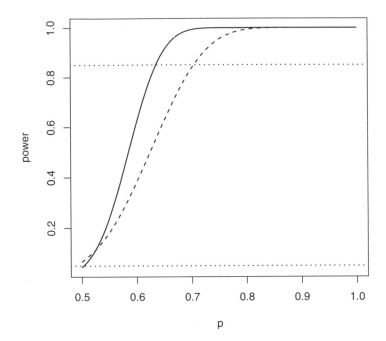

**Fig. 1.3.** Power functions

Statistical tests, regardless of the sample size, the type of the collected measurements, and the form of the test statistic, are usually chosen to have a given significance level. Therefore, the discrimination between different testing procedures cannot be based on the properties of the test statistic under the null distribution alone. Instead, the features of the test statistic under the alternative hypothesis are used in the evaluation of statistical tests. A central feature is the *statistical power* of the test, which is the probability of rejecting the null hypothesis. This probability can be computed for any distribution permitted by the model. When the distribution is one of those specified by $H_1$, high power is preferred in the comparison between alternative tests.

Consider again the binomial example. We have assumed that the probability $\pi$ of identity by descent (IBD) under the alternative hypothesis is larger than one-half and that it reflects the strength of the genetic effect. (See Chap. 9 for a detailed discussion.) Hence, the power is equal to the probability that a binomial random variable with the probability of success $\pi > 1/2$ exceeds the threshold. The threshold is set with respect to $\pi = 1/2$ for a sample of size 100. A question of potential interest is how much power would we lose if we would collect a smaller sample, say of size 36?

An answer can be found with the aid of the *power function*. The plot of the power function for a sample of size 100 (*solid line*) and a sample of size 36 (*broken line*) can be found in Fig. 1.3. This figure is generated by a simple R code, which we will present below. The power function represents the statistical power as a function of the parameters that determine the distribution of the test statistic under the alternative hypothesis. In the case we consider the parameter is the probability of success. Observe that when the parameter approaches the value of 0.5 – its value under the null hypothesis – then the power function converges to the significance level of the test. (The target significance of 0.05 is indicated in the figure by the lower *dashed line*.) On the other hand, the larger the parameter $p$ the closer the power is to its maximal value of one. Typically, the larger the sample size the higher the power. In general, this can be observed also in the current plot of the power functions. (Can you give an explanation for the fact that the order is reversed in this figure for smaller values of success probabilities?) An assessment of the loss in power can be carried out, for example by the identification of the minimal IBD probability that can be detected with at least a given power, say 85%. Refer to the upper *dashed line*. If the sample size is 100, then that minimal probability is about 0.6. If the sample size is 36, then the probability is closer to 0.7. The decision which sample size to prefer would ideally depend on the relative importance of detecting smaller genetic effects compared to the increase in cost and workload associated with the recruitment and assessment of a larger sample.

Let us consider the code that generated Fig. 1.3:

```
> n <- 100
> x <- 1.645*sqrt(n/4)+n/2
> p <- seq(0.5,1,length=100)
> plot(p,1-pbinom(x,n,p),type="l",ylab="power")
> n <- 36
> x <- 1.645*sqrt(n/4)+n/2
> lines(p,1-pbinom(x,n,p),lty=2)
> abline(h=c(0.05,0.85),lty=3)
```

The value of x is the threshold of the test in the scale of the binomial random variable. The vector p is a sequence of probabilities of success in the range $[1/2, 1]$. In this example we set the length of the sequence instead of setting the size of the increment. Consider the application of the function "plot" in the fourth line. The $y$ values are generated with the aid of the function "pbinom", which computes the cumulative distribution function of the binomial distribution. The first argument to this function is a real number (or a vector of real numbers) and the output is the probability that a binomial random variable obtains a value less than or equal to the first argument. The second and third arguments specify the parameters of the binomial distribution. Observe that the probability of success is entered as a vector. As a result, the output of the function is also a vector in which each component corresponds to the cumulative probability with respect to the probability of success in the component of the input vector. The argument "type" in the function "plot" was introduced earlier. The argument "ylab" in that function sets a title for the $y$ axis.

The next three lines of code are merely a repeat of the first lines with the new sample size and with the function "lines" applied in order to add the new power function to the plot. In the last line the function "abline" is used in order to add the dashed lines. This is a lower-level plotting function for the addition of parametric straight lines. The argument "h" is used in order to add horizontal lines. The line type, dotted in this case, is set by the argument "lty".

For statisticians, increasing the sample size, and hence the amount of data available for the analysis, is a blessing. It enables a more reliable characterization of subtle signals. However, increasing the amount of data comes with a price. Obtaining it is usually more expensive and processing it is more laborious. One would expect that the statistical analysis will also be more complicated. Surprisingly, this need not be the case. On many occasions the distribution of the statistics used in the inference is relatively simple when the sample is large. That fact was hinted when we introduced the central limit theorem, but it has a broader implications. Indeed, the statistical theory that deals with large samples is much more complete and uniform compared to the theory for small samples.

Let us introduce the concepts behind the theory of large samples with reference to Fig. 1.4. This figure contains three plots which describe distributions of a test statistic in various situations. Again, the code that generates these plots will be given later. The first plot corresponds to standardized test statistic when the sample size is 36 and the other plots correspond to a test statistic when the sample size is 100.

Examine the first plot. The density function of the test statistic is plotted under two scenarios. In *black* is the distribution under the null hypothesis (probability of success is equal to one-half) and in *gray* is the distribution under a parameter value from the alternative (probability of success equal 0.75 in this example). The *dashed line* corresponds to the threshold for rejection. The null hypothesis is rejected if the value of the test statistic is to the right of the *dashed line*. About 5% of the *black* distribution is to the right of that line. Most, but not all, of the *gray* distribution is in the rejection region.

Let the sample increase to 100. Two different approaches for the assessment of the testing procedure may be considered. In the first approach the statistical model under the alternative is kept fixed (corresponding, in this case, to keeping $\pi = 0.75$ as the probability of success and to the second plot). In this case the statistical power converges rapidly to one as the sample size increases, because the two distributions become clearly separated. In a different approach, which seems unnecessarily complicated here but will be very useful when the underlying model is itself more complicated and one cannot find a simple mathematical expression for the power, one modifies the statistical model when the sample size changes. The modification is such that

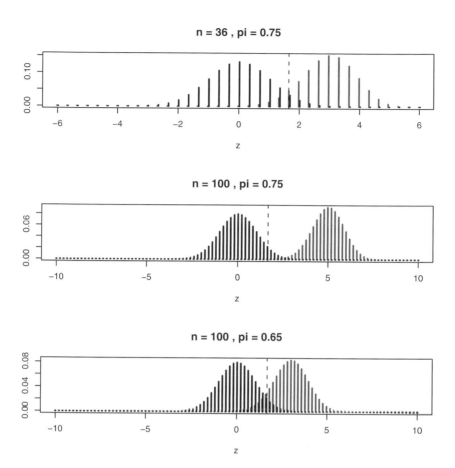

**Fig. 1.4.** Local, and non-local alternatives

the probability of rejection under the alternative hypothesis remains essentially constant for all (large) $n$. In our numerical example the modification resulted in changing of the probability of success to 0.65.

This second approach, where one considers weaker signals in response to an increase in the sample size, is called the approach of *local alternatives*. The consideration of weaker signals is not an artificial construction. The natural habitat of statistics is the border region between the trivial and the impossible. To put it differently, if for a given inferential problem an ever increasing amount of information is available, then eventually the true answer will emerge to everybody's eyes with no need to carry out sophisticated analysis. On the other hand, if too few observations are available, then a valid conclusion cannot be drawn, no matter how clever the analysis. In both cases, statistical analysis is either not needed or cannot help to produce a solution.

Only when the picture is not clear, but also not hopeless, can insights provided by statistical analysis be of help. The approach of local alternative may be interpreted as a formulation of this general observation by focusing attention on the borderline region.

An essential component in the mathematical theory of local alternatives and large samples is convergence of the problem to an appropriate limit that concerns inference on the mean of a normally distributed random variable with a fixed variance. Consider the binomial setting as an example. The expectation of the test statistics, for a given alternative probability of success $\pi$, is given by:

$$\mathrm{E}(Z_n) = \frac{\mathrm{E}(\sum_{i=1}^{n} J_i) - n/2}{(n/4)^{1/2}} = \frac{n\pi - n/2}{(n/4)^{1/2}} = 2\,n^{1/2}(\pi - 1/2)\ .$$

The variance of the test statistic is given by:

$$\mathrm{var}(Z_n) = \frac{\mathrm{var}(\sum_{i=1}^{n} J_i)}{n/4} = \frac{n\pi(1-\pi)}{n/4} = 4\pi(1-\pi)\ .$$

The local alternative, as a function of the sample size $n$, is selected in a manner that keeps the expectation of the test statistic fixed. This means that as $n$ increases, we assume that $\pi$ converges to $1/2$ at the rate of $1/n^{1/2}$. It follows that the variance of the test statistic, $4\pi(1-\pi)$, converges to one, which is its value under the null hypothesis (the limiting case $\pi = 1/2$). Local alternatives will produce an approximation of the power function of the given test by the power function for an appropriate normal "mean shift" problem.

Mathematically, this means we assume that as $n \to \infty$, $\pi$ is a function of $n$ that satisfies $n^{1/2}(\pi - 1/2) \to \mu$ for some value $\mu \geq 0$. Then the power, which equals $\Pr(Z_n \geq z) \approx \Pr(Z_n - \mathrm{E}(Z_n) \geq z - \mu/\sigma)$ converges to $1 - \Phi(z - \mu/\sigma)$, where $z$ denotes the significance threshold defined by $1 - \Phi(z) = 0.05$ and $\sigma^2 = 1/4$, the variance of the binomial distribution in the limiting case $\pi = 1/2$. For a numerical example consider again the case $n = 100$ and $\pi = 0.65$. Then $\mu = 1.5$, so the power is approximately $1 - \Phi(1.645 - 1.5/0.5) = 0.91$, which is also the value obtained by evaluating the exact probability that a binomial random variable exceeds 58. The value of the approach based on local alternatives will become apparent when we deal with substantially more complicated problems in Chaps. 4 and 6.

Let us conclude this section by the presentation of the code that produced the three plots in Fig. 1.4. The first plot was created by running:

```
> n1 <- 36
> x <- 0:n1
> p1 <- 0.75
> z <- (x-n1/2)/sqrt(n1/4)
> plot(z-0.02,dbinom(x,n1,p1),type="h",col=gray(0.5),
+      main=paste("n =",n1,", pi =",p1),xlab="z",ylab="")
> lines(z+0.02,dbinom(x,n1,0.5),type="h")
> abline(v=1.645,lty=2)
```

Some points should be made: The expression "0:n1", which appears in the second line, is a shorter way of writing "seq(from=0,to=n1,by=1)". It produces, in this case, the set of values that the binomial random variable may have. Consider next the application of the "plot" function. Observe that the function "dbinom", which produces the binomial density, was applied. The argument "h" in "type" produces vertical, histogram-like, lines. Colors are set with the aid of the argument "col". The simplest way to assign values to this argument is by introducing the color's name. For example, one may write "col="red"". Alternatively, color number codes can be used. For technical reasons we will produce in this book only black and white plots. Hence we will apply the function "gray", which produces levels of gray. The input to the function is the level – a number between zero and one. The argument "main" produces the title of the plot. It takes as values character strings. Such strings may be entered directly or as outputs of string-manipulating functions. The former approach is applied in the argument "xlab="z"", which adds the title to the $x$ axis. For the main title the function "paste" is used. This function takes as input character strings and pastes them together. If the input is numerical it is translated, first to a character string representation of the number. In this example the output of the function is

```
> paste("n =",n1,", p =",p1)
[1] "n = 36 , p = 0.75"
```

which is a character string.

The second plot is produced by the code:

```
> n2 <- 100
> x <- 0:n2
> z <- (x-n2/2)/sqrt(n2/4)
> plot(z-0.02,dbinom(x,n2,p1),type="h",col=gray(0.5),
+       main=paste("n =",n2,", pi =",p1),xlab="z",ylab="")
> lines(z+0.02,dbinom(x,n2,0.5),type="h")
> abline(v=1.645,lty=2)
```

and the third by the code:

```
> xi <- sqrt(n1)*(2*p1-1)
> p.local <- 0.5 + xi/sqrt(4*n2)
> plot(z-0.02,dbinom(x,n2,p.local),type="h",col=gray(0.5),
+       main=paste("n =",n2,", pi =",p.local),xlab="z",ylab="")
> lines(z+0.02,dbinom(x,n2,0.5),type="h")
> abline(v=1.645,lty=2)
```

Observe that we have changed the number of observations to 100. In order to compute the appropriate value of the probability of success under the local alternative scenario, we first compute the expectation of the test statistic in the original sample. The new probability of success is the one that will produce the same expectation but for the new sample size.

## 1.7 Correlation and Regression

In most scientific experiments not one but several variables are measured. In particular, we shall be primarily interested in the joint analysis of genotype and phenotype. It is of interest to quantify and assess the relationships between variables. A popular summary statistic for the quantification of pairwise relationships is the covariance. An alternative is the correlation, which is a standardized covariance.

Imagine that we are given a sample of $n$ unrelated individuals, who express a quantitative phenotype of interest. Consider a target autosomal gene and assume that the phenotype and the alleles of the gene are measured. This produces one phenotypic measurement and a combination of the two alleles for each subject in the sample. For simplicity, we suppose that the gene has only two possible alleles, one denoted the *wild type* and the other the *variant* or *mutant type*. Use $y_i$ to symbolize the value of the phenotype for subject $i$ and $x_i$ to be the number of variant alleles for the same subject. Observe that $y_i$ takes numerical values and that $x_i$ may take the values 0, 1, or 2. The empirical covariance between the two sets of measurements, here the phenotypic and genotypic measurements, is defined by:

$$\frac{1}{n}\sum_{i=1}^{n}(y_i - \bar{y})(x_i - \bar{x}) \, ,$$

where $\bar{y}$ and $\bar{x}$ are the average values of the $y$'s and the $x$'s, respectively. The correlation between the two measurements is obtained by dividing the above by the product of the standard deviations of the two sequences. It takes the form:

$$\hat{\rho} = \frac{\frac{1}{n}\sum_{i=1}^{n}(y_i - \bar{y})(x_i - \bar{x})}{[\frac{1}{n}\sum_{i=1}^{n}(y_i - \bar{y})^2 \frac{1}{n}\sum_{i=1}^{n}(x_i - \bar{x})^2]^{1/2}} \, .$$

The correlation coefficient quantifies linear dependence between two measurements. It has values in the range between -1 and 1. A value of 1 corresponds to an exact linear relation with a positive slope. Likewise, a value of -1 corresponds to an exact linear relation with a negative slope. When the value of the coefficient is equal to zero, we say that the two measurements are uncorrelated. Independence between variables implies lack of correlation in the sense that the correlation coefficient will tend to take values close to zero.

We are also interested in the population covariance and correlation. The population covariance of $x$ and $y$ is defined as

$$\mathrm{cov}(x, y) = \mathrm{E}[(y - \mathrm{E}(y))(x - \mathrm{E}(x))] \, ,$$

where $\mathrm{E}(y)$ is the mean value of $y$ and $\mathrm{E}(x)$ is the mean value of $x$. The correlation is the covariance of the variables scaled to have unit variance:

$$\mathrm{cor}(x, y) = \mathrm{cov}(x, y)/(\sigma_y \sigma_x) \, ,$$

where $\sigma_y$ is the standard deviation of $y$ and $\sigma_x$ is the standard deviation of $x$. As the sample size $n$ increases, the empirical covariance and correlation converge to the population covariance and correlation. Independence between measurements implies that the population correlation coefficient equals zero.

The covariance and correlation parameters are closely related to the statistical model of linear regression, which may be used as a model for the relation between the two measurements. Regression models regard one of the variables as the dependent variable and the other as the independent, or explanatory, variable. According to the model, the conditional expectation of the dependent variable, given the value of the independent variable, is a linear function of the independent variable. Typically, one also assumes that the *residuals* (defined below) are independent of the explanatory variable.

In the genetic example, linear regression is also referred to as an *additive model*. This model assumes the relation

$$y_i = \mu + \alpha x_i + e_i \ .$$

The parameter $\mu$ is the expected value of the phenotype among wild-type homozygotes and the parameter $\alpha$ is the additive effect of one variant allele on the phenotype. The residual $e_i$ is a zero-mean random variable, which is assumed to be independent of $x_i$. For the simulations given below, since $x_i$ can assume the values 0, 1, or 2, we assume that $x_i$ is binomially distributed with parameters $n = 2$ and $p$; and we take the distribution of the residuals to be normal and denote their variance by $\sigma_e^2$.

In the case where $\alpha = 0$ we see that $x$ and $y$ are independent. This can be interpreted as saying that the gene has no effect on the trait. In order to test for such independence one may formulate the problem in terms of the hypotheses $H_0 : \alpha = 0$ versus $H_1 : \alpha \neq 0$, and look for an appropriate test statistic.

The quantity $Z = n^{1/2}\hat{\rho}$ can be used as a test statistic, although now the alternative is two-sided since $\alpha$ could be positive or negative. By a more general version of the central limit theorem than the one stated above, one finds that its null distribution is approximately normal with mean 0 and variance 1. Large absolute values of $Z$, which correspond to values of $\hat{\rho}$ substantially different from zero, are an indication for a non-zero value for the slope parameter $\alpha$. Consequently, for the conventional 5% significance level, the null hypothesis of independence between the phenotype and the genotype measurements will be rejected when $Z$ is either less than the 0.025-quantile or greater than the 0.975-quantile of the normal distribution.

Equivalently, one can look at the square of the test statistic and again reject the null hypothesis for large values. The distribution of the square of a standard normal distribution is called the chi-square distribution on one degree of freedom.

A number of commonly occurring statistics involved in tests of more than one parameter lead to approximate distributions in the form

$$U = Z_1^2 + \cdots + Z_d^2 \,,$$

where the $Z_i$ are independent $N(0,1)$ random variables. In this case $U$ is said to have a chi-square distribution on $d$ degrees of freedom. An example with $d = 2$ would arise in the current context if we wanted to test the hypothesis $\mu = 0$ and $\alpha = 0$. We will see examples in Chaps. 4 and 5.

Now we return to examine the statistical properties of the test for the slope parameter in the regression model. Consider a sample of size $n = 100$. We generate the test statistic under the null assumption of independence between $y$ and $x$:

```
> n <- 100
> U <- numeric(length=10^4)
> for (i in 1:length(U))
+ {
+     x <- rbinom(n,2,0.3)
+     y <- rnorm(n,15,3)
+     U[i] <- n*cor(x,y)^2
+ }
> mean(U > qchisq(0.95,1))
[1] 0.0527
```

Observe the use of the "`for`" loop. The indexing variable, to the left of the word "`in`" inside the brackets, sequentially obtains as a value the components of the vector to the right of that word. The expression following the brackets is evaluated for each value of the indexing variable. Several expressions may be grouped together by placing them between curly brackets. In this example we compute the test statistic 10,000 times. In each iteration x values are generated according to the binomial distribution. The phenotypes are generated according to the normal distribution with mean equal to 15 and standard deviation of 3. The test statistic is formed by squaring the correlation coefficient and multiplying the result by the sample size. The result is stored in the vector U. The function `qchisq` computes the quantile of the chi-square distribution. The simulated significance level is approximately equal to the target value of 5%.

A more comprehensive description of the distribution of the test statistic under the null distribution is given in Fig. 1.5. The *solid black line* represents the theoretical cumulative distribution function of the chi-square distribution, which is appropriate since we are computing the square of test statistic. The *solid gray line* represents the empirical distribution function of the simulated U statistic. The *broken black line* represents the empirical distribution function of the statistic when it is simulated under the alternative hypothesis. Note that the simulated *solid gray line* and the theoretical *solid black line* are practically identical. These two lines were formed using the code:

```
> u <- seq(0,30,length=100)
> plot(u,pchisq(u,1),type="l",ylab="probability")
```

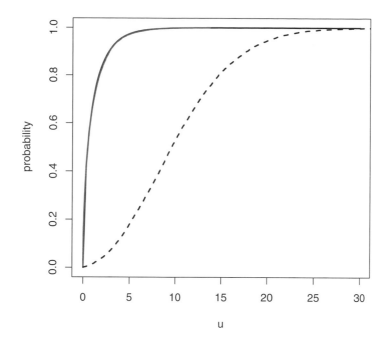

**Fig. 1.5.** Distribution of the correlation test statistic

```
> p <- (1:length(U))/length(U)
> U <- sort(U)
> lines(U,p,col=gray(0.5))
```

Observe that we have not used the built-in function "ecdf" for plotting the empirical distribution function. Instead, we wrote our own code, which uses the fact that the empirical distribution function increases in steps of size equal to $1/n$, where $n$ is the number of observations. The steps occur at the observed values. In order to make the plot we sorted the values of the vector U with the function "sort". The resulting vector serves as $x$ values. The $y$ values are a monotone vector in the range $(0, 1]$ of the same length as U, which is stored in an object named p.

Under the alternative hypothesis the value of $\alpha$ is different from zero. In the following example we take it to be equal to 1.5. The simulation of U under the alternative distribution is carried out exactly like the simulation under the null. The only difference is in the expectations of the y values, which are taken to be a functions of the x values. This is implemented by using a different "mean" value as an argument in the function "rnorm". The resulting power is about 0.89. The plot of the empirical distribution of the test statistic under the given alternative is added as a *broken line* to the plot:

```
> for (i in 1:length(U))
+ {
+     x <- rbinom(n,2,0.3)
+     y <- rnorm(n,15+1.5*x,3)
+     U[i] <- n*cor(x,y)^2
+ }
> mean(U > qchisq(0.95,1))
[1] 0.8898
> U = sort(U)
> lines(U,p,lty=2)
```

## 1.8 Stochastic Processes

A characteristic of modern gene mapping is the availability of a large number of genetic markers distributed across the genome, with the possibility of forming a different test statistic for each marker. The statistical complication resulting from that fact is the need to analyze simultaneously several correlated measurements taken over markers residing on the same chromosome. A standard approach for the joint analysis of correlated random variables regards them as a stochastic processes. In this book we will consider mainly two types of processes: *Gaussian processes* and *Markov processes*.

A (discrete time) stochastic process refers to a sequence of random variables, which are typically correlated. It is convenient to index the random variables by points on the real line and denote them by $t$, whether or not they can be interpreted as "time". Often, but not always, these points can be taken to be the set of integers or some other collection of regularly spaced set of points. The distribution of the stochastic process is understood to be the joint distribution of its components, denoted, for example, $X_t$. This distribution, however, can be partially characterized by the distribution of subsets of the components. Of special interest are the marginal distributions of each component $X_t$ separately and the joint distribution of pairs of components $X_t, X_{t'}$.

An important family of stochastic processes are the Gaussian processes. These processes are defined by the requirement that each finite subset of variables has a joint Gaussian, or multivariate normal, distribution. The multivariate normal distribution is a natural generalization of the normal distribution in the sense that each component has a marginal normal distribution. (More generally, each linear combination of components possesses a normal distribution.) The joint distribution of a Gaussian process is completely determined by the marginal expectation of its components, $E(X_t)$, and by the covariance between any pair of components, $cov(X_t, X_{t'})$. It is convenient to write the expectation in the form of a vector, called the *mean vector* and covariances in the form of a matrix, called the *covariance matrix*. The discussion of Gaussian processes will be continued in Chap. 5, where we begin a more systematic discussion of the statistics of gene mapping in the presence of multiple markers.

Considering Gaussian processes, which require only the specification of pairwise joint distributions, is one way of reducing the complexity of the joint distribution of a stochastic process. An alternative is to consider sequential dependencies. The simplest and most popular model of sequential dependence is the Markovian model. According to this model, given the value of a component, $X_t$, the process to the left of the component is independent of the process to the right of it. The distribution of such process can be described by the distribution of the first element in the process (called the *initial distribution* and denoted by $\pi$) and the transition probabilities, which describe the distribution of a subsequent component, given the value of the current one.

Of special interest to us will be Markov chains taking only a finite set of values, known as *states*. In this case, the conditional distributions that describe the movement of the chain from one state to another may be conveniently represented in a form of *probability transition matrices*, which we denote by $T_t$. Each row in such a matrix corresponds to the current state of the process, while the column describes the conditional distribution of the state in the next step, given the current value. It follows that each row of a transition matrix contains nonnegative entries that sum to one. If the matrices are the same for all $t$, the Markov chain is said to have *stationary transition probabilities*.

Given the marginal distribution of the current state of the process, written as a row vector, the marginal distribution of the states one step later are given by multiplying this row vector on the right by the transition matrix. In the particular case of a Markov chain with stationary transition probabilities where these marginal distributions are the same, i.e., the distribution of $X_t$ is the same for all $t$, we say that the distribution is *stationary*. Symbolically, the marginal distribution $\pi$ of the process is a stationary distribution if it satisfies the equation: $\pi'T = \pi'$. A simple set of conditions that are met in all the examples that we will consider, will establish the existance and uniqueness of the stationary distribution $\pi$ as a function of the transition matrix. It will follow that the joint distribution of a finite state stationary Markov chain is fully determined by its transition matrix.

## 1.9 Likelihood-Based Inference

Tools for making statistical inference can be derived in a variety of ways. A standard practice, which is supported by solid theoretical foundations, is a likelihood-based approach. In simple problems of the type discussed above, the likelihood-based approaches agree with the common sense approaches we have adopted. In more complex problems it may not be so obvious what exactly the "common sense" approach is.

The likelihood function is the probability density function of the observations, interpreted as a function of the parameters that determine the distribution. In many cases it is more convenient to consider the logarithm of the likelihood function, denoted the *log-likelihood* function. Either of these functions may be used in order to identify good estimates of unknown parameters or in order to construct efficient test statistics.

For example, if the observation has a binomial distribution, then the log-likelihood function takes the form:

$$\ell(p) = \log \binom{n}{X} + X \log p + (n - X) \log(1 - p) .$$

It is considered as a function of $p$, the probability of a success. The argument $X$ is the observed number of successes in the trial. An estimate of the unknown parameter, as a function of the observation, may be obtained by maximizing the likelihood function with respect to the parameter. The result is the *maximum likelihood estimate* (MLE). Equivalently, the log-likelihood function may be used. The maximizer can be found by equating the derivative of the log-likelihood to zero:

$$\dot{\ell}(p) = \frac{X}{p} - \frac{n - X}{1 - p} = \frac{X - np}{p(1 - p)} = 0 .$$

The MLE in this case turns out to be $\hat{p} = X/n$, the empirical frequency of successes in the sample. Note that we have used the "dot" notation to represent derivatives.

The log-likelihood function and its derivatives are a good source for finding efficient test statistics. The motivation for this proposition is the Neyman-Pearson Lemma, which states that the likelihood ratio is the most powerful statistic for testing a simple null distribution, e.g., $p = p_0$ against a simple alternative, say $p = p_1 \neq p_0$ The likelihood ratio statistic is computed as the ratio of the likelihood function at the given alternative divided by the likelihood at the given null value. Alternatively, the difference of the log-likelihoods, $\ell(p_1) - \ell(p_0)$, can be used. The null hypothesis is rejected when the statistic exceeds a threshold, which is set by the null distribution of the statistic to control the significance level.

For a composite hypotheses involving many values of $p$ it is not clear which parameter values to use in the log-likelihoods that form the difference. One approach, called the *generalized likelihood ratio test* (GLRT), is to use maximum likelihood estimates. One of the log-likelihood functions is maximized over the entire space of parameters. This corresponds to plugging the MLE into the log-likelihood function. The other log-likelihood is maximized over the subset of parameters that form the null hypothesis. This corresponds to plugging into the function the MLE for a sub-model that corresponds to the null hypothesis. It can be shown that under suitable regularity conditions the null distribution of twice the log-likelihood difference is approximately chi-square when the sample size is large. The number of degrees of freedom for the chi-square distribution is the difference between the dimension of the space of parameters and the dimension of the sub-space formed by the null hypothesis. For example, the GLRT for the hypothesis $H_0 : p = 1/2$ versus $H_1 : p \neq 1/2$ in the binomial case is:

$$2(\ell(\hat{p}) - \ell(1/2)) = 2\,n\left[\hat{p}\log\frac{\hat{p}}{1/2} + (1-\hat{p})\log\frac{1-\hat{p}}{1/2}\right].$$

One degree of freedom is assigned to the asymptotic chi-square distribution. This follows from the fact a single parameter is used, which corresponds to dimension one. The null hypothesis contains only one point – a subspace of dimension zero. The difference in the dimensions equals one.

The following example is artificial, but it arises naturally in Chap. 4 as an approximation to a more complex problem. Suppose $y_1$ and $y_2$ are independently and normally distributed, with $y_i$ having mean value $\mu_i$ and variance $\sigma_i^2$. Assume that the $\sigma$'s are known and we want to test the null hypothesis that $\mu_1 = \mu_2 = 0$. The log-likelihood function, except for numerical constants that play no role, is

$$\ell(\mu_1, \mu_2) = -(y_1 - \mu_1)^2/(2\sigma_1^2) - (y_2 - \mu_2)^2/(2\sigma_2^2).$$

By differentiating with respect to $\mu_i$ for $i = 1, 2$ and setting the derivatives equal to 0, we see that the maximum likelihood estimators for $\mu_1$ and $\mu_2$ are simply $y_1$ and $y_2$, respectively. Hence twice the log-likelihood ratio statistic is

$$(y_1/\sigma_1)^2 + (y_2/\sigma_2)^2.$$

Under the null hypothesis that $\mu_1 = \mu_2 = 0$, each of the terms in this sum is the square of a standard normal random variable, and hence the sum is a chi-square random variable with two degrees of freedom. Now,

```
> qchisq(0.95,2)
[1] 5.991465
```

is the 95th quantile for a chi-square random variable with two degrees of freedom, which is very close to 6. So the null hypothesis would be rejected at the 0.05 level if the expression given above exceeds 6.

The rest of this section is devoted to a third approach, which involves the *score statistic*. This test statistic uses the first and second derivatives of the log-likelihood function. In the multi-parameter setting it uses the gradient vector and the Hessian matrix of the function, i.e., the vector of partial derivatives and the matrix of mixed partial second derivatives. Here we discuss the relatively simple case of a scalar parameter, which we denote by $p$.

Given the likelihood function $\ell(p)$, the first derivative $\dot{\ell}(p)$ is called the efficient score. The second derivative is $\ddot{\ell}(p)$. The negative of its expected value is called the Fisher information and is denoted by $I(p)$. The score statistic for testing the null hypothesis $p = p_0$ against a one-sided alternative is based on $\dot{\ell}(p_0)$. It turns out that when the null hypothesis is true, this statistic is approximately normally distributed with mean value 0 and variance $I(p_0)$. Hence, a significance threshold can be obtained for the standardized statistic $\dot{\ell}(p_0)/[I(p_0)]^{1/2}$ from the standard normal distribution.

Let us consider the binomial model for the last time. The second derivative of the log-likelihood function is given by:

$$\ddot{\ell}(p) = -\frac{X}{p^2} - \frac{n - X}{(1 - p)^2} .$$

Using $p_0 = 1/2$ we obtain:

$$\dot{\ell}(1/2) = 4(X - n/2) , \quad I(1/2) = -\mathrm{E}[\ddot{\ell}(0.5)] = 4n .$$

As a result we get:

$$[\dot{\ell}(1/2)]/[I(1/2)]^{1/2} = \frac{X - n/2}{(n/4)^{1/2}} ,$$

which is the same statistic discussed above.

## 1.10 Properties of Expectations and Variances

The following rules for manipulating expectations, variances, and covariances are used repeatedly in the text. Let $y_1, y_2, \ldots, y_n$ be random variables and $a_1, a_2, \ldots, a_n$ constants. Then

$$\mathrm{E}\left( \sum_{i=1}^{n} a_i y_i \right) = \sum_{i=1}^{n} a_i \mathrm{E}(y_i) , \tag{1.1}$$

and

$$\mathrm{var}\left( \sum_{i=1}^{n} a_i y_i \right) = \sum_{i=1}^{n} a_i^2 \mathrm{var}(y_i) + \sum_{i \neq j} a_i a_j \mathrm{cov}(y_i, y_j) . \tag{1.2}$$

If $y_i$ and $y_j$ are independent, then $\mathrm{cov}(y_i, y_j) = 0$. In particular, if $y_1, \ldots, y_n$ are independent, then the preceding formula simplifies to

$$\mathrm{var}\left( \sum_{i=1}^{n} a_i y_i \right) = \sum_{i=1}^{n} a_i^2 \mathrm{var}(y_i) . \tag{1.3}$$

A special case of particular interest is independent $y_i$ with $\mu = \mathrm{E}(y_i)$, $\sigma^2 = \mathrm{var}(y_i)$, and $a_i = 1/n$ for all $i$. Then $\mathrm{E}(n^{-1} \sum_{i=1}^{n} y_i) = \mu$ and $\mathrm{var}(n^{-1} \sum_{i=1}^{n} y_i) = \sigma^2/n$. These formulas are perhaps the most basic in all statistics, because they express quantitatively the possibility of reducing the variability of individual, noisy measurements of a quantity $\mu$ by using averages of independently repeated measurements.

As an example, suppose that $X$ is binomially distributed with parameters $n$ and $p$. Such an $X$ can be represented as the sum $Y_1 + \cdots + Y_n$ of the independent Bernoulli trials, where each $Y_i$ is 1 or 0 according as the $i$th trial is success or failure. It is easy to see that $E(Y_i) = p$ and $\text{var}(Y_i) = p(1-p)$. It follows immediately that $E(X) = np$, $\text{var}(X) = np(1-p)$, $E(X/n) = p$, $\text{var}(X/n) = p(1-p)/n$. These quantities can be computed directly from the binomial distribution by writing, for example, $E(X) = \sum_{k=0}^{n} k \binom{n}{k} p^k (1-p)^{n-k}$ and simplifying this expression by algebraic manipulations. This is a more complicated route even in this simple problem, and it is often unavailable in more complex problems.

Finally, we note for later reference that if $x$ and $y$ are two random variables, then

$$E[xy] = E[x E(y|x)] , \qquad (1.4)$$

where $E(y|x)$ denotes the conditional expectation of $y$ give the value of $x$. Similarly,

$$\text{var}(y) = E[\text{var}(y|x)] + \text{var}[E(y|x)] . \qquad (1.5)$$

## 1.11 Bibliographical Comments

Among the many elementary texts of statistics, one that stands out is Rice [64]. A more advanced text, which contains a thorough discussion of likelihood methods and of local alternatives, is Cox and Hinkley [15]. An excellent text combining statistical analysis and computation is Venables and Ripley [87].

For probability, including Markov chains, Feller [29] is excellent. Another very good basic text is Ross [68].

## Problems

In the following exercises you are asked to simulate test statistics in various settings.

**1.1.** The first problem deals with the situation described in Sect. 1.5. Consider two binomial random variables $X_1$ and $X_2$ such that $X_1 \sim B(n_1, p_1)$ and $X_2 \sim B(n_2, p_2)$. One is interested in testing the null hypothesis $H_0 : p_1 = p_2$ versus the alternative $H_1 : p_1 \neq p_2$. A standard test considers the difference between the sample proportions $\hat{p}_1 = X_1/n_1$ and $\hat{p}_2 = X_2/n_2$, which are standardized with the aid of the pooled sample proportion $\hat{p} = (X_1 + X_2)/(n_1 + n_2)$. The resulting test statistic is given in the form

$$Z = \left( \frac{n_1 n_2}{n_1 + n_2} \right)^{1/2} \frac{\hat{p}_1 - \hat{p}_2}{[\hat{p}(1-\hat{p})]^{1/2}} ,$$

which has approximately a standard normal distribution under the null hypothesis when $n_1$ and $n_2$ are large.

(a) Investigate, using simulations, the distribution of this test statistic under the null hypothesis. Consider various values of the parameters $n_1$, $n_2$, and $p = p_1 = p_2$. A famous rule of thumb to determine what it means for $n_1$ and $n_2$ to be large requires that $\min\{n_1 p, n_1(1-p), n_2 p, n_2(1-p)\} \geq 5$. Evaluate this rule of thumb.

(b) Investigate numerically the power properties of this test as a function of the difference $p_1 - p_2$.

(c) Investigate numerically, for fixed values of $p_1 \neq p_2$, the power of the test as a function of the sample size. Specifically, for a given $n$, let $n_1 = \beta n$ and $n_2 = (1 - \beta)n$, $0 < \beta < 1$. What value of $\beta$ would maximize the statistical power?

**1.2.** Consider the following generalization of the regression model that was introduced in Sect. 1.7:

$$y = \mu + \alpha_1 x_1 + \alpha_2 x_2 + e.$$

(This model would be appropriate for simultaneously testing for a relation between a pair of genes on different chromosomes and the phenotype $y$. See Chap. 2.) Again, assume that each gene has two alleles and denote by $x_1$ and $x_2$ the count of variant alleles in the first gene and in the second gene, respectively. Assume that $x_1 \sim B(2, p_1)$ and $x_2 \sim B(2, p_2)$ and that the two are independent. A reasonable test statistic may take the form

$$U = n\hat{\rho}_1^2 + n\hat{\rho}_2^2 \, ,$$

where $\hat{\rho}_i$ is the correlation coefficient between $y$ and $x_i$, for $i = 1, 2$. Under the null hypothesis of no relation with either gene, the asymptotic distribution of the given statistic is chi-square on two degrees of freedom.

(a) Investigate, using simulations, the distribution of this test statistic under the null hypothesis and compare it to the theoretical asymptotic distribution. Consider various values of the sample size $n$ and of the allele frequencies, $p_1$, and $p_2$.

(b) Investigate numerically the power of this test for $p_1 = p_2 = 1/2$, $n = 200$, and different values of $\alpha_1$ and $\alpha_2$.

(c) Compare the statistical properties, i.e., rejection threshold and power function, of the statistic considered above with the statistic for a single gene that is described in Sect. 1.7. Describe conditions under which you would prefer one or the other of the tests.

# Introduction to Experimental Genetics

Variability in observed phenotypes may result from a variety of factors – inherited as well as environmental. The blend of all these influences gives rise to the unique being every living creature is. Still, the main role of science is to identify the rules which unite seemingly unrelated phenomena. The role of genetics is no different. Its first, and most important, task is to identify the major factors that give rise to different phenotypical characteristics. Once these major factors have been identified, the investigation can be carried on in order to identify genes having secondary effects.

The basic strategy in experimental sciences is to isolate the phenomena being investigated and study them under controlled conditions. Ideally, an experiment will involve measurement of the phenomena, taken at various levels of one or a small number of potentially influencing factors with the levels of other factors kept fixed. Thence, observed variation in the phenomena can be attributed to the variation in that single explanatory factor. Experimental genetics applies this approach. Effort is made to ensure uniform environmental conditions, so one hopes that environmental effects on the phenotype are minimized. Moreover, the use of an experimental cross of inbred strains (see below) potentially reduces the variability in the genetic background, and thus facilitates the dissection of the genetic factors that give rise to the observed phenotypic variation. (This simplification brings with it the disadvantage that there may well be genetic variability in nature that is not represented in a given set of inbred strains.)

In reality, the ideal experiment is almost never feasible. Environmental condition may vary slightly from one stage of the experiment to the next. Pure strains may be contaminated. Some factors may not be possible to isolate from others, thus forcing the investigation of several factors jointly. And on top of that, measurement errors may introduce unwanted error into the system. Consequently, precautionary measures need to be taken. These may include the use of experimental controls, taking repeated measurements, and reproducing the results in other laboratories. The analysis of the outcomes of the experiments must take into account the potential effect of unplanned factors. This is the role of statistical models and statistical inference.

Models from statistical genetics are used in order to incorporate under one roof the investigated effects of the genetic factors, measurement errors, and a whole range of influencing factors that may have not been taken into account during the planning of the experiment. The analysis of the outcomes of the experiment, and the conclusions, are based on applying the principles of statistical inference to these models. The realization that eventually the outcomes of the experiment will be investigated with statistical tools should have an effect on the way the experiment is planned. A thoughtful design of an experiment may increase its statistical efficiency. A poor design may be wasteful, or may not produce enough information to attain its stated goal.

Our discussion concentrates on mouse genetics, but the methods are valid more generally. The mouse model was selected due to its importance in relation to human genetics. The rest of this chapter provides some background in specific statistical models and experimental designs for genetic mapping based on crosses of inbred strains. Because we are not able to do breeding experiments with humans, human genetics involves more complex statistical considerations that will be investigated in later chapters. The first section below outlines some basic facts regarding the mouse model and its genetics. The material in this section is mainly borrowed from the excellent textbook by Lee Silver [77]. (See the link http://www.informatics.jax.org/silver/ for an online version of the book.) The second section deals with a statistical model of the relation between genetic factors and the observed phenotype. Like reality, statistical models can become quite complex. However, even simple models can provide insight and lead to useful analysis. The model that will be discussed here is as simple as such models can get. Thus, it is perfect for the introduction of the basic concepts. Yet, even in its simplest form, the model is frequently used by researchers – a testimony to its value. Some simulations are conducted in accordance with the model in order to demonstrate the effect of genetic factors on the distribution of the phenotype. In the third section some common experimental designs in mice are described. The merits and drawbacks of the different designs will be revisited in later chapters after going into the details of the statistical tools and their properties. Finally, a short and non-technical description of the genetic markers that are commonly used today is provided.

## 2.1 The Mouse Model

Already in the early days of modern genetics, the house mouse was identified as a perfect model for genetic investigation in mammals. The house mouse is a small, easy to handle animal. It is relatively inexpensive to maintain, it breeds easily, and it has a high rate of reproduction. A significant portion of biological research is aimed at understanding ourselves as human beings. Although many features of human biology at the cell and molecular levels are shared across all living things, the more advanced behavioral and other characteristics of human beings are shared only in a limited fashion with other species or are unique to humans. In this vein, the importance of mice in genetic studies was first recognized in the intertwined biomedical fields of immunology and cancer research, for which a mammalian model was essential. Today, specially developed mouse strains serve as models for many human traits, e.g., obesity or diabetes.

The mouse genome, like the genomes of most mammals, contains about three billion base-pairs (bp). It is organized in 19 pairs of homologous autosomal chromosomes, compared to the 22 pairs in human, and a single pair of sex chromosomes. Yet, there is a large amount of homology between the mouse and the human genes. Large chunks of DNA, 10–20 million bp in length, remained intact during evolution, and are practically identical in both species. As a matter of fact, the entire human genome can be, more or less, recovered by cutting the mouse genome into 130–170 pieces and reassembling them in a different order.

At the genetic level, processes of meioses, mating, and reproduction in the two species are similar. In particular, in mice like in human, autosomal chromosomes may experience crossovers during meiosis. This process of recombination mixes up the genetic material that is passed on from the parent to the offspring. Owing to recombination, the genetic contribution of the parent is a random mosaic of segments originating from the two grandparents. However, the rate of crossovers per base pair during meiosis in mouse is about half the rate in human.

The founding father of mouse research was W. E. Castle, who (intentionally) brought the mouse into his laboratory at the beginning of the 20th century. Subsequently, he, and his many students, began developing genetically homogeneous inbred lines of mice. These pure inbred lines became a very valuable resource and the key to the success of the mouse model in genetics. Genetically homogeneous strains provide the means to control for the effect of genetic factors. Moreover, since all mice of the same line are genetically identical, results of experiments carried out in a specific laboratory can be compared to results of experiments from other laboratories. A major source of such genetically pure strains is the Jackson Laboratory in Bar Harbor, Maine. This laboratory was founded in 1929 by C. Little, a student of Castle's, with the aim of promoting mouse research; it serves to this day as a center of mouse research.

Genetically homogeneous inbred strains are created by a process of successive brother–sister mating. Random drift in finite inbred populations eventually results in the fixation of a given locus, namely in the extinction of all other alleles. Once a locus is not polymorphic, it remains so in all subsequent generations. With additional brother–sister mating, the genomes in the population become less and less polymorphic. The formal definition of an inbred strain requires at least 20 generations of strict brother–sister mating. Some of the classical inbred strains have a history of more than 100 generations of inbreeding. See Chap. 3 for a more systematic discussion of inbreeding.

### 2.1.1 A Quantitative Trait Locus (QTL)

We consider measurement of a continuous trait, such as body weight or blood pressure, in a population of mice. Denote the level of the measurement for a randomly selected mouse by $y$. One usually observes that such measurements show variability across the population. Some of this variability may be attributed to genetic factors. The task is to model the overall genetic contribution, and the genetic contribution of each specific locus, to the overall variance.

We assume that the phenotypic value $y$ is a simple summation of the mouse's genotype and environment influences:

$$y = m + G + E , \tag{2.1}$$

where $m$ is a constant, $G$ denotes the effect of all genes contributing to the phenotype, and $E$ denotes the environmental effects. By writing $m$ separately we can assume without loss of generality that both $G$ and $E$ have mean value 0, so $m$ is in fact the average phenotypic value. We also make the critical assumption that $G$ and $E$ are uncorrelated. The variance of $G$, say $\sigma_G^2$, is called the genetic variance; the variance of $y$, $\sigma_y^2$ is the phenotypic variance. From our assumption that $G$ and $E$ are uncorrelated, it follow that $\sigma_y^2$ is the sum of $\sigma_G^2$ and the variance of $E$, $\sigma_E^2$. An important quantity is the heritability $H^2$ defined to be the ratio $\sigma_G^2/\sigma_y^2$ of the genotypic to phenotypic variance. It measures the percentage of variation in the phenotype that has a genetic origin.

We now want to make more specific assumptions about the contribution to $G$ arising from one genetic locus. Consider a given polymorphic genetic locus, specifically, a bi-allelic locus with its two alleles denoted by $A$ and $a$. The genotype at the given locus is one of the three types: $AA$, $Aa$, or $aa$. Consequently, the underlying population can also be subdivided into three subclasses according to these three types. The model we propose may assign different average measurements for each of these subclasses. However, the variance of the measurement is assumed to be unchanged across the three subclasses. Let $x_M$ be the number of $A$ alleles (0 or 1) inherited from the mother and $x_F$ the number inherited from the father. Observe that $x = x_M + x_F$, the total number of $A$ alleles in the genotype, can take on the values 0, 1, or 2. Consider a model of the form

$$y = \mu + \alpha(x_M + x_F) + \delta |x_M - x_F| + e , \tag{2.2}$$

where $e$ is a zero mean random deviate. Note that $|x_M - x_F| = 1$ if, and only if, $x = 1$, namely that the mouse is *heterozygous*. The term $\mu$ is the mean level of $y$ for an *aa-homozygote*. The mean level for an $AA$-homozygote is $\mu + 2\alpha$, and the mean level for an heterozygote is $\mu + \alpha + \delta$. The locus is said to be *additive* if $\delta = 0$, since in this case each $A$ allele adds the amount $\alpha$ to the average phenotype. If $\delta = \alpha$, the allele $A$ is said to be *dominant*; it is called *recessive* if $\alpha = -\delta$. These terms are consistent with the usage for qualitative traits. For a dominant locus, a single $A$ allele produces the full genetic effect; for a recessive locus a single $A$ allele produces no effect, while two $A$ alleles do produce an effect. Note that $A$ is dominant if and only if $a$ is recessive.

The term $x = x_{\mathrm{M}} + x_{\mathrm{F}}$ is the number of $A$ alleles in a randomly selected mouse. Both $x_{\mathrm{M}}$ and $x_{\mathrm{F}}$ are Bernoulli random variables. We assume here that they are independent with the same probability $p$ to take the value 1 (indicating an $A$ allele). Then the distribution of $x$ is binomial, namely $\Pr(x = 2) = p^2$, $\Pr(x = 1) = 2\,p(1-p)$, and $\Pr(x = 0) = (1-p)^2$, where $p$ is the frequency of allele $A$ in the population. It follows that the mean value of $x$ is $2\,p^2 + 2\,p(1-p) + 0 = 2\,p$. Therefore, the overall mean of the phenotype is $m = \mu + 2\,p\,\alpha + 2\,p(1-p)\delta$. (The assumption that $x_{\mathrm{M}}$ and $x_{\mathrm{F}}$ are independent Bernoulli variables can be derived from the notion of *Hardy-Weinberg Equilibrium*, which will be discussed in the next chapter.)

The residual $e$ incorporates all remaining factors that contribute to the variability. Such factors can include the genetic contribution from loci other than the one we investigate, as well as environmental factors. We assume that $e$ is *uncorrelated* with the other terms on the right-hand side of (2.2). This assumption may not be satisfied. In most cases it rules out the possibility of other genes on the same chromosome that contribute to the trait. Such genes would be *linked* to and hence correlated with the investigated locus. (See the discussion of recombination in Chap. 3.) In the special case where the locus explicitly modeled in (2.2) is the *only* genetic locus contributing to the trait, then $c$ in (2.2) is the same as $E$ in (2.1); and since $G$ has mean 0, it can be written explicitly as $G = \alpha(x - 2\,p) + \delta[|x_{\mathrm{M}} - x_{\mathrm{F}}| - 2\,p(1-p)]$.

Often in what follows we assume also that $e$ is normally distributed, although this assumption is not strictly necessary. Because of the central limit theorem, it would be a reasonable assumption if $e$ is made up of a sum of approximately independent small effects, either genetic or environmental. It would not be satisfied, however, if there is another major gene having a substantial effect in addition to the one modeled explicitly in (2.2). We will return to this point below.

By simple, but somewhat tedious algebra (see Prob. 2.4 below) one can rewrite the model (2.2) in the form:

$$y = m + \{\alpha + (1 - 2\,p)\delta\} \times [(x_{\mathrm{M}} - p) + (x_{\mathrm{F}} - p)] - \{2\,\delta\} \times [(x_{\mathrm{M}} - p)(x_{\mathrm{F}} - p)] + e\,, \quad (2.3)$$

and show that $[(x_{\mathrm{M}} - p) + (x_{\mathrm{F}} - p)]$ and $[(x_{\mathrm{M}} - p)(x_{\mathrm{F}} - p)]$ are uncorrelated by virtue of the assumption that $x_{\mathrm{M}}$ and $x_{\mathrm{F}}$ are independent. Since by assumption $e$ is also uncorrelated with these terms, the variance of $y$ can be written as the sum of three terms:

$$\sigma_y^2 = \sigma_{\mathrm{A}}^2 + \sigma_{\mathrm{D}}^2 + \sigma_e^2\,, \quad (2.4)$$

where

$$\sigma_A^2 = 2\,p(1-p)[\alpha + (1-2p)\delta]^2\,, \quad \sigma_D^2 = 4\,p^2(1-p)^2\delta^2\,, \quad \sigma_e^2 = \mathrm{var}(e)\,. \quad (2.5)$$

The term $\sigma_A^2$ is called the locus specific *additive variance*, while $\sigma_D^2$ is called the locus specific *dominance variance*. Note the potential confusion with the terminology introduced above. The allele $A$ is additive if and only if the dominance variance is 0. If $A$ is dominant *or* recessive, the dominance variance is positive. In the very important special case of an *intercross*, defined below in Sect. 2.2, $p = 1/2$, so $\sigma_A^2 = \alpha^2/2$, and $\sigma_D^2 = \delta^2/4$.

A measure of the importance of the locus under study is the *locus specific heritability*, denoted by $h^2$ and defined to be the ratio $(\sigma_A^2 + \sigma_D^2)/\sigma_y^2$. In the special case that only a single gene contributes to the trait $h^2 = H^2$, the heritability of the trait.

### 2.1.2 Simulation of Phenotypes

Let us explore the effect of the different model parameters on the distribution of the phenotype in a population. For the exploration we use R, which was introduced in the previous chapter. Start with formation of the vector of mean values, for the case where $\mu = 5$, $\alpha = 1$, and $\delta = -1$ (i.e., a recessive model). For each animal the mean expression level is equal either to 5 or to 7, depending on the animal's genotype:

```
> n <- 6; p <- 0.5; x <- rbinom(n,2,p)
> mu <- 5; alpha <- 1; delta <- -1
> mu + alpha*x + delta*(x==1)
[1] 7 5 5 7 5 5
```

A normal residual term is added to the mean:

```
> sig <- 0.5
> mu + alpha*x + delta*(x==1) + sig*rnorm(n)
[1] 8.042733 5.351743 4.859863 7.133684 4.671937 5.558227
```

The default application of the function "rnorm" generates independent standard normal random variables. Multiplying by the standard deviation and adding the mean transforms the distribution to having the given mean and standard deviation. An alternative way of obtaining the same distribution is by using the "mean" and "sd" parameters of the function:

```
> rnorm(n, mu + alpha*x + delta*(x==1), sig)
[1] 8.312484 5.351396 4.635130 6.936316 4.829702 5.295540
```

Yet, a third way would be to have:

```
> mu + alpha*x + delta*(x==1) + rnorm(n, sd=sig)
[1] 6.982847 6.354027 5.177322 6.852138 5.111498 4.644922
```

Observe that we must introduce the standard deviation with the "par_name = par_value" argument assignment format since the parameter "sd" is not the second parameter of the function "rnorm".

Let us put some real action into the story by simulating a population of 100,000 animals according to the given genetic model, under two different scenarios for non-genetic variability:

```
> n <- 10^5; p <- 0.5; x <- rbinom(n,2,p)
> mu <- 5; alpha <- 1; delta <- -1; sig <- 0.5
> y <- mu + alpha*x + delta*(x==1) + sig*rnorm(n)
> h2 <- var(alpha*x + delta*(x==1))/var(y)
> plot(density(y),main=
+    paste("A recessive model:\n h^2 = ",round(h2,3),sep=""))
```

This produces the density function on the left in Fig. 2.1. The density on the right is produced by:

```
> mu <- 5; alpha <- 1; delta <- -1; sig <- 1
> y <- mu + alpha*x + delta*(x==1) + sig*rnorm(n)
> h2 <- var(alpha*x + delta*(x==1))/var(y)
> plot(density(y),main=
+    paste("A recessive model:\n h^2 = ",round(h2,3),sep=""))
```

The object "h2" stores the value of $h^2$, i.e., the fraction of the genetic variance for the modeled gene within the total variance. The function "density" produces an estimate of the density of the population, based on the values stored in the vector y. The resulting object is then plotted with the function "plot".

The title for the figure is set with the argument "main" of the function "plot". The assignment of this argument is a character string. Character strings are entered using either double (") or single (') quotes. The function "paste" takes an arbitrary number of arguments and concatenates them one by one into character strings, with a separating string determined by the argument "sep". Numbers are converted into character strings. The symbol "\n" enters a line break.

The distributions of the phenotypes in both populations are given in Fig. 2.1. Both pictures represent a mixture of two normal distributions. The mean of each of the normal distributions that form the mixture is the same (5 and 7), and the fraction is the same (3/4 and 1/4). The only difference between the two cases is the magnitude of the variance attributed to non-genetic factors (1/4 versus 1). Note that this small change is enough to change a bimodal distribution into a distribution with a single mode. In fact, a casual glance at the second density function in Fig. 2.1 suggests that it is roughly normal, although a more careful inspection shows that it is slightly skewed to the right.

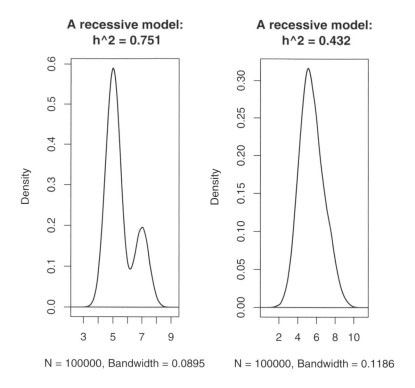

**Fig. 2.1.** Distributions of QTL phenotypes

## 2.2 Segregation of the Trait in Crosses

Crosses between inbred strains are the bread and butter of experimental genetics. Inbred strains are specially developed to be homozygous at all loci throughout the genome, in contrast to outbred populations which are often polymorphic. Standard approaches for dissecting an heritable trait in experimental genetics often involve crossing inbred strains. The genetic contribution to a trait cannot be demonstrated by looking at individuals from a single inbred strain alone, since in principle all members of the same strain are genetically identical and the observed phenotypical variability among them has to be attributed to environmental factors. Therefore, in order to map genetic factors one must select at least two different strains, which show different levels of expression of the phenotype under the same environmental conditions. In many cases one can screen the commercially available inbred strains for an appropriate pair of strains, which are phenotypically distinct. Careful crosses between the selected strains produce the population that is used for the genetic mapping.

The two most popular protocols for crossing inbred strains are the *back-cross* and the *intercross*, which we describe below. However, before going into the details of the two different breeding schemes, let us introduce the standard terminology. Practically all breeding experiments start with an *outcross*, a mating between two animals or strains considered unrelated to each other. Specifically, we consider here an outcross between two distinct inbred strains that show a phenotypic difference. The resulting offspring are called the first filial generation, or $F_1$. Consider any genetic locus where the two strains differ. Recall that inbred strains are homozygous at all loci. Say that the genotype of one of the inbred strains at a given locus is $AA$, and the genotype of the other inbred strain is $aa$. It follows that the genotype of the $F_1$ generation at the polymorphic site must be $Aa$, since each parent passes on one of its alleles to the offspring. Consequently, $F_1$ mice have a fixed genetic composition – all are heterozygous at all polymorphic loci.

A backcross is obtained by mating the $F_1$ offspring with mice from either one of the original inbred strains. Note that there are two possible types of backcross, depending on the choice of the inbred strain for the cross, denoted here by the $AA \times Aa$ backcross and the $aa \times Aa$ backcross. (As a matter of fact, one can further divide the backcross breeding scheme based on the sex of the $F_1$ mice in the cross. This may be important if a sex-linked trait or imprinting is considered. However, we ignore these possibilities in the sequel.)

Consider the $aa \times Aa$ backcross design in the context of the simple QTL model described above in (2.2). Assume the mother is inbred and the father is the $F_1$. Then $x_M$ is always equal zero, but $x_F$ may be zero or one. The backcross offspring are either $aa$ homozygous, which corresponds to $x_F = 0$, or $Aa$ heterozygous, which corresponds to $x_F = 1$. The probability of each of the values is $1/2$, with corresponding phenotypic mean levels of $\mu$ and $\mu+\alpha+\delta$. The resulting regression model is given by

$$y = \mu + (\alpha + \delta)x_F + e , \qquad (2.6)$$

with a similar equation for the $AA \times Aa$ backcross. The offspring are either $AA$ homozygous or $Aa$ heterozygous. In this case, the variable $1 - x_F$ has a Binomial(1, 0.5) distribution, and the regression model takes the form:

$$y = (\mu + 2\alpha) + (\delta - \alpha)(1 - x_F) + e . \qquad (2.7)$$

Both models (2.6) and (2.7) have a similar form. However, their statistical properties may be quite different depending on the relations between $\alpha$ and $\delta$. Since a Bernoulli random variable with $p = 1/2$ has variance $p(1 - p) = 1/4$, the variance component associated with the genetic factor is $(\alpha + \delta)^2/4$ for the first model and $(\alpha - \delta)^2/4$ for the second. Consequently, the locus specific heritability is larger for the first model if $\alpha$ and $\delta$ have the same sign, and vice versa if they have opposite signs. For the additive model ($\delta = 0$) the ratios are equal.

The intercross is a result of the mating of an $F_1$ male and an $F_1$ female. In terms of the notation, we will refer to the intercross as the $Aa \times Aa$ cross. The term $F_2$ may also be used. (Subsequent generations of mating are denoted $F_n$, where $n$ is the number of generations since the initial outcross.) The offspring of the $Aa \times Aa$ intercross can have any one of three genotypes. The distribution of the genotypes follows the ratios 1:2:1, thus $x$ has a binomial distribution, $B(2, 1/2)$. It was shown above (Equation (2.5) and the following sentences) that the variance component associated with the given locus is $\alpha^2/2 + \delta^2/4$.

We turn now, with the aid of a small simulation study, to a demonstration of the segregation of the trait in a cross. The simulation will generate segregation of the alleles from parents to offspring and the resulting expression of the phenotype in the offspring. Recall that a parent carries two alleles at a given locus, one inherited from the grandfather and the other inherited from the grandmother. Only one of these two alleles will be passed on to the parent's offspring. According to Mendel's first law of segregation each of the alleles has an equal chance to be passed on. For example, if the parent is $Aa$ at the given locus, then it will pass with equal probabilities either the allele $A$ or the allele $a$. Of course, if the parent's genotype is $AA$ ($aa$) it will pass on the allele $A$ (respectively $a$) with certainty.

Assume the parent is an $F_1$ mouse:

```
> n <- 9
> pat <- rep("A",n)
> mat <- rep("a",n)
> pat
[1] "A" "A" "A" "A" "A" "A" "A" "A" "A"
> mat
[1] "a" "a" "a" "a" "a" "a" "a" "a" "a"
> mode(pat)
[1] "character"
```

The function "rep" produces a vector with n repeats of its first argument. Since the first argument is a character string, the result is a vector of character strings. The content of a regular vector can either be numerical, logical ("TRUE" or "FALSE"), character strings, raw bytes, or complex numbers.

```
> from.mat <- rbinom(n,1,0.5)
> offspring <- pat
> from.mat==1
[1] FALSE FALSE  TRUE  TRUE FALSE  TRUE FALSE FALSE  TRUE
> offspring[from.mat==1]
[1] "A" "A" "A" "A"
> mat[from.mat==1]
[1] "a" "a" "a" "a"
> offspring[from.mat==1] <- mat[from.mat==1]
> offspring
[1] "A" "A" "a" "a" "A" "a" "A" "A" "a"
```

The vector "from.mat" is a numerical vector of zeros and ones while the vector "from.mat==1" is a logical vector. The index of a vector is given within square brackets. A logical vector can be used in order to select a part of a vector – the part associated with a "TRUE" values for the indexing vector. Compare, for example the indexing vector produced by the third expression above with the indexed vectors given in the subsequent expressions. Finally, note that the assignment in the sixth expression produces a vector of character strings for the segregated alleles according to the randomization produced in the first expression. An alternative approach for selecting a part of a vector is by using a vector of integers for indexing. Using a minus sign will produce the complementary vector. Examine the expressions below. (Recall that the binary operation "a:b" produces the vector of integers between a and b. See the the description of the function "seq".)

```
> 2:6
[1] 2 3 4 5 6
> offspring[2:6]
[1] "A" "a" "a" "A" "a"
> offspring[-(2:6)]
[1] "A" "A" "A" "a"
```

Return to the simulation. We will need to repeat the process of segregation of alleles from a parent to its offspring several times. It is convenient to have a function that conducts this task, instead of writing the same lines of code over and over again. Below we create the appropriate function and store it in an object called "meiosis". Observe the format of a function. It starts with the reserved word "function", followed by a list of its arguments enclosed in parentheses. Next comes the expression that the function applies to the arguments. Instead, one can put between the curly brackets a sequence of expressions to manipulate the arguments. The output of the function is the value of the expression, or it may also be set by the function "return". In our example, if the arguments of the function are the two alleles of the given parent, then the output is the random allele segregated to the offspring. The function "length" determines the length of a vector.

```
> meiosis <- function(GF,GM)
+ {
+     from.GM <- rbinom(length(GF),1,0.5)
+     GS <- GF
+     GS[from.GM==1] <- GM[from.GM==1]
+     return(GS)
+ }
> meiosis(pat,mat)
[1] "a" "a" "a" "A" "a" "a" "a" "a" "a" "A"
```

Another special type of vector is called "list". Regular vectors must have all their components of the same type (numerical, logical, character, bites, or complex). A list, on the other hand, may have any type of object as its component.

```
> n <- 10^5
> model <- list(mu=5,alpha=1,delta=-1,sigma=0.5,allele="A")
> pheno1 <- rnorm(n,model$mu,model$sigma)
> pheno2 <- rnorm(n,model$mu + 2*model$alpha,model$sigma)
> a <- rep("a",n)
> A <- rep("A",n)
> IB1 <- list(pat=a,mat=a,pheno=pheno1)
> IB2 <- list(pat=A,mat=A,pheno=pheno2)
```

The list "model" contains the parametrs of the genetic model. Names are assigned to the components of the vector. One alternative for referring to a component of a vector is by its name: "vector.name$component.name". (Or, one may use the format: "vector.name["component.name"]".) The vectors "pheno1" and "pheno2" store the generated phenotypes of the two inbred lines. Finally, we store the genotype and phenotype information of the two inbred lines as lists, titled "IB1" and "IB2", respectively.

The function "cross" applies the function "meiosis" in order to simulate a cross between two mice. The first two input arguments are lists, "fa" and "mo", which contain the genetic information of the two parents. The third argument is a list with the details of the genetic model. (The format "argument=argument.value" may be used in order to assign a default value to the argument. The default value is used unless another value is specifically assigned.) The output of the function is a list with the genetic and phenotypic information of the offspring. Note that the offspring's "pat" genotype is an allele from the father's genotype and the offspring's "mat" genotype is an allele from the mother's genotype. The object x is a vector of integers (0, 1, or 2), m is the vector of the offspring's mean phenotype, and y is the vector of expressed phenotypes.

```
> cross <- function(fa,mo,model)
+ {
+     pat <- meiosis(fa$pat,fa$mat)
+     mat <- meiosis(mo$pat,mo$mat)
+     x <- (pat==model$allele)+(mat==model$allele)
+     m <- model$mu + x*model$alpha + (x==1)*model$delta
+     y <- m + model$sigma*rnorm(length(x))
+     return(list(pat=pat,mat=mat,pheno=y))
+ }
```

Note that the function "cross" may use the object "meiosis" even though this object is not implicitly passed as one of the arguments. In general, any existing object may be used inside a function. This is a useful property, but may cause unexpected side effects if applied carelessly.

We apply the function "cross" in order to create the $F_1$, intercross, and the two types of backcross:

```
> F1 <- cross(IB1,IB2,model)
> BC1 <- cross(IB1,F1,model)
> BC2 <- cross(IB2,F1,model)
> F2 <- cross(F1,F1,model)
```

Since we would like to make several plots of the same format, we create the function "plot.cross" in order to facilitate plotting. The input arguments are the name of the cross, which appears in the title of the plot, and a cross list. The output of the function is null. The side effect is the appropriate plot. The expression that produces the plot is very similar to those that were used in the previous plots.

```
> plot.cross <- function(cross.name,cross)
+ {
+       plot(density(cross$pheno),main=paste(cross.name,
+           ": mean = ",round(mean(cross$pheno),2),
+           ", sd = ",round(sd(cross$pheno),2),sep=""))
+ }
```

The following expressions produce the plots in Fig. 2.2.

```
> op <- par(mfrow=c(2,3))
> plot.cross("IB1",IB1)
> plot.cross("IB2",IB2)
> plot.cross("F1",F1)
> plot.cross("BC1",BC1)
> plot.cross("BC2",BC2)
> plot.cross("F2",F2)
> par(op)
```

The function "par" is used in order to set the parameters for high level plotting. We save the current setting of plotting in the object "op" and set the plotting region to contain six plots in two rows and three columns. After plotting, the default setting is restored.

Examine the six plots in Fig. 2.2. The first two plots describe the distribution of the phenotype in the two pure inbred strains. Note that the distribution follows the bell shape of the normal distribution. The two distributions look the same, but they have different means. The distribution of the phenotype among the $F_1$ mice is identical to the distribution among the inbred strain with the low expression, since the genetic effect is recessive. The picture is the same for the BC1 $= aa \times Aa$ backcross. All animals in this cross have at most a single $A$ allele. Genetic variability emerges in the last two plots. Observe that the standard deviation of the phenotype increases from 0.5 to 1 in the case of the intercross and to 1.12 in the case of the backcross. Note that the distribution is no longer normal, but rather a mixture of normals.

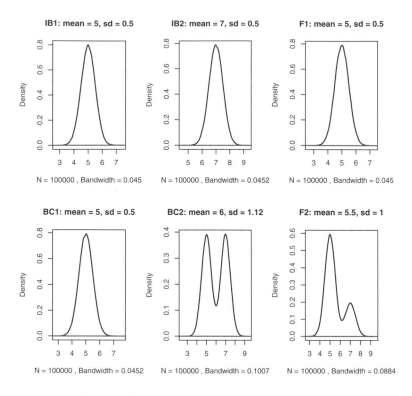

**Fig. 2.2.** Segregation of the phenotype in crosses

The mixture frequencies are $(1/2, 1/2)$ for the backcross and are $(3/4, 1/4)$ for the intercross.

The investigation of the backcross and the intercross will proceed throughout most of the part of the book that deals with experimental genetics. Other crossing designs will be considered occasionally. In particular, we will deal with *recombinant inbred strains*, which are important resources in genetic mapping. These inbred strains are created by the formation in parallel of several $F_2$ mating pairs, followed by several generations of inbreeding within each pair. The result is a set of inbred strains, all originating from the initial cross of the two inbred strains. In Chap. 3 we will explain in more detail the genetic population dynamics of the formation of inbred strains and their genetic properties.

## 2.3 Molecular Genetic Markers

The emergence of modern experimental genetics goes hand in hand with the discovery and the development of the technology for genotyping molecular genetic markers. These markers are specific variations in the sequence of the

DNA molecule. The ability to measure these variations enables the researcher to trace the segregation of segments of genetic material from one generation to the next. The information gathered this way, together with the phenotypic data, is the input used for the statistical analysis. We consider here the two most important types of variations in the context of gene mapping: the *Single Nucleotide Polymorphism* (SNP) markers and the *Simple Sequence Repeat* (SSR) markers.

### 2.3.1 The SNP Markers

The SNP, as the name suggests, involves variation in the nucleotide at a specific locus. While some in a population of a given species may have, for example, the base *A* at a given locus on a particular strand of DNA, others may have *T* at the very same location in the DNA sequence. Although, in principle, SNPs may have four distinct alleles, in practice they are typically bi-allelic. Another, less frequent type of a SNP, is a deletion, i.e., the absence of a given base pair.

The SNPs are the most abundant form of variation found in the genomes of mammals. More than a million such variations have been mapped in the last few years in the human genome. A similar effort, currently under way for the mouse genome and for other genomes, is sure to produce similar numbers for use in experimental genetics in the near future.

Current technologies for genotyping involve multiplying the region of the variation using the *Polymerase Chain Reaction* (PCR). Via this simple but miraculous laboratory process, a tube that originally contained a relatively small number of very long and complex genomic DNA molecules, ends up containing a huge number of copies of a small segment of the molecule about the point of variation. Thus, instead of having to determine the color of a needle in a stack of hay, the problem becomes that of determining the color of the needle in a stack of identical needles.

Modern technologies involve various methods of partial sequencing of the polymorphic locus, different approaches of tagging the locus, and signaling to an electro-optic or other type of a detection device. New technologies emerge almost daily, pushing down the price and increasing the rate of the genotyping of SNPs.

### 2.3.2 The SSR Markers

Undoubtedly, SNPs will have a significant role in the future of experimental genetics, including the mouse model. However, until that day comes, SSR markers are still the most important form of variation for mapping in experimental genetics (and in family-based linkage analysis in humans).

An SSR – more commonly known under the nickname "microsatellite" – is a genomic element that consists of a mono-, bi-, tri-, or tetrameric sequence repeated (hence the name) a number of times that varies from one individual to another. These elements are very polymorphic. They also tend to mutate relatively rapidly. However, this is not a major concern in experimental genetics and in linkage analysis, which typically involve tracing the segregation process only a limited number of generations.

The alleles of an SSR marker are determined by the length of the SSR element. Thus, applying PCR to segments containing the SSR element will yield products that vary in length according to number of tandem repeats in the alleles. Separating these products by length, either by the use of gel electrophoresis, or by the use of more sophisticated capillary-based systems, gives a direct read of the different alleles. Tens of thousands of such SSR markers have been identified and mapped. Commercial kits for their genotyping make this tool handy for even the most modest genetic laboratories.

## 2.4 Bibliographical Comments

The regression model of this chapter originated in the pioneering paper of Fisher [30]. A systematic development of a general version of this model is given by Kempthorne [44]. Clearly written expositions appear in Falconer and Mackay [27], Crow and Kimura [16], Lander and Botstein [47], and Lynch and Walsh [51], among others.

## Problems

For the following problems, assume that $e$ has a standard normal distribution.

**2.1.** Simulate the distribution of $y$ for a backcross design. Do the same for an intercross design. Consider various levels of $\alpha$ and $\delta$.

**2.2.** Consider the two major genes additive model:

$$y = \mu + \alpha_1 x_1 + \alpha_2 x_2 + e , \tag{2.8}$$

where $x_i$ denote the number of $A_i$ alleles at locus $i$ ($i = 1, 2$). Assume the genes corresponding to $x_1$ and $x_2$ lie on two different chromosomes, so by Mendel's laws $x_1$ and $x_2$ are independent.

(a) Investigate the distribution of the phenotype for various values of the model parameters (including the probabilities $p_i$ of the allele $A_i$).
(b) Assume that the indicated genes are the only genes contributing to the trait, so $e$ can be regarded as the environmental effect $E$. Find expressions for $\sigma_G^2$ and $\sigma_y^2$. (It will be helpful to rewrite the model as in (2.3).)
(c) Assume in addition that the parental strains are inbred. How would you estimate $H^2$ if you know the phenotypes of samples from both parental strains and from the intercross?

(d) Extend the model to include $k$ independent genes, $k \geq 2$. Assume that $\alpha_i = \alpha$, for all $i$ and that the parental strains are inbred. Can you figure out a way to estimate $k$ and $\alpha$ from phenotypic data involving both parental strains and the intercross progeny?

**2.3.** Show that (2.3) follows from (2.2). Verify that the two terms involving $x_M$ and $x_F$ on the right-hand side of (2.3) are uncorrelated. Hence verify (2.4) and (2.5). Hint: To facilitate algebraic manipulations, it may be helpful to observe that $|x_M - x_F| = x_M + x_F - 2\,x_M x_F$.

**2.4.** The model (2.3) is customarily written in the somewhat different form

$$y = m + \tilde{\alpha}(x - 2\,p) + \tilde{\delta}[I_{\{x=1\}} - 2\,p(1-p) - (1-2\,p)(x - 2\,p)] + e \; ,$$

where $\tilde{\alpha} = \alpha + (1 - 2\,p)\delta$ and $\tilde{\delta} = -2\,\delta$. Show that this is the same as (2.3). The form given in (2.3) seems slightly easier to manipulate computationally and illustrates that what geneticists call "dominance" is exactly what statisticians call "interaction," in this case the interaction of the allele inherited from the mother with that inherited from the father.

**2.5.** Generalize the model of Prob. 2.2 to include interaction between loci, as follows: Starting from $y = \mu + \alpha_1 x_1 + \alpha_2 x_2 + \gamma x_1 x_2 + e$, where as above the two loci are assumed to lie on different chromosomes, re-write the model in the form:

$$y = m + \tilde{\alpha}_1(x_1 - 2\,p_1) + \tilde{\alpha}_2(x_2 - 2\,p_2) + \gamma(x_1 - 2\,p_1)(x_2 - 2\,p_2) + e \; .$$

What are $\tilde{\alpha}_1$ and $\tilde{\alpha}_2$? Find an expression for $\sigma_y^2$. The term $\gamma^2 p_1(1-p_1)p_2(1-p_2)$ is called the interaction variance, or more precisely the additive-additive interaction variance to distinguish this form of interaction from other possibilities.

# 3

# Fundamentals of Genetics: Inbreeding, Recombination, Random Mating, and Identity by Descent

This chapter introduces some concepts of population genetics that are used in various places later in the book. In addition it allows us to expand our knowledge of R functions and to practice conditional probability arguments, which are used here to analyze the genetic makeup of the present generation by conditioning on that of the parental generation. The reader who has a working knowledge of R may wish to skip Sect. 3.1 and the latter part of Sect. 3.2 that refers to Sect. 3.1. The first part of Sect. 3.2 (including the programs "meiosis.rec" and "cross.rec") introduces the concept of recombination, which plays a very important role throughout the book. It can be read independently of Sect. 3.1. Sections 3.3 and 3.4 are important for the third part of the book that deals with human genetics. This chapter contains some relatively difficult mathematical material, which is marked by an asterisk (*) and can be omitted.

## 3.1 Inbreeding

Inbreeding plays two important, somewhat distinct roles in experimental genetics: first as the starting place for the breeding experiments leading to back crosses and intercrosses, and second as a resource for genetic mapping through recombinant inbreds.

An inbred strain can be created by repeated inbreeding of a small population. The most extreme examples are self-fertilizing plants such as Arabidopsis that can reproduce by themselves. A population of size one can be maintained by selfing for many generations. A heterozygous locus, say $Aa$, will eventually become either $AA$ or $aa$ and remain the same (fixed) thereafter. In mammals selfing is not an option and the population size must be at least of size two – a male and a female. However, by repeated brother–sister mating it is in principle possible to maintain a population of this minimal size. Again, each locus will eventually reach homozygosity throughout the population and will remain in that state thereafter. In much of the rest of this chapter we will assume that inbreeding is taking place in a population of size two. In practice, the stock size during inbreeding is kept larger in order to reduce the chance of the stock becoming extinct. The study of inbreeding in a larger population will be left as an exercise.

*Recombinant inbred* strains (RI) are a special type of inbred strains. They are constructed by an outcross of two established inbred parental strains. From the second generation of $F_2$ animals, which are not genetically identical, a set of pairs is selected. Each selected pair is used in order to create a new inbred strain via repeated brother–sister mating. Eventually, a set of inbred strains is created, which is called a *recombinant inbred set*. Each strain in a given set is genetically homogeneous within, but is genetically different from the other strains in the set and from the original parental strains. Thus, at each locus for which the original strains differ, the new recombinant inbred strain is homozygous for one of the two alleles. However, the parental source of the allele may vary from one strain to the next within the recombinant inbred set, and at different loci the source of the allele may be different parental strains. For example, if the original parental strains are $A/B$ and $a/b$ homozygous at two loci (i.e., one strain is $AA$ at one locus and $BB$ at another, while the other strain is $aa$ at the first locus and $bb$ at the other), recombinant inbreds can be $A/B$, $A/b$, $a/B$, or $a/b$ homozygous at those two loci. There are $2^3 = 8$ possibilities with three loci, $2^4 = 16$ with four, etc.

The name of the RI set is formed by combining the names of the two parental inbred strains. For example, if the $F_1$ mice are created by mating an A female and a C57BL/6J (B6) male, then the resulting recombinant inbred set is denoted AXB. If, on the other hand, the female is a B6 and the male is A then the set is called BXA.

Recombinant inbred sets can be used for mapping traits. At each locus the allele is equally likely to have come from each of the parental strains. In the notation of the previous chapter, the genotype is either $AA$ or $aa$ with probability $1/2$ each. The representation (2.3) can be re-written (Prob. 3.1) in the simpler form $y = m + \alpha(x - 1) + e$, where $x$ is 0 or 2 with probability $1/2$. Hence the genetic variance component for a recombinant inbred set at the given locus is $\alpha^2$. As we will see in later chapters, mapping is conducted by correlating expressed phenotype with genotypes. A major advantage of recombinant inbred sets, and an important motivation for their establishment, is that genotypes are fixed for the entire set. Consequently, the experiment for mapping a trait with recombinant inbreds is conducted by phenotyping mice from the different strains that form the set. Genotyping is unnecessary for the commercially available sets, since their genotypes have already been determined. In addition, since mice from the same strain are genetically identical, one can average the phenotypes of a number of different mice from the same strain, which has the effect of decreasing the environmental variance relative to the genetic variance.

In order to understand the dynamics of inbreeding, in the following sub-section we will investigate the process of forming a recombinant inbred strain and expand the discussion of the segregation of the phenotype to include the case of a recombinant inbred strain. In the next subsection, we take a closer look at a mathematical analysis of inbreeding.

### 3.1.1 Segregation of the Phenotype in a Recombinant Inbred Strain

We begin with a small simulation of brother–sister inbreeding and segregation of the trait after inbreeding. The simulation is built using the functions and the objects that were created in Sect. 2.2 to simulate the segregation of the phenotype in a backcross and in an intercross. The only difference is that we replace the recessive model used there ($\delta = -\alpha$) with an additive model ($\delta = 0$):

```
> N <- 10^5
> model <- list(mu=5,alpha=1,delta=0,sigma=0.5,allele="A")
> pheno1 <- rnorm(N,model$mu,model$sigma)
> pheno2 <- rnorm(N,model$mu + 2*model$alpha,model$sigma)
> a <- rep("a",N)
> A <- rep("A",N)
> IB1 <- list(pat=a,mat=a,pheno=pheno1)
> IB2 <- list(pat=A,mat=A,pheno=pheno2)
> F1 <- cross(IB1,IB2,model)
> BC1 <- cross(IB1,F1,model)
> BC2 <- cross(IB2,F1,model)
> F2 <- cross(F1,F1,model)
```

Next we create a set of inbred strains. Two new elements are added: the "for" loop and the "c" function. The "for" loop is one of several flow control functions that may be used for standard programming. Its format is "for (var in seq) expression". It allows us to repeat brother–sister mating for (in this case) 20 generations. The function "c" concatenates its argument to form a single vector. In our case, the vector of the paternal allele for the last male offspring and the vector of the paternal allele for the last female offspring are concatenated to form a single vector for the paternal allele of the recombinant inbred strain. Likewise, a single vector for the maternal allele of the recombinant inbred strain is formed.

```
> Fn.fa <- cross(F1,F1,model)
> Fn.mo <- cross(F1,F1,model)
> for (g in 2:20)
+ {
+     New.fa <- cross(Fn.fa,Fn.mo,model)
+     New.mo <- cross(Fn.fa,Fn.mo,model)
```

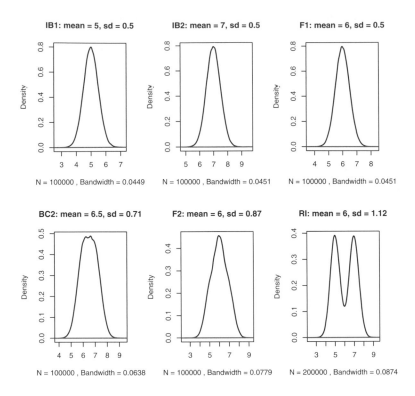

**Fig. 3.1.** Segregation of the phenotype after inbreeding

```
+       Fn.fa <- New.fa; Fn.mo <- New.mo
+ }
> RI <- list(pat=c(New.fa$pat,New.mo$pat),
+          mat=c(New.fa$mat,New.mo$mat),
+          pheno=c(New.fa$pheno,New.mo$pheno))
> op <- par(mfrow=c(2,3))
> plot.cross("IB1",IB1)
> plot.cross("IB2",IB2)
> plot.cross("F1",F1)
> plot.cross("BC2",BC2)
> plot.cross("F2",F2)
> plot.cross("RI",RI)
> par(op)
```

Consider the distributions in Fig. 3.1. It can be seen that the distribution of the recombinant inbred strain has the largest variability. Since the residual variance was equal in all three cases, this increased variability, reflected in the very apparent bimodality, suggests that recombinant inbred strains may be efficient resources for gene mapping. Observe also that the two distinct modes of about the same height are in agreement with the distribution of the backcross under the recessive model (Fig. 2.2). This density is consistent with the assumed model of equally likely fixation in one of the two homozygous states. However, regarding the actual extinction of heterozygosity, one may ask: "Are we sure to have complete fixation after 20 generations of inbreeding?" A closer look at the data reveals an answer. Counting the number of heterozygous animals in the simulated example gives 2,045 out of the total of 200,000, or about 1%. In other words, the probability of fixation by the 20th generation of brother–sister mating is in the vicinity of 99%.

```
> x <- (RI$pat=="A") + (RI$mat=="A")
> table(x)
x
    0    1    2
99771 2045 98184
> table(x)/sum(table(x))
x
       0        1        2
0.498855 0.010225 0.490920
```

In order to produce the counts we used the function "table". This function creates a table of the levels of frequency counts for object of the type "factor", a notion that will be explained at a later stage. In our case, however, x may obtain the levels 0, 1, and 2 and "table" counts the observed frequency of each level. The function "sum" forms a summation of the components of a numerical vector. When the output of "table" serves as the input of "sum", the result is the total frequency. Consequently, the last expression in the above produces the relative frequencies of the levels of x. (As a matter of fact, the output of "table" is not a numerical vector, but rather a slightly more complex object. Nonetheless, when assigned to a function that expects numerical vectors as input, then the more complex object is treated as such.)

*Remark 3.1.* It is interesting to compare the backcross and intercross panels in Fig. 3.1 with those obtained under a recessive model in Fig. 2.2. There is still evidence for two modes in the BC panel of Fig. 3.1, but the two underlying distributions are much less clearly separated, because the genetic contribution to the overall variance is much smaller. Also the IC panel looks roughly like a single normal distribution, although it is actually a mixture of three substantially overlapping normal distributions, since now each of the three possible genotypes has its own expected phenotype whereas for the recessive model in Fig. 2.2 there are only two distinct expected phenotypes.

*Remark 3.2.* An inbred strain is in principle fixed at *all* loci. The preceding simulation showed that with 20 generations of brother–sister mating the probability of fixation at any particular locus is about 0.99, but this means that about 1% of loci may not be fixed. In practice the percentage of heterozygous loci may be even larger, since at some loci heterozygotes may be more viable and reproduce more successfully than homozygotes. Although our discussion of statistical methods to study recombinant inbreds will assume that there is fixation at all loci, the methods must in practice be slightly modified to allow for the possibility that this is not the case for some loci.

### 3.1.2 Dynamic of Inbreeding at a Single Locus

As we saw above, by the 20th generation of inbreeding there is a very large probability of fixation. Our next goal is to understand the dynamics of that probability, as we move from one generation to the next. We start by repeating the previous experiment. This time, however, we save the proportion of heterozygotes in each generation:

```
> Fn.fa <- cross(F1,F1,model)
> Fn.mo <- cross(F1,F1,model)
> hetero.average <- mean(c(Fn.fa$pat != Fn.fa$mat,
+                          Fn.mo$pat != Fn.mo$mat))
> for (g in 2:20)
+ {
+     New.fa <- cross(Fn.fa,Fn.mo,model)
+     New.mo <- cross(Fn.fa,Fn.mo,model)
+     Fn.fa <- New.fa; Fn.mo <- New.mo
+     hetero.average <- c(hetero.average,
+         mean(c(Fn.fa$pat != Fn.fa$mat,
+         Fn.mo$pat != Fn.mo$mat)))
+ }
> plot(hetero.average,type="l",xlab="generations")
```

Note that the expression "Fn.fa$pat != Fn.fa$mat" produces a logical vector ("TRUE" if the animal is heterozygous and "FALSE" otherwise). The same holds true for the other expression, which is concatenated by the function "c" to form a single logical vector. The function "mean" expects a numerical vector as an input and computes its average value. Before computing the mean, the logical vector is converted into a numerical vector by setting "TRUE" equal to one and "FALSE" equal to zero. The result is the relative frequency of "TRUE"s, i.e., the relative frequency of heterozygous mice.

A plot is produced by the "plot" function, which contains only the *solid black line* in Fig. 3.2. (The other lines and the legend are added to the plot by subsequent R commands given below.) Observe that the probability of heterozygosity decays exponentially as a function of the number of brother–sister mating generations. In our case, the input to "plot" is a numerical vector, which serves as the $y$ values. Unless specified otherwise, the vector's indices are used as $x$ values. The default plot is a scatter plot of points. Selecting the argument "type="l"" connects consecutive points with lines.

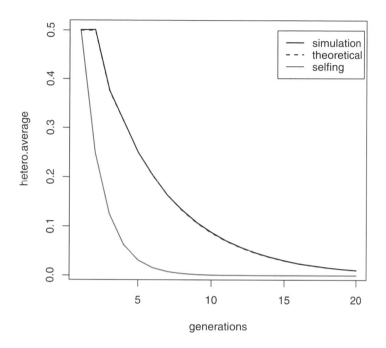

**Fig. 3.2.** Reduction of heterozygosity by inbreeding

The control over the title for the x-axis is gained by assigning as a value to the argument "xlab" and appropriate character string.

We now turn to a mathematical analysis of inbreeding. Since brother–sister inbreeding is rather complicated, we start with the simpler case of inbreeding in a self-fertilizing plant. The principal ingredients of the argument are very important in population genetics and will reappear later in this text. However, our later applications are only a few, so it is not necessary for the reader to follow all mathematical details.

Suppose, then, that we consider a self-fertilizing plant and a locus at which that plant is heterozygous. We are interested in the probability, $H_g$, that it is still heterozygous after $g$ generations of self-fertilization. We have assumed that $H_0 = 1$. The key to our analysis is to write $H_g$ in terms of $H_{g-1}$. The equation is $H_g = \frac{1}{2}H_{g-1}$, which follows from the observation that in order for the plant to be heterozygous after $g$ generations it must be heterozygous after $g - 1$ (because we are neglecting the possibility of mutation) and whichever allele at the given locus is in the egg, the other allele must occur in the sperm. This latter event occurs with probability 1/2. By iterating this basic relation, we have $H_g = \frac{1}{2}H_{g-1} = (\frac{1}{2})^2 H_{g-2} = \cdots (\frac{1}{2})^g H_0 = (\frac{1}{2})^g$ since $H_0 = 1$.

We compute the probability of heterozygosity in each of the 20 generations of self-fertilization and add the line to the plot in Fig. 3.2:

```
> H <- 0.5^(1:20)
> lines(H,col=gray(0.5))
```

Although the decay is also exponential, the dynamics of heterozygosity in brother–sister mating are different. Note that H is a numerical vector of length 20. The function "lines" is a low-level plotting function that adds lines to existing plots. The argument "col" sets the line's color, a *gray line* in this case.

One can analyze the probability of heterozygosity in brother–sister mating using the same tools as in the case of self-fertilization. In this case, however, one must consider heterozygosity two generations back as well. It turns out that the recursive formula for heterozygosity is given by the equation:

$$H_g = \frac{1}{2}H_{g-1} + \frac{1}{4}H_{g-2} \,. \tag{3.1}$$

We do not prove this formula here (cf. Prob. 3.2). Instead, we apply an approach that is analogous to the argument given above for selfing. It tracks the genotypes in the population in each generation and uses techniques based on Markov chains for the computation of probabilities. Although requiring more mathematical analysis than (3.1), it appears to be relatively easily adapted to deal with more complex problems.

The population is of size two in each generation. Each mouse in the population may either be in state 0, 1, or 2, depending on its genotype. Denote by 0-0 and 2-2 the states where both mice are homozygous for the same allele, and note that these are *absorbing states*. Once such a state is reached, the process will remain indefinitely in the same state. The other states, denoted here by 1-0, 2-0, 1-1, and 2-1 are *transient*. Fixation occurs when the process reaches an absorbing state. At initiation, all the probability is assigned to the state 1-1. Given the state of the population at a given generation, the distribution over the states in the next generation can be described by a transition probability matrix. The distribution in generation $g$ is given by multiplying the distribution in generation $g - 1$ by the probability transition matrix.

Denote by $Q$ the sub-matrix of transition probabilities between transient states and let $\pi_0'$ be a row vector giving an initial distribution over these states. We are interested in the case $\pi_0' = (0,0,1,0)$, which corresponds to starting in the state 1-1. The distribution over the transient states after $g$ generations is the vector $\pi_0'Q^g$, where $Q^g$ corresponds to the multiplication of the matrix $Q$ by itself $g$ times. The probability of heterozygosity after $g$ generations is $\pi_0'Q^gv$, where $v' = (1/2,0,1,1/2)$, since $1/2$ of the animals in the state 1-0 are heterozygous, none of the animals in 2-0 is heterozygous, etc. In order to make the analysis more concrete let us put the computer to work:

```
> hetero.states <- paste(c(1,2,1,2),"-",c(0,0,1,1),sep="")
```

```
> hetero.states
[1] "1-0" "2-0" "1-1" "2-1"
> Q <- matrix(0,4,4)
> Q
     [,1] [,2] [,3] [,4]
[1,]    0    0    0    0
[2,]    0    0    0    0
[3,]    0    0    0    0
[4,]    0    0    0    0
> rownames(Q) <- colnames(Q) <- hetero.states
> # Transition probabilities
> Q["1-0",c("1-0","1-1")] <- c(0.5,0.25)
> Q["2-0","1-1"] <- 1
> Q["1-1",] <- c(0.25,0.125,0.25,0.25)
> Q["2-1",c("1-1","2-1")] <- c(0.25,0.5)
> Q
      1-0   2-0   1-1  2-1
1-0 0.50 0.000 0.25 0.00
2-0 0.00 0.000 1.00 0.00
1-1 0.25 0.125 0.25 0.25
2-1 0.00 0.000 0.25 0.50
```

Let us explain: Q is an object of the class "matrix". It was created with the aid of the function "matrix". Elements of the matrix are identified by the double indexing format "[,]", left for rows and right for columns. Sub-matrices can be assigned by appropriate selection of indices, with rules that follow similar patterns as the rules used for indexing vectors. As in vectors, names can be used for indexing. The first row of the matrix Q corresponds to the probabilities of transition from the state 1-0 to any of the four transient states. For example, if one parent has the genotype 0 and the other the genotype 1, then there is a 50% chance that one of the offspring will have genotype 0 and the other genotype 1. There is a 25% chance that both will have the genotype 1 and a 25% that both will have the genotype 0. The last case produces an absorbing state. This computation gives rise to the first row of the matrix Q. Similar computations will lead to the other rows of the matrix.

Note that there is a line that starts with the symbol "#". R ignores any text that appears on the same line after that symbol. This allows the addition of informative remarks whenever they are needed.

```
> initial <- c(0,0,1,0)
> het.count <- c(0.5,0,1,0.5)
> hetero.prob <- NULL
> QQ <- Q
> for (g in 1:20)
+ {
+       hetero.prob[g] <- initial %*% QQ %*% het.count
```

```
+      QQ <- QQ %*% Q
+ }
> lines(hetero.prob,lty=2)
> legend(14.5,0.5,
+      legend=c("simulation","theoretical","selfing"),
+      lty=c(1,2,1),col=gray(c(0,0,0.5)))
```

The vector "initial" gives the starting state; it was denoted by $\pi_0$ above. The vector "het.count", denoted above by $v$, is used in order to compute the probability of heterozygosity from the distribution of transient states. The vector "hetero.prob" stores the computed probabilities. Initially this vector is assigned a null value. This creates an object that can be identified and manipulated by R, but contains no information. Information is accumulated in each iteration of the "for" loop. Note that the final length of "hetero.prob" need not be preassigned. The system automatically expands the object upon request. Finally, observe that the binary operation "%*%" corresponds to matrix multiplication, unlike the operation "*", which is applied term-by-term. (Try both "Q * Q" and "Q %*% Q" to see the difference between the two operations.)

The results of the theoretical computation are added to the plot in Fig 3.2 with the aid of the function "lines". Note the good agreement between the simulation and the theoretical computation. For clarity, a legend is added with the low-level plotting function "legend". The first two arguments of the function are used in order to determine the location of the upper-left corner of the legend box in Fig 3.2; and the other arguments set the text, the type, and the color of the lines to appear inside the box. The line type must be provided. In the current case the first and the third lines are of the same type: "lty = 1", which corresponds to a solid line. The second line is of type 2, which correspond to a broken line. Observe that since there are three lines in the legend, the argument should be a vector of length three. If all lines were of the same type, then one could create a vector with repeated components by using the function "rep". The first argument to that function is the object to be repeated. If the second argument is an integer, then the output is a vector which is formed by repeating the object the given number of times. The function "rep" is very handy in order to create sequences, in particular when it is used in combination with the function "seq".

## 3.2 The Recombination Fraction

*Crossovers* between chromosomal segments may occur during meiosis. These crossovers cause the gamete that passes to an offspring to be a mosaic of genetic material originating from the two grandparents. A *recombination event* between two loci occurs whenever the number of crossovers is odd. In this case, the genetic material at one locus is from one grandparent, whereas the genetic material at the other locus is from the other grandparent. The probability of a recombination is known as the recombination fraction and is denoted by $\theta$. The value of $\theta$ depends on the relative position of the two loci. The closer they are on a chromosome, the closer the recombination frequency is to zero. At the other extreme, for loci on different chromosomes, the recombination frequency is $1/2$, which is equivalent to independent assortment of the chromosomes. It may be shown under some very general assumptions that $\theta$ can never be greater than $1/2$.

Consider two loci that are polymorphic with respect to the two parental pure inbred strains. Denote by $a$ the allele of one strain at one locus, and by $b$ the allele of the same strain at the other locus. Similarly, for the other strain, denote by $A$ and $B$ the alleles at the two loci. Within each inbred strain the loci are not polymorphic. In short, we can say that the first strain is $a/b$ homozygote and the second strain is $A/B$ homozygote. Regardless of what crossovers may occur during the formation of the outcross, the gamete that passes on to the $F_1$ mouse from the first strain is always $a/b$, and the gamete that passes on from the second strain is always $A/B$. Consequently, the $F_1$ mouse must be $(A/B, a/b)$ heterozygote.

The gamete that is passed on from an $F_1$ parent to its offspring can be any one of the four possible types. If a recombination does not occur during meiosis, then the gamete may be either $A/B$ or $a/b$ with equal probability. If a recombination does occur, then the gamete may be $A/b$ or $a/B$, again with equal probabilities. We call gametes of the second kind *recombinant* gametes. The probability of recombinant gametes in a given cross is an essential parameter for the assessment of the cross as a resource for QTL mapping. The aim in this section is to evaluate this parameter for the backcross, intercross, and a recombinant inbred strain.

The evaluation of the fraction of recombinants for the backcross and for the intercross is straightforward. For the backcross one of the gametes is inherited from an inbred parent and cannot be recombinant. The other is inherited from the $F_1$ and is recombinant with probability $\theta$. Consequently, the probability that a random gamete taken from a backcross population is recombinant equals $\theta/2$. For the intercross both gametes come from an $F_1$ parent. The probability of random gamete to be recombinant is exactly $\theta$. The situation for the recombinant inbred is more complex. Before attempting an exact mathematical analysis let us examine the fraction of recombinants by simulation.

The functions "meiosis" and "cross" can be used in order to simulate a recombinant inbred strain. However, in the original programs we considered only a single locus. Modifications are needed in order to track two loci instead of one. Below is an edited version of the original function. In the original version of the function "meiosis" the objects "GF" and "GM" were vectors. In the new version these two objects are matrices with a column dimension of two. The first column represents the QTL, and the second represents another locus, which may be linked to the QTL. The recombination fraction between the two loci is added as a new argument to the function:

```
> meiosis.rec <- function(GF,GM,rec.frac)
+ {
+       N <- nrow(GF)
+       GS <- GF
+       from.GM <- rbinom(N,1,0.5)
+       GS[from.GM==1,1] <- GM[from.GM==1,1]
+       rec <- rbinom(N,1,rec.frac)
+       from.GM <- from.GM*(1-rec) + (1-from.GM)*rec
+       GS[from.GM==1,2] <- GM[from.GM==1,2]
+       return(GS)
+ }
```

In the code for the new function "meiosis.rec", the function "nrow" returns the number of rows in a matrix. The segregation of the first locus uses exactly the same algorithm as before. The segregation of a maternal allele at the second locus depends on the segregation at the first locus and on the recombination process. A maternal segregation occurs at the second locus if such segregation occurs in the first locus and if there is no recombination, or if there is a paternal segregation at the first locus and a recombination does occur.

The function cross.rec is practically identical to the function "cross". The only modification, which is really needed only for safety reasons, is that the recombination fraction is explicitly passed to the function "meiosis.rec".

```
> cross.rec <- function(fa,mo,QTL.model,rec.frac)
+ {
+     pat <- meiosis.rec(fa$pat,fa$mat,rec.frac)
+     mat <- meiosis.rec(mo$pat,mo$mat,rec.frac)
+     x <- (pat[,1]==QTL.model$allele) +
+              (mat[,1]==QTL.model$allele)
+     m <- QTL.model$mu + x*QTL.model$alpha +
+              (x==1)*QTL.model$delta
+     y <- m + QTL.model$sigma*rnorm(length(x))
+     return(list(pat=pat,mat=mat,pheno=y))
+ }
```

For convenience we write a new function "rec.count" that returns the frequency of recombinant gametes in a cross. This function exploits again the function "table", which returns in this case the cross table of the frequencies of the different combinations of the levels of "loc1" and "loc2". In our case, the output of the "table" function can be treated as a $2 \times 2$ matrix.

```
> rec.count <- function(Fn)
+ {
+     loc1 <- c(Fn$pat[,1],Fn$mat[,1])
```

```
+       loc2 <- c(Fn$pat[,2],Fn$mat[,2])
+       cross.tab <- table(loc1,loc2)
+       theta <- (cross.tab[2,1]+cross.tab[1,2])/sum(cross.tab)
+       return(theta)
+ }
```

Finally, we run the simulation and make the plot:

```
> N <- 10^5
> a <- rep("a",N)
> b <- rep("b",N)
> A <- rep("A",N)
> B <- rep("B",N)
> IB1 <- list(pat=cbind(a,b),mat=cbind(a,b),pheno=pheno1)
> IB2 <- list(pat=cbind(A,B),mat=cbind(A,B),pheno=pheno2)
> F1 <- cross.rec(IB1,IB2,model,0)
> theta.BC <- theta.F2 <- theta.RI = NULL
> rec.frac <- seq(0,0.5,by=0.05)
+       for (g in 2:20)
+       {
+           New.fa <- cross.rec(Fn.fa,Fn.mo,model,theta)
+           New.mo <- cross.rec(Fn.fa,Fn.mo,model,theta)
+           Fn.fa <- New.fa; Fn.mo <- New.mo
+       }
+       RI <- list(pat=rbind(New.fa$pat,New.mo$pat),
+             mat=rbind(New.fa$mat,New.mo$mat),
+             pheno=c(New.fa$pheno,New.mo$pheno))
+       theta.RI <- c(theta.RI, rec.count(RI))
+ }
> plot(rec.frac,theta.RI,type="l",xlab="theta",
+       ylab="rec. fraction")
> lines(rec.frac,theta.F2,lty=2)
> lines(rec.frac,theta.BC,lty=3)
> legend(0,0.5,legend=c("RI","F2","BC"),lty=1:3)
```

Examine the three lines in Fig. 3.3. As expected, the backcross and intercross yield straight lines with slopes $1/2$ and 1, respectively. The fraction of recombinant gametes in the recombinant inbred strain is larger and is not linear. The derivative of the line for recombinant inbred strain at zero is about equal to four. The greater rate of recombination for recombinant inbreds arises from the much larger number of meioses involved.

## * A Mathematical Derivation of the Recombination Fraction for Recombinant Inbreds

The probability in question can be derived by consideration of the different genotypes in the population and examination of a related Markov chain.

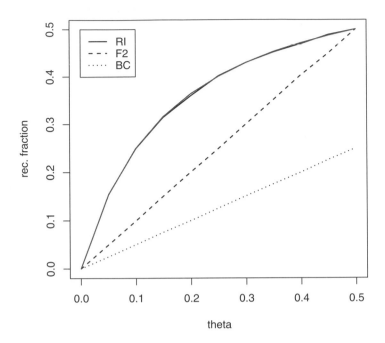

**Fig. 3.3.** The recombination fraction in RI, F2, and BC

Here we outline an alternative approach, which is developed in Kimura [45]. The beauty of this approach is that it allows substantial simplification in the analysis by exploiting the many symmetries that are present.

Consider, in particular, three probabilities, computed for the pair of male and female mice at the $g$th generation of inbreeding:

$C_g =$ The probability that a randomly selected gamete is of the type $A/b$.

$S_g =$ The probability that a randomly selected gamete in a randomly selected mouse carries the allele $A$ and, at the same time, the other gamete of that mouse carries the allele $b$ (at the other locus).

$T_g =$ The probability that a random gametes in a randomly selected mouse carries the allele $A$ and a random gamete in another mouse carries the allele $b$ (again, at the other locus).

Of course, we are interested in $C_g$, which is equal to one-half the probability of a recombinant gamete. However, keeping track of the other two probabilities will allow us to write down the recursive relations:

$$C_g = (1 - \theta)C_{g-1} + \theta S_{g-1} \tag{3.2}$$

$$S_g = T_{g-1} \tag{3.3}$$

$$T_g = 0.5\,T_{g-1} + 0.5 \times (0.5\,C_{g-1} + 0.5\,S_{g-1})\,. \tag{3.4}$$

The first relation follows from the fact that the random gamete is inherited from an animal in the previous generation. On the one hand, if no recombination takes place, then the segregated gamete is $A/b$ if and only if it appears in the parent (and is the one selected). On the other hand, if a recombination does take place, then the gamete $A/b$ occurs if the event described in the definition of $S_g$ holds for the parent. The second relation follows simply by the fact that the two gametes of the offspring are a random sample of the gametes of the parents. Recombination is not relevant in this computation, since we are considering only the marginal frequencies of each of the two different loci in the parents. Consider, last, relation (3.4). The first locus for one offspring and the second locus for the other offspring are either inherited from different parents or from the same parent with equal probabilities. In the former case we get that the probability of the event in question is identical to the same probability for the parents. In the later case, however, there are two possibilities. Either both originate from the same gamete or from the two homologous gametes. Again, each possibility has probability of $1/2$. The conditional probability of the event when they originate from the same gamete is $C_{g-1}$ and the conditional probability when they emerge from homologous gametes is $S_{g-1}$. Note that recombination is again irrelevant, since only marginal probabilities are of concern.

The solution of the recursion (3.2–3.4) requires some extra work. Let $u_g = (C_q, S_q, T_q)'$ and observe the relation

$$u_g = Q u_{g-1}, \tag{3.5}$$

where

$$Q = \begin{pmatrix} 1-\theta & \theta & 0 \\ 0 & 0 & 1 \\ 0.25 & 0.25 & 0.5 \end{pmatrix}.$$

The matrix $Q$ is a transition probability matrix, since its rows sum to 1. Hence it can be regarded as defining a Markov chain. From the theory of Markov chains it is ensured to have a stationary distribution $\pi = (\pi_1, \pi_2, \pi_3)'$ with the property that $\pi' = \pi'Q$. Solving this system of equations together with the condition that $\sum_i \pi_i = 1$ produces:

$$\pi' = (1, 2\theta, 4\theta)/(1 + 6\theta)\,.$$

By (3.5) and stationarity we get $\pi'u_g = \pi'Qu_{g-1} = \pi'u_{g-1} = \cdots = \pi'u_0$, for all $g$. All entries to the vector $u_g$ converge to the same quantity, $C_\infty$, since the population is becoming more and more inbred. Since $C_0 = 0$, $S_0 = 0.5$, and $T_0 = 0.25$ we have $u_0' = (0, 1/2, 1/4)$, and hence

$$C_\infty = \lim_{g \to \infty} \pi'u_g = \pi'u_1 = 2\theta/(1 + 6\theta)\,.$$

Since a recombinant is either $A/b$ or $a/B$, and these are equally likely, we can obtain the exact formula for the fraction of recombination in a recombinant inbred strain: $4\theta/(1+6\theta)$. Observe that the derivative of this fraction at $\theta = 0$ is indeed equal to four. Our work will be completed when we add the exact curve for a recombinant inbred strain to Fig. 3.3:

```
> lines(rec.frac,4*rec.frac/(1+6*rec.frac),lty=gray(0.5))
```

Observe, again with reassurance, the excellent agreement between the simulated and the theoretically computed curves.

## 3.3 Random Mating

This section and the next are important for Chaps. 8–13 dealing with gene mapping in natural populations.

In contrast to the situation in the previous sections, where there was a finite, usually very small, mating population, in this section we assume an infinitely large population. Random mating corresponds to the case where the choice of the mating partner is not influenced by the genotypes under consideration. The backcross and intercross are examples of non-random mating. On the other hand, mating in wild populations in general, and in human genetics in particular, is often modeled by random mating.

Random mating enhances the mixing of the genetic material in the population and drives the genotypes toward steady-state conditions. Two important steady states are *Hardy-Weinberg equilibrium* and *linkage equilibrium*. Hardy-Weinberg equilibrium refers to the joint distribution of the two homologous copies of an autosomal allele. In a steady state these two copies are independent and identically distributed. Linkage equilibrium, on the other hand, refers to the joint distribution at two or more loci. In linkage equilibrium the distribution of alleles at different polymorphic loci are independent of each other.

In order to motivate the idea of Hardy-Weinberg equilibrium consider a single bi-allelic locus with alleles $A$ and $a$. The genotype of a random individual can be $AA$, $Aa$, or $aa$. Under Hardy-Weinberg equilibrium the frequency of these genotypes are $p_A^2$, $2p_A(1 - p_A)$, and $(1 - p_A)^2$, respectively, where $p_A$ denotes the fraction of allele $A$ in the population and $1 - p_A$ the fraction of allele $a$. These frequencies of genotypes will emerge in offspring of random mating. Indeed, the probability that the father will contribute the allele $A$ to his offspring is $p_A$. The probability of inheriting allele $A$ from the mother is also $p_A$. Random mating corresponds to independence of the two contributions, which taken together produce the indicated binomial probabilities. Note also that when the fractions of the genotypes $AA$, $Aa$, and $aa$ are $p_A^2$, $2p_A(1-p_A)$, and $(1 - p_A)^2$, the frequencies of the alleles $A$ and $a$ are, respectively, $[2p_A^2 + 2p_A(1-p_A)]/2 = p_A(p_A+1-p_A) = p_A$ and $1-p_A$. Hence the allelic frequencies of the children's generation are the same as in the parents generation, so the Hardy-Weinberg frequencies apply to the genotypes of the grandchildren, etc. This is in contrast to the finite populations we considered earlier in this chapter, where allele frequencies change from one generation to the next and heterozygosity eventually disappears with repeated random mating.

A statistical test of Hardy-Weinberg equilibrium based on a sample of unrelated individuals may be obtained as an application of a chi-square test to the frequency table of genotypes. Hence, for example, if a bi-allelic locus is considered, then the frequency table is composed of three cells, one for each genotype. The observed cell counts are compared to the expected counts. The latter are obtained using the observed allele frequencies and the assumed independence. The distance between the two tables is measured with a chi-square statistic, with one degree of freedom in this case. The power of this test is poor if the locus is actually multi-allelic. In Prob. 3.8 an alternative test is suggested based on the notion of relatedness introduced in the next section.

Unlike Hardy-Weinberg equilibrium, linkage equilibrium does not occur in one generation of random mating. In order to illustrate convergence to linkage equilibrium consider two bi-allelic loci inherited from the same parent. Denote, as before, the alleles of the first locus by $A$ and $a$ and use $B$ and $b$ to denote the alleles of the second locus. Let $\theta$ be the recombination fraction between the two loci. Let us examine the probability of the *haplotype* $A/B$ in a given generation, namely the relative frequency in that generation of pairs of loci inherited from the same parent with allele $A$ at the first locus and allele $B$ at the second. Call this probability $p_{AB}$. Under linkage equilibrium this probability equals the product of the marginal probabilities $p_A p_B$. In general, one may consider the difference $D = p_{AB} - p_A p_B$ as a measure of linkage disequilibrium in the population. Consider the level of linkage disequilibrium in the next generation following random mating. The haplotype $A/B$ will emerge as a result of one of two possibilities. In the case where recombination does not occur in the parent, the haplotype appears in an offspring if it was inherited from the parent. The probability of inheritance is $p_{AB}$. If recombination in the parent does take place, then the given haplotype will emerge if the allele in the parent at the first locus of one homologous chromosome is $A$ and the allele at the second locus of the other homolog is $B$. The probability of the first event is $p_A$. The probability of the second event is $p_B$. Under random mating, hence Hardy-Weinberg equilibrium, the two loci inherited from the different grandparents are independent. It follows that the probability in the case of recombination is $p_A p_B$. Combining these arguments, we see that the disequilibrium in the next generation is equal to:

$$\tilde{D} = (1-\theta)p_{AB} + \theta p_A p_B - p_A p_B = (1-\theta) \times (p_{AB} - p_A p_B) = (1-\theta)D .$$

By recursion we find that after $g$ generations the level of linkage disequilibrium shrinks to $D_g = (1-\theta)^g \times D_0$. Hence under random mating there is relatively fast convergence to linkage equilibrium; but unlike Hardy-Weinberg equilibrium, linkage equilibrium does not occur in a single generation, even when the two loci are on different chromosomes.

## 3.4 Inbreeding in Infinite Populations and Identity by Descent

In finite populations random mating will occasionally produce mating between relatives. Here we consider mating between relatives in infinite populations. Two individuals are related if they have a common ancestor. Siblings and half-siblings have at least one common parent; first cousins have common grandparents, etc. If two individuals have a common ancestor, then it is possible that at a given locus they have inherited the same allele from that ancestor. Such an allele is said to be inherited identical by descent (IBD). The coefficient of relatedness of two individuals is defined to be the probability that at a given locus a randomly selected allele from one of the individuals is identical by descent with a randomly selected allele at the same locus in the other individual. For example, two siblings, whose parents are unrelated, have two common ancestors, their mother and their father. If we select a random allele from one of the siblings, there is probability $1/2$ it was inherited from their mother and probability $1/2$ it was inherited from their father. In either case, if we select a random allele from the other sibling, there is a $1/2$ chance it was inherited from the same parent, and if so, there is then a $1/2$ chance it is the same allele. Hence the coefficient of relatedness of the siblings is $(1/2) \times (1/2) = 1/4$. If two relatives mate, the coefficient of inbreeding of their offspring is by definition the probability of relatedness of the parents, i.e., the probability that at a given locus the two alleles in that offspring are identical by descent.

The following equations modify Hardy-Weinberg equilibrium to accommodate inbreeding. Assume that in a population that is otherwise in Hardy-Weinberg equilibrium, mating occurs between two relatives having a coefficient of relatedness $F$. Then at a locus with alleles $A$ and $a$, a child can have a genotype $AA$ because (i) it inherits the $A$ allele from one parent and the same allele IBD from the other parent, which occurs with probability $Fp_A$, or (ii) it inherits the allele $A$ independently (not IBD) from both parents, which happens with probability $(1-F)p_A^2$. Adding these two terms together, we find that the probability of the genotype $AA$ is $p_{AA} = p_A^2 + Fp_A(1 - p_A)$. A similar formula holds for the genotype $aa$. For the genotype $Aa$, the alleles cannot be inherited IBD, since they are different, so $p_{Aa} = 2p_A(1 - p_A)(1 - F)$. A consequence of inbreeding is an increase in homozygosity compared to random mating.

## 3.5 Bibliographical Comments

Excellent references for the fundamentals of theoretical population genetics are Crow and Kimura [16] and Ewens [25].

Classical papers that deal with the computation of recombination fractions in recombinant inbred strains are Haldane and Waddington [35] and Wright [90]. In the text we outline an alternative approach, which with minor variations is taken from Kimura [45]. While Kimura's argument is by no means simple, the reader who compares his paper to its predecessors will see how ingenious his argument is.

## Problems

**3.1.** Consider the representation (2.3) with the conditions that $x_M = x_F$ (to achieve homozygosity) and $p = 1/2$. By observing that the factor multiplying $2\,\delta$ is always $1/4$ show that (2.3) reduces to the representation of $y$ for recombinant inbreds given in the text, although the value of $m$ is different (how?).

**3.2.** Demonstrate the generalization of formula (3.1) given below for a randomly mating population of size $N$. Hints: We assume that the population is a constant size of $N$ individuals ($2N$ genes) and that generations are discrete. We also assume that each individual can be male or female; a more elaborate argument shows that this has a negligible effect in a population that is half male and half female. In addition to $H_g$, the probability that two homologous genes in the same individual are different, let $G_g$ denote the probability that at a given locus a randomly selected allele from one individual and that from another individual are different. This immediately implies that $H_g = G_{g-1}$. Now derive a formula for $G_g$ in terms of $H_{g-1}$ and $G_{g-1}$ by considering two mutually exclusive possibilities: (i) the alleles in question have the same parental origin in generation $g-1$ or (ii) they come from two different parents. For case (i) the probability that the alleles originate with the same person is $1/N$, and in order for them to be different, they must be from different chromosomes (probability $1/2$) and the parent must not be homozygous (probability $H_{g-1}$). Put these together with the somewhat easier analysis of case (ii) to obtain the recursion

$$H_g = \left(1 - \frac{1}{N}\right) H_{g-1} + \frac{1}{2N} H_{g-2} \,. \tag{3.6}$$

Modify the program for brother–sister mating and plot the outcome for $N = 50$. What is the most striking difference from the plot given in the text for $N = 2$? Explain why this makes intuitive sense.

**3.3.** Solve (3.6) by assuming that $H_g = H_0 r^g$ for a suitable value of $r$. What is $r$? Show that $r \approx (1 - 1/2N)$ when $N$ is large. For large $N$, roughly how many generations are required for all individuals to become homozygous?

**3.4.** The theory of Markov chains is helpful in thinking about recombination. For example, in a backcross of $Aa$ to $aa$, suppose there are markers on a chromosome with a recombination fraction between consecutive markers of $\theta$. Let $x(s)$ denote the number (0 or 1) of $A_s$ alleles in an offspring at the marker $s$. Consider an adjacent locus $t$, where the alleles are $A_t$ and $a_t$, and let $x(t)$ denote the number of $A_t$ alleles. Let $p_{ij} = \Pr\{x(t) = j | x(s) = i\}$. Then $p_{00} = 1 - \theta$, $p_{01} = \theta$, $p_{10} = \theta$, and $p_{11} = 1 - \theta$ can be thought of as a Markov chain transition matrix for the process that records the number of $A$ alleles at consecutive marker loci along the chromosome. (This point of view will be particularly helpful in Chaps. 5–7.) Find the corresponding $3 \times 3$ matrix for an intercross, where the genotypes are $aa$, $Aa$, and $AA$. The first and third rows are comparatively easy, since, for example, $p_{00}$, the probability that $x(t) = 0$ given that $x(s) = 0$, is the probability that there is no recombination between $s$ and $t$ in the two parental meioses. This equals $(1 - \theta)^2$. The second row is trickier. For example consider $p_{11}$, the conditional probability that $x(t) = 1$ given that $x(s) = 1$. This event occurs if and only if between the markers $s$ and $t$ there are either no recombinations *or* two recombinations in the parental meioses. This has probability $\theta^2 + (1 - \theta)^2$.

**3.5.** What is the coefficient of relatedness of two half-siblings, of an aunt and her niece, of two first cousins, of a grandmother and her grandchild? In a child whose parents are first cousins, what is the probability that the child's alleles are IBD at a given autosomal locus?

**3.6.** Modify the argument given in the text for brother–sister mating to find the recombination for a recombinant inbred line created by repeated self-fertilization. *Hint*: There are two recursions, for $C_g$ and $S_g$, to find and analyze.

**3.7.** Use simulations in order to investigate the decay of linkage disequilibrium in a randomly mating infinite population that is discussed in Sect. 3.3. In principle, one can apply the R code that was used in Sect. 3.2 to investigate the recombination fraction following brother–sister mating. In order to modify the program to accommodate random mating one can randomly reorder the parents in each generation before mating. The function "`sample`", which in its default application produces a random permutation of its input, may by handy in that regard. (Hint: Let "`1:N`" be the row index of a matrix `A`. Then "`A[sample(1:N),]`" is the same matrix with rows permuted randomly.)

**3.8.** This problem is concerned with a test for Hardy-Weinberg that is motivated by the departures from equilibrium suggested by the discussion of inbreeding. Suppose alleles at a given locus $i = 1, \ldots, k$ have frequencies $p_i$. Denote genotypes by *ordered* pairs $(i, j)$, which under Hardy-Weinberg equilibrium have frequency $p_i p_j$. (In reality we do not observe the ordered genotype, but the mathematical notation is simplified if we pretend that we do.) Suppose we have a sample of $n$ genotypes, and $X_{ij}$ is the number of $(i, j)$

genotypes. Consider the inbreeding model that the probability of genotype $(i, j)$ is given by $p_{ii} = p_i^2 + F p_i(1 - p_i)$, while for $i \neq j$, $p_{ij} = p_i p_j(1 - F)$.

(a) Identify the log likelihood function, $\ell(F, p_1, \ldots, p_k)$.

(b) Show that the efficient score $\partial \ell / \partial F$ evaluated at $F = 0$ is given by $\sum_i X_{ii}/p_i - n$.

(c) Show that the variance of the efficient score when $F = 0$ is $n(k - 1)$. What is its expectation when $F > 0$?

(d) We cannot use the efficient score directly as a test statistic because the parameters $p_1, \ldots, p_n$ will usually be unknown. Show that reasonable estimators of the $p_i$ when Hardy-Weinberg equilibrium holds are $\hat{p}_i = (2 n)^{-1}[\sum_j X_{ij} + X_{ii}]$. It may be shown, although the argument is still more complicated, that when Hardy-Weinberg equilibrium holds, the statistic $[\sum_i X_{ii}/\hat{p}_i - n]/(n(k - 1))^{1/2}$ has approximately a standard normal distribution when $n$ is large, and hence can be used as a test of Hardy-Weinberg equilibrium. Suppose $k = 5$. How large a sample size $n$ would be required to have 90% power to detect a value of $F = 0.01$, when the significance level is set to 0.05?

# Part II

# Experimental Genetics

# 4

# Testing for Linkage with a Single Marker

In this chapter we introduce the basic principles behind testing a genetic marker for linkage to a quantitative phenotype. We split the presentation into two parts. In the first part we deal with the case of a marker located at a *quantitative trait locus* (QTL). The parameters that determine the statistical properties of the test are the parameters of the regression model that relates the observed phenotype to the QTL as presented in Chap. 2. Later in this chapter we extend the investigation to the case of a marker in the vicinity of, but not necessarily on top of, the QTL, so the recombination fraction also plays a role. Fortunately the effect of recombination can be separated from the effect of the QTL. From Chaps. 5 onward we discuss the case, which is the common practice today, of testing a set of markers for linkage.

The backcross design serves as our main platform for discussing the general principles. At the end of this chapter, though more briefly, we discuss the intercross design and the use of recombinant inbred sets.

## 4.1 Tests for the Presence of a QTL

Consider the regression model that was presented previously in (2.2):

$$y = \mu + \alpha(x_\text{F} + x_\text{M}) + \delta|x_\text{F} - x_\text{M}| + e .$$

In order to simplify the discussion, we will assume that the genotypic information at the given locus is available. This is the case if the QTL itself, or a marker completely linked to it, is genotyped. In the next section this assumption will be relaxed.

### 4.1.1 Testing a QTL in the Backcross Design

In a typical backcross experiment, aimed at dissecting the genetic component of a given trait, two inbred strains that show a large separation in their phenotypic values are selected. An outcross of these two strains produces the $F_1$

generation, which is then crossed with one of the two original inbred strains in order to produce the backcross. The choice of which of the two original inbred strains to use at this stage depends on the variability of the phenotype of the resulting backcross. Unless constrained by other reasons, one should choose the cross that produces the larger amount of phenotypic variability $\sigma_y^2$. This larger phenotypic variability usually means larger heritability $H^2$, which in turn suggests a better trait specific heritability $h^2$, hence (as shown below) a larger chance of detecting QTLs. (See also Prob. 4.5.)

As we have seen previously, the genotype of a backcross at a given locus can be either a heterozygote, or the appropriate homozygote, depending on the type of backcross. Each of the two possibilities occurs with probability $1/2$. Without loss of generality, we can denote these two possibilities by $Aa$ and $aa$ and assume that the $F_1$ mouse was the father.

The regression model for the backcross design takes the form

$$y = \mu + (\alpha + \delta)x_{\mathrm{F}} + e\,, \tag{4.1}$$

where $x_{\mathrm{F}}$, denoted henceforth by $x$, is equal to one when the genotype is $Aa$, and it is equal to zero when the genotype is $aa$. Since the variance of $x$ is $1/4$, the phenotypic variance is $\sigma_y^2 = (\alpha + \delta)^2/4 + \sigma_e^2$.

The information collected in a genetic experiment is composed of a measurement of the phenotype and the determination of the genotype for a collection of mice from the backcross. Assume that $n$ mice are phenotyped and genotyped. Thus, the raw data are $\{(y_i, x_i) : i = 1, \ldots, n\}$, where $i$ is the index of the mouse. The interrelation between $y_i$ and $x_i$ is given in (4.1). The information $(y_i, x_i)$ for the $i$th mouse is stochastically independent of the information $(y_j, x_j)$ from the $j$th mouse. This statement is based on the assumptions that the environmental conditions are held fixed across the experiment and there is no social interaction between the mice that correlates with the expression of the phenotype. Mendel's first law will ensure that the genotypes of different mice are independent. Consequently, the independence property holds.

The phenotypic variance $\sigma_y^2$ can be estimated by the sample variance

$$\hat{\sigma}_y^2 = \frac{1}{n}\sum_{i=1}^{n}(y_i - \bar{y})^2,$$

where $\bar{y} = n^{-1}\sum_{i=1}^{n} y_i$ is the sample mean. In what follows, it will be helpful to regard $\sigma_y^2$ as given; its value, although it must be estimated from the data, does not depend on the truth or falsity of the hypotheses discussed next.

The data are to be used in order to assess the contribution of the given locus to the trait. The first step in such an assessment is trying to answer the question: "Is there a contribution at all?" This question can be addressed via statistical hypothesistesting. The scientific null hypothesis is that the marker locus under consideration is not a QTL (and is not linked to a QTL). One way to specify this hypothesis mathematically is $H_0 : \alpha + \delta = 0$. The alternative to this null hypothesis is the hypothesis that there is a contribution from the given locus. This alternative can be formulated as: $H_1 : \alpha + \delta \neq 0$.

Another way to interpret these two hypotheses is in terms of the component of the variance associated with the genetic contribution of the locus. Under the null hypothesis there is no such contribution to the variance of the phenotype. Under the alternative, on the other hand, there is a positive contribution. The null hypothesis can be specified mathematically by $\sigma_e^2 = \sigma_y^2$. The corresponding alternative hypothesis is $\sigma_e^2 < \sigma_y^2$.

We must be careful in the mathematical formulation of the null hypothesis not to lose sight of the scientific null hypothesis. Although in Chap. 2 we assumed that $e$ might contain both genetic and environmental effects, and the genetic effects in principle could be substantial, it is tempting (as we have in several calculations) to regard (2.2) as a model with one major QTL and a normally distributed residual effect $e$. In this case the null hypothesis $\alpha + \delta = 0$ would mean there is *no* major QTL. This is much more than we want to hypothesize, which is only the weaker statement that if there are major QTL, they are on different chromosomes from the genomic location we are currently studying.

Let us introduce some notation in order to write down a formula for the test statistic. Let

$$\hat{p}_{\text{Aa}} = \frac{1}{n} \sum_{i=1}^{n} x_i = \frac{n_{\text{Aa}}}{n}$$

be the proportion of heterozygous mice in our sample. The integer $n_{\text{Aa}}$ is the number of heterozygous mice in the sample. Let $n_{\text{aa}} = n - n_{\text{Aa}}$ be the number of homozygous mice. Split the sample according to the genotype at the investigated locus. Denote by $\bar{y}_{\text{Aa}}$ the mean response among the $n_{\text{Aa}}$ mice with genotype $Aa$. Similarly, $\bar{y}_{\text{aa}}$ is the mean response among the $n_{\text{aa}}$ mice with genotype $aa$.

Next we derive a test of the null hypothesis, and give it an intuitive justification. (A more formal justification for this test is given in Subsect. 4.1.3 below.) The difference $\bar{y}_{\text{Aa}} - \bar{y}_{\text{aa}}$ is a natural measure of the phenotypic difference between heterozygous and homozygous mice. Under the null hypothesis, the expected phenotypic difference in the population is equal to zero. Under the alternative, this difference is different from zero. Consequently, a significantly large absolute difference in the sample, between the heterozygous phenotypic mean and the homozygous phenotypic mean, is an indication that the null hypothesis is wrong and should be rejected. For the proposed test a cutoff level should be provided. When the absolute value of the genotype-specific difference in phenotypic means exceeds the cutoff level, then the null hypothesis is rejected. Alternatively, when the absolute value of the difference is less than the cutoff level, the null hypothesis is not rejected.

The cutoff level is based on the null sampling distribution of $\bar{y}_{\text{Aa}} - \bar{y}_{\text{aa}}$. Conditional on the number of homozygous and heterozygous mice, the null mean of this statistic is zero, and its null variance is

$$\sigma_y^2(1/n_{\mathrm{Aa}} + 1/n_{\mathrm{aa}}) = \sigma_y^2/[n\hat{p}_{\mathrm{Aa}}(1 - \hat{p}_{\mathrm{Aa}})] \ .$$

Therefore, by a version of the central limit theorem, the standardized random variable

$$\frac{(\bar{y}_{\mathrm{Aa}} - \bar{y}_{\mathrm{aa}}) \cdot [n\hat{p}_{\mathrm{Aa}}(1 - \hat{p}_{\mathrm{Aa}})]^{1/2}}{\sigma_y}$$

has approximately a standard normal distribution. The standard deviation $\sigma_y$ is an unknown quantity. However, $\hat{\sigma}_y$ can be used to estimate this quantity; and for reasonably sized samples, substituting an unknown parameter by an estimate will leave the distribution, more or less, the same. As a result,

$$Z = \frac{(\bar{y}_{\mathrm{Aa}} - \bar{y}_{\mathrm{aa}}) \cdot [n\hat{p}_{\mathrm{Aa}}(1 - \hat{p}_{\mathrm{Aa}})]^{1/2}}{\hat{\sigma}_y} \tag{4.2}$$

is approximately normal. The test statistic is $|Z|$. The 5%-significance level can be taken to be 1.96, the 5%-cutoff level for a two-sided test for the standard normal distribution. Such a level will ensure that, on the average, only one in every 20 experiments *testing a single marker* will falsely claim association with the phenotype when, in fact, such association is absent.

The function "Z.BC" computes the test statistic for a backcross trial. The input is a list containing the genetic and phenotypic data from the trial. The output is the value of $Z$ based on information in the second column of the genetic matrices. (In the simulations below we will assume that the first column represents the QTL and the second represents the marker.)

```
> Z.BC <- function(BC)
+ {
+     x <- BC$pat[,2] != BC$mat[,2]
+     y <- BC$pheno
+     n <- length(x)
+     p.hetero <- mean(x); p.homo <- 1-p.hetero
+     if((p.hetero > 0) & (p.hetero < 1))
+     {
+         y.hetero <- mean(y[x]); y.homo <- mean(y[!x])
+         Z <- (y.hetero-y.homo)*sqrt(n*p.hetero*p.homo)/sd(y)
+     } else Z <- 0
+     return(Z)
+ }
```

Note that the computed x is actually the indicator of heterozygosity, and not the count of a given allele. In the extreme situation where no genetic or phenotypic variability is present in the sample, the naïve use of the formula in (4.2) may lead to an error. Variability that ensures appropriateness of the test is examined and the proper value of the test statistic is determined with the aid of the function "if" which takes the structure: "if (*cond*) *exp1* else *exp2*". A "TRUE" value in the logical expression "*cond*" leads to the evaluation of the expression "*exp1*". Alternatively, a "FALSE" value leads to the evaluation of the other expression. (The "else" need not be included, in which case no alternative action will be taken if the logical condition is false.)

We would like to explore the distribution of the test statistic $Z$ under the null assumption of no correlation between the investigated locus and the trait phenotype. The function "BC.sim" simulates the test statistic for given size of a backcross sample, for a given model of the QTL. It uses the function "cross.rec" from Sect. 3.2. (Recall that this function depends on the recombination fraction between the marker and the QTL. For the present we are interested in only the recombination fraction $1/2$, which indicates that the null hypothesis is true. Later we consider a recombination fraction of 0, so the QTL and marker effectively coincide, and then intermediate values.)

```
> BC.sim <- function(sample.size,model,rec.frac,n.iter)
+ {
+      a <- rep("a",sample.size)
+      b <- rep("b",sample.size)
+      A <- rep("A",sample.size)
+      B <- rep("B",sample.size)
+      IB <- list(pat=cbind(a,b),mat=cbind(a,b),pheno=NULL)
+      F1 <- list(pat=cbind(A,B),mat=cbind(a,b),pheno=NULL)
+      Z <- NULL
+      for (i in 1:n.iter)
+      {
+          BC <- cross.rec(IB,F1,model,rec.frac)
+          Z <- c(Z,Z.BC(BC))
+      }
+      return(Z)
+ }
```

Let us consider two sample sizes, sample.size = 20 or 200, and two different methods of generating the null distribution. In the first case there is no major QTL anywhere. The distribution of the phenotype in this case is normal. We take the recombination frequency to equal $1/2$. (The value of the recombination fraction in this case is irrelevant.) In the second case we assume the presence of a major QTL on an unlinked chromosome. Like the first case, the genotype at the marker is unlinked to the phenotype, so the recombination fraction is again $1/2$. The difference from the first case is that the distribution of the phenotype is not normal, but a mixture of two normals.

```
> model.0 <- list(mu=0,alpha=0,delta=0,sigma=1,allele="A")
> model.1 <- list(mu=0,alpha=0.75,delta=0.75,sigma=1,
+                 allele="A")
> n.iter <- 10^4
> Z <- matrix(nrow=n.iter,ncol=4)
```

```
> Z[,1] <- BC.sim(20,model.0,0.5,n.iter)
> Z[,2] <- BC.sim(200,model.0,0.5,n.iter)
> Z[,3] <- BC.sim(20,model.1,0.5,n.iter)
> Z[,4] <- BC.sim(200,model.1,0.5,n.iter)
> colnames(Z) <- paste("n=",c(20,200,20,200),",a+d=",
+                       c(0,0,1.5,1.5),sep="")
```

Note that the results are stored in a matrix form, where each sample size and null distribution gets a column. We conveniently assigned the columns names in order to keep track of the matrix's content.

Next we plot the distribution of the $Z$ statistics. The function "seq" produces numerical vectors of equally spaced numbers between the first and second argument. The function "dnorm" gives the normal density. The first application of the function "plot" yields the theoretical density of the test statistic under null model. The four densities are added via a repeated application of the low-level function "lines". The legend serves for clarity.

```
> z <- seq(-3.5,3.5,by=0.01)
> plot(z,dnorm(z),type="l",xlab="Z",ylab="Density")
> for (i in 1:4) lines(density(Z[,i]),lty=i+1)
> legend(-3.7,0.41,legend=c("Theoretical",colnames(Z)),
+    lty=1:5)
```

Observe that the theoretical and the four simulated distributions in Fig. 4.1. To the naked eye the distribution of the test statistic looks quite robust and independent of both the sample size and the form by which the null distribution at the marker was introduced. Monte Carlo estimates of the actual probability of rejecting the null hypothesis for the four simulated statistics results in values which are not too different from the theoretical value of 0.05:

```
> apply(Z,2,function(x) mean(abs(x)>=1.96))
  n=20,a+d=0   n=200,a+d=0   n=20,a+d=1.5 n=200,a+d=1.5
     0.0497        0.0497        0.0475        0.0490
```

In the computation we used the very handy function "apply", which applies a specified function to each of the rows or columns of a matrix. The first argument of "apply"" is the input matrix. If the second argument is set to "2", as in the example, then the function in the third argument is applied to each of the columns. Setting the value of the argument to "1" will apply it to the rows. If the function to be applied contains only a single and a simple expression, then it is convenient to introduce it directly. Otherwise, one may save the function as an object, and then use the name of the object in order to refer to the function.

### 4.1.2 The Noncentrality Parameter

The significance level of the test, and the determination of the cutoff level, are computed under the assumptions of the null hypothesis. Other statistical

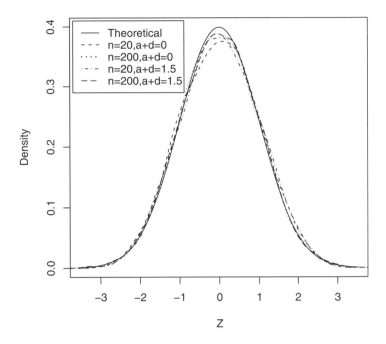

**Fig. 4.1.** Densities of the Z statistic under the null distribution

properties, such as the statistical power of the test, the selection of the sample size, and the construction of confidence intervals, rely on the stochastic behavior of $Z$ under the alternative hypothesis as well. In this subsection we will investigate this alternative distribution. We would also like to reintroduce the important notion of local alternatives, which was described in Chap. 1, and consider it in the given context of the backcross design. The main product of this exercise will be the derivation of the noncentrality parameter – the asymptotic mean of the test statistic under the alternative – which determines the main features of the distribution of the test statistic under the alternative.

### The Distribution of $Z$ Under the Alternative

Under the alternative distribution, according to our model, the expectation of $\bar{y}_{aa}$ is $\mu$ and the expectation of $\bar{y}_{Aa}$ is $\mu + \alpha + \delta$. Therefore, the expectation of the difference $\bar{y}_{Aa} - \bar{y}_{aa}$ is equal to $\alpha + \delta$, which can be positive or negative. The variance of that difference under the alternative can be computed the same way it has been computed under the null distribution.

One can ignore for practical purposes, unless the sample size is very small, the fact that the sample variance $\hat{\sigma}_y^2$ is being used in the standardization of the test statistic $Z$, instead of the actual variance $\sigma_y^2$. Making that simplification, one gets that the variance of $Z$ is equal to one, not only under the null distribution but also under the alternative. In order to obtain this simplification we have also ignored the sample variability of the genotypes. However, once the sample is reasonably large, $\hat{p}_{Aa}$ approximates quite well the expected proportion of heterozygotes, which is $1/2$.

Setting $\hat{p}_{Aa} = 1/2$, we see from (4.2) that the expectation of the test statistic $Z$ under the alternative, which we will henceforth call the noncentrality parameter, is given approximately by:

$$\mathrm{E}(Z) \approx \frac{n^{1/2}(\alpha + \delta)/2}{\sigma_y} = \xi \ . \tag{4.3}$$

Note also that by the central limit theorem the distribution of $Z$ is approximately normal, not only under the null hypothesis but also under the alternative. The standard deviation of the standardized test statistic remains constant. One can identify, therefore, that the way an association of the locus with the trait affects the distribution of the test statistic is by shifting its location parameter. Instead of being centered about zero at the absence of association, it is centered about a new mean when an association is present. The expression for the mean given in (4.3) allows us to compute approximately the *power* of the test, i.e., the probability of rejecting the null hypothesis, computed under the alternative distribution:

$$\Pr(Z \geq z_0) + \Pr(Z \leq z_0) \approx \Pr(Z - \xi \geq z_0 - \xi) \ .$$

This probability, using the normal approximation, is about equal to 0.85 when $\xi$ is such that $z_0 - \xi \approx -1$, which for $z_0 \approx 2$ would mean $\xi \approx 3$.

## Local Alternatives

Examining the expression for the mean in (4.3) we see that the sample size $n$ enters as a multiplicative factor in the numerator. As the sample size increases, the absolute value of the numerator, thus the noncentrality parameter, increases. Consequently, the distribution of the test statistic under the alternative diverges from its distribution under the null. This divergence makes the distinction between the null and the alternative hypothesis easier and enhances the power of statistical inference. The improvement in the power due to increased sample size can express itself in several ways. One way is the increase in the probability of rejecting the null hypothesis for a fixed alternative. This probability typically approaches one quite rapidly. Another way is the ability to deal with alternatives which are closer to the null hypothesis. This second way is the route taken in the approach of local alternatives.

A weaker signal corresponds to a smaller genetic term in the regression model. For the backcross design, this means that the sum $\alpha + \delta$ is closer to 0. In order to get a simple, but useful nonetheless, asymptotic expression one can assume the sum converges to zero proportionally to $1/n^{1/2}$. More specifically, let

$$\alpha + \delta = \alpha_n + \delta_n = \alpha_0/n^{1/2} + \delta_0/n^{1/2} \, ,$$

for some constants $\alpha_0$ and $\delta_0$. Then by (4.3) the noncentrality parameter can be rewritten as

$$\mathrm{E}(Z) \approx \frac{\alpha_0 + \delta_0}{2\sigma_y} = \xi \, .$$

Thus, the expectation of the test statistic converges to some limit as the sample size increases.

By the preceding analysis we have approximated the original testing problem by a normal shift testing problem, i.e., a problem that involves testing for shift in the mean of a normally distributed population with a fixed variance. It is also interesting to note that when we consider local alternatives, we find that the distribution of the test statistic is (approximately) independent of the sample size. This is a major simplification, since now we have one less element to worry about in the investigation of statistical properties. The sample size can always be introduced back by substituting the original parameters for the alternative local parameters. Thus, in any place where $\alpha_0$ appears, we can substitute $\alpha n^{1/2}$. Similarly, we can replace $\delta_0$ by $\delta n^{1/2}$.

The phenomenon of convergence to a normal-shift problem, which was demonstrated in the context of the backcross design, is actually a quite general phenomenon. Inferential procedures constructed for the normal shift problem can be translated into inferential procedures in the context of local alternatives in the original problem. The statistical properties of the translated procedures will resemble those of their normal shift counterparts. In particular, optimality properties possessed by specific inferential procedures in the normal shift context will approximately hold for the translated versions of these procedures. This allows for a very rich and widely applicable statistical theory. We will introduce some of these application in other parts of this book.

As an example, which parallels an example given in Sect. 1.6, we compare an experiment with a sample size of 20 to an experiment with a sample size of 200. While the genetic effect is identical in both cases, the statistical properties of $Z$ are very different:

```
> Z <- matrix(nrow=n.iter,ncol=2)
> colnames(Z) <- c("null","Alternative")
> Z[,1] <- BC.sim(20,model.0,0,n.iter)
> Z[,2] <- BC.sim(20,model.1,0,n.iter)
> apply(Z,2,mean)
       null Alternative
0.009614507 2.615090543
> apply(Z,2,sd)
```

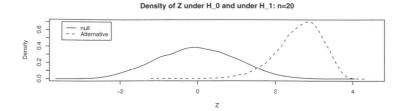

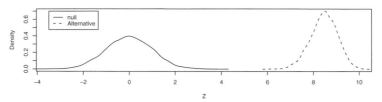

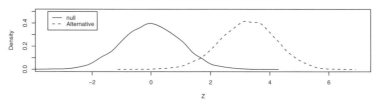

**Fig. 4.2.** Local, and non-local alternatives

```
       null Alternative
  1.0179658   0.5957992
> apply(Z,2,function(x) mean(abs(x) >= 1.96))
       null Alternative
     0.0495     0.8681
> plot(c(-3.5,4.5),c(0,0.7),type="n",xlab="Z",ylab="Density")
> for (i in 1:2) lines(density(Z[,i]),lty=i)
> legend(-3.5,0.7,legend = colnames(Z),lty=1:2)
> title("Density of Z under H_0 and under H_1: n=20")
> Z[,1] <- BC.sim(200,model.0,0,n.iter)
> Z[,2] <- BC.sim(200,model.1,0,n.iter)
> apply(Z,2,mean)
       null Alternative
-0.01035956  8.46298170
> apply(Z,2,sd)
       null Alternative
  1.0067297   0.5825284
> apply(Z,2,function(x) mean(abs(x) >= 1.96))
```

```
                  null Alternative
            0.0523        1.0000
> plot(c(-3.5,10),c(0,0.7),type="n",xlab="Z",ylab="Density")
> for (i in 1:2) lines(density(Z[,i]),lty=i)
> legend(-3.5,0.7,legend = colnames(Z),lty=1:2)
> title("Density of Z under H_0 and under H_1: n=200")
```

However, if the genetic effect for the larger trial is smaller, then the statistical differences are much more subtle, and the normal shift structure clearly emerges:

```
> a0 <- 0.75*sqrt(20)
> a0
[1] 3.354102
> d0 <- a0
> model.local <- list(mu=0,alpha=a0/sqrt(200),
+                     delta=d0/sqrt(200),sigma=1,allele ="A")
> Z[,2] <- BC.sim(200,model.local,0,n.iter)
> apply(Z,2,mean)
          null Alternative
-0.01035956   3.25015769
> apply(Z,2,sd)
          null Alternative
   1.0067297    0.9332931
> apply(Z,2,function(x) mean(abs(x) >= 1.96))
          null Alternative
      0.0523       0.9122
> plot(c(-3.5,7),c(0,0.5),type="n",xlab="Z",ylab="Density")
> for (i in 1:2) lines(density(Z[,i]),lty=i)
> legend(-3.5,0.5,legend = colnames(Z),lty=1:2)
> title("Density of Z under H_0 and under local H_1: n=200")
```

Observe that one can apply the function "plot" in order to create an empty plot. Setting the argument "type="n"" ensures that the two points that serve to determine the range of the axis are not plotted. Lines, legends, and titles may later be added with the aid of low-level functions.

*Remark 4.1.* An important theoretical use of local alternatives is to help us interpret the noncentrality parameters of different statistics in terms of sample sizes. Suppose two statistics $T_1, T_2$ are both approximately standard normal under the null hypothesis and for a local alternative are approximately normal with variance one and means $\xi_1, \xi_2$, respectively. The statistic with the larger value of $\xi_i$ has greater power, but can we give a quantitative statement about how much better that statistic is? The noncentrality parameters for samples of size $n$ will often be proportional to $n^{1/2}$, so $\xi_i = n^{1/2}\mu_i$, where $\mu_1, \mu_2$ are noncentrality parameters for samples of size one. Suppose, to be specific, that $\mu_1 > \mu_2 > 0$. Now we change the point of view and ask how much larger a sample size we would need if we want to use $T_2$ and yet have the same noncentrality parameter as $T_1$ has for a given sample size. To address this question, suppose now that the test statistics $T_1$ and $T_2$ are actually based on two different sample sizes, $n_1$ and $n_2$, so $\xi_i = n_i^{1/2}\mu_i$. The two noncentrality parameters would be the same if $n_1^{1/2}\mu_1 = n_2^{1/2}\mu_2$, or equivalently $n_2 = n_1(\mu_1/\mu_2)^2$. Hence to compensate for the smaller value $\mu_2$, we would have to have a sample size that is larger by the square of the ratio of $\mu_1$ to $\mu_2$. If $\mu_1 = 1.4\mu_2$, $n_2$ would have to be approximately twice as large as $n_1$ in order that $T_2$ would have the same noncentrality parameter as $T_1$. This is sometimes expressed by saying that $T_2$ is only 50% as efficient as $T_1$.

### 4.1.3 The BC Test as a Test of the Regression Coefficient

For the discussion of an intercross given below and in later chapters it will be helpful to view (4.1) as a linear regression model. A regression statistic for testing the null hypothesis that the marker and trait are unlinked is

$$Z = \frac{\sum_{i=1}^{n}(y_i - \bar{y})(x_i - 1/2)}{[\sum_{i=1}^{n}(x_i - \bar{x})^2]^{1/2}\hat{\sigma}_y} \ .$$

(See Sect. 1.7.) Some algebra shows is this expression exactly equals (4.2) given above.

When the sample size is large, $\bar{y} \approx m = \mu + (\alpha + \delta)/2$, $\sum(x_i - \bar{x})^2 \approx n/4$ and $\hat{\sigma}_y \approx \sigma_y$, so

$$Z \approx \frac{\sum_{i=1}^{n}(y_i - m)(x_i - 1/2)}{(n/4)^{1/2}\sigma_y} \ . \tag{4.4}$$

This approximation will be particularly useful in Chaps. 5 and 6 when we discuss the use of many markers to search the genome for those markers that are linked to the QTL.

## 4.2 The Effect of Recombination

In the previous section we identified the form of the test for checking the relation between the QTL and the phenotype. The statistical characteristics of the test statistic were spelled out in the context of large samples and local alternatives. The parameter that was highlighted in this context was the noncentrality parameter. We wrote a formula for this parameter in terms of the local regression coefficient and the standard deviation of the phenotype. This single parameter essentially determines the distribution of the test statistic for a marker perfectly linked to the QTL. In this section we would like to extend that analysis to markers that are only partially linked to the functional QTL. Again, we will use the theory of local alternatives in order to approximate the distribution of the test statistic in a manner that does not depend on the sample size.

Let us consider a genetic marker near a functional polymorphism. If that marker is genotyped for a sample of backcross mice, then one may compute the correlation between the alleles of the marker and the measured expression levels of the phenotype. This correlation measures the relation between the marker and the trait. By itself, the marker does not affect the expression level of the trait. However, since the marker is correlated with the functional polymorphism, and that functional polymorphism is in turn correlated with the expression level, then we may be able to observe correlation between the marker and the phenotype. Obviously, the strength of the correlation, or linkage, between the functional polymorphism and the genotyped marker is central in the determination of the chances of establishing a relation between the marker and the trait. This correlation is determined by the *recombination fraction*, which allows us to determine the noncentrality parameter of a statistic for testing the linkage between the linked marker and the trait. The recombination fraction was described in the previous chapter. We start by reviewing that concept.

### 4.2.1 The Recombination Fraction

To continue here the discussion of recombination that we began in Sect. 3.2, consider a backcross and two loci that are polymorphic with respect to the two pure inbred strains used for the backcross. The first strain is an $a/b$ homozygote and the second strain is an $A/B$ homozygote. Regardless of which crossovers may occur during the formation of the outcross, the $F_1$ mouse must be an $(A/B, a/b)$ heterozygote. When the $F_1$ mouse is backcrossed with the inbred $a/b$ mouse, then the gametes of the resulting offsprings can be $(a/b, a/b)$ or $(A/B, a/b)$ in the case of no recombination. Alternatively, if a recombination does occur, the possible pairs of gametes are $(a/B, a/b)$ or $(A/b, a/b)$. The probability of each pair of gametes is $(1-\theta)/2$ in the former case and $\theta/2$ in the latter. The observed genotype in the case of no recombination in the $F_1$ meiosis will be either $aa$ and $bb$ or $Aa$ and $Bb$. In the case of recombination, the observed genotype will be either $aa$ and $Bb$, or $Aa$ and $bb$. Consequently, the presence of a recombination can be determined from the observed genotypes. It is convenient to summarize the distribution of the four possibilities in a tabular form. See Table 4.1 for such a summary.

Knowing the genotype at the marker, the genotypic status at the QTL can be partly predicted. Specifically, if the genotype at the marker is $bb$, then the genotype at the QTL is $aa$ with probability $1-\theta$, and it is $Aa$ with probability $\theta$. Similarly, if the genotype at the marker is $Bb$ then the genotype at the QTL is $Aa$ with probability $1 - \theta$, and it is $aa$ with probability $\theta$.

**Table 4.1.** Distribution of observed genotypes at a pair of linked markers in the backcross design

| Genotype | $aa$ | $Aa$ | total |
|:---:|:---:|:---:|:---:|
| $bb$ | $(1-\theta)/2$ | $\theta/2$ | $1/2$ |
| $Bb$ | $\theta/2$ | $(1-\theta)/2$ | $1/2$ |
| total | $1/2$ | $1/2$ | $1$ |

### 4.2.2 The Distribution of a Test Statistic at a Linked Marker

In Sect. 4.1 we developed a statistic to test for a relation between the genotype at a given locus and the phenotype. We then investigated the distribution of this test statistic, assuming that the genotyped locus is perfectly linked to a QTL. Typically, however, it is unlikely that one will *a* priori know the exact location of the QTL. Under a more realistic scenario, one would genotype a marker that is only partially linked to the QTL, rather than perfectly linked. Nonetheless, a test statistic can be computed based on the genotypic information at the marker and the measured phenotypic data. Denote the QTL by $\tau$. In order to distinguish between the marker and the QTL we will denote the location of the marker by $t$.

Using the same notation as in the definition of the test statistic (4.2), but denoting this time the alleles of the marker by the capital and the small case versions of the letter $b$, we get that the test statistic for the marker is:

$$Z_t = \frac{(\bar{y}_{\mathrm{Bb}} - \bar{y}_{\mathrm{bb}}) \cdot [n\hat{p}_{\mathrm{Bb}}(1 - \hat{p}_{\mathrm{Bb}})]^{1/2}}{\hat{\sigma}_y} \,. \tag{4.5}$$

Under the null assumption, neither the QTL nor the given marker have any measurable correlation with the phenotype. Thus, just like before, the expected value of this test statistic is zero. Furthermore, under the null hypothesis, the distribution of the test statistic for the QTL is identical to the distribution of the test statistic for the marker. In particular, using the same two-sided threshold of 1.96 will produce a test with the approximate significance level of 5%.

Under the alternative hypothesis the picture is more complex. Still, under local alternatives, the distribution of $Z_t$ is approximately normal in large samples, and its variance is approximately equal to one. Thus, the statistical properties of the test at the marker are determined by the noncentrality parameter, or the expectation of the test statistic. We turn to the examination of this parameter.

The critical computation in the determination of the noncentrality parameter involves the expectation of the difference $\bar{y}_{\mathrm{Bb}} - \bar{y}_{\mathrm{bb}}$. It is useful to express the terms in this difference in a slightly more elaborate form. Indeed, note that the two terms are averages computed over subsets of mice, subsets determined by the genotype of the mice at the locus of the marker. Each of these averages can be represented as a weighted average of averages, where these new averages are computed over a partition of these subsets. Before getting completely lost, let us look at a specific representation.

The subset of mice with genotype $Bb$ can be sub-divided into two groups according to the genotype of the QTL: one group indexed by $(Aa, Bb)$ and the other indexed by $(aa, Bb)$. Denote the average phenotype expression over the group $(Aa, Bb)$ by $\bar{y}_{\mathrm{Aa,Bb}}$, and the average phenotype expression over the group $(aa, Bb)$ by $\bar{y}_{\mathrm{aa,Bb}}$. Next, denote the proportion of the group $(Aa, Bb)$ among the $Bb$-mice by $\hat{p}_{\mathrm{Aa:Bb}}$, and the proportion of the $(aa, Bb)$ among the $Bb$-mice by $\hat{p}_{\mathrm{aa:Bb}}$. The representation we have in mind is:

$$\bar{y}_{\mathrm{Bb}} = \hat{p}_{\mathrm{Aa:Bb}} \times \bar{y}_{\mathrm{Aa,Bb}} + \hat{p}_{\mathrm{aa:Bb}} \times \bar{y}_{\mathrm{aa,Bb}}.$$

Let us carry out the computation by conditioning on the genotypic information at both loci. The average $\bar{y}_{\mathrm{Aa,Bb}}$ is computed for a sub-collection of $Aa$-type of mice. According to our model, the expected expression level for such mice is $\mu + (\alpha + \delta)$. Likewise, the average $\bar{y}_{\mathrm{aa,Bb}}$ is computed for $aa$-type of mice. Their expected expression level is $\mu$. It follows that the expectation of $\bar{y}_{\mathrm{Bb}}$ is equal to:

$$\mathrm{E}(\bar{y}_{\mathrm{Bb}}) = \mu + (\alpha + \delta) \times \mathrm{E}(\hat{p}_{\mathrm{Aa:Bb}}) \ .$$

Essentially, the same type of representation and computation gives for the average $\bar{y}_{\mathrm{bb}}$ the result:

$$\mathrm{E}(\bar{y}_{\mathrm{bb}}) = \mu + (\alpha + \delta) \times \mathrm{E}(\hat{p}_{\mathrm{Aa:bb}}) \ ,$$

where $\hat{p}_{\mathrm{Aa:bb}}$ is the proportion of the $Aa$-mice among the $bb$-mice. It follows that

$$\mathrm{E}(\bar{y}_{\mathrm{Bb}} - \bar{y}_{\mathrm{bb}}) = (\alpha + \delta) \times \mathrm{E}(\hat{p}_{\mathrm{Aa:Bb}} - \hat{p}_{\mathrm{Aa:bb}}) \ .$$

The sample fractions $\hat{p}_{\mathrm{Aa:Bb}}$ and $\hat{p}_{\mathrm{Aa:bb}}$ should be close to their population values when the sample size is large. These population values are $1 - \theta$ and $\theta$, respectively. We can conclude, therefore, that

$$\mathrm{E}(\bar{y}_{\mathrm{Bb}} - \bar{y}_{\mathrm{bb}}) \approx (\alpha + \delta) \times (1 - 2\theta) \ .$$

As we can see, the expectation of the difference at the marker is proportional to the expectation of the difference at the QTL (which is $\alpha + \delta$). Subsequently, the noncentrality parameter at the marker is essentially proportional to the noncentrality parameter at the QTL. In particular, if local alternatives are considered, then the limiting expectation of the test statistic is proportional to the limiting expectation of the test statistic at the QTL:

$$\mathrm{E}(Z_t) \approx (1 - 2\theta) \times \frac{n^{1/2}(\alpha + \delta)}{2\sigma_y} = (1 - 2\theta) \times \xi \ , \tag{4.6}$$

where $\xi$ is the expected value of the test statistic considered in Sect. 4.1, i.e., the statistic that correlates the genotype of the QTL itself with the observed phenotype.

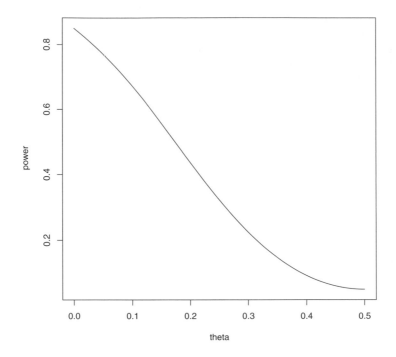

**Fig. 4.3.** The statistical power as a function of the recombination fraction.

Let us examine the reduction in the asymptotic power of the test statistic as one moves away from the QTL. We take the noncentrality parameter at the QTL to be equal to 3, which produces a statistical power of about 85% for a marker completely linked to the QTL. As the recombination fraction increases, the power decreases. Finally, in the case of no linkage ($\theta = 0.5$) the probability is equal to the significance level of 0.05. The power function is presented in Fig. 4.3:

```
> th <- seq(0,0.5,0.01)
> ncp <- 3
> r <- 1-2*th
> p <- pnorm(-1.96, ncp*r)+ 1 - pnorm(1.96,ncp*r)
> plot(th,p,type="l",ylab="power")
```

### 4.2.3 *Covariance of (4.4) at Different Marker Loci

We can use similar arguments to compute the covariance of (4.4) at two different loci $s$ and $t$ on an unlinked chromosome. This will be very useful in the following chapter, where we consider the simultaneous use of many markers to map genes of unknown genomic location, which requires that we understand the joint distributions of (4.4) at different marker positions on the genome. Indeed, let $x(s)$ and $x(t)$ denote the indicators of heterozygosity at markers $s$ and $t$, respectively. Then

$$
\begin{aligned}
\operatorname{cov}(x(s), x(t)) &= \mathrm{E}[x(s)x(t)] - \mathrm{E}[x(s)]\mathrm{E}[x(t)] \\
&= \Pr[x(s) = 1]\mathrm{E}[x(t)|x(s) = 1] - 1/4 \\
&= (1/2)(1 - \theta) - 1/4 = (1 - 2\theta)/4 .
\end{aligned}
$$

The crucial step in this chain of equalities is the observation that if we are given $x(s) = 1$, then $x(t) = 1$ if and only if there is no recombination between $s$ and $t$, which has probability $1 - \theta$. Since $\operatorname{var}[x(s)] = \operatorname{var}[x(t)] = 1/2$, we have

$$
\operatorname{cor}[x(s), x(t)] = 1 - 2\theta.
$$

Denote by $Z_s$ and $Z_t$ the statistic $Z$ in (4.2) at the marker loci $s$ and $t$, respectively. From (4.4) we see that $Z_t$ is a linear function of the $x_i(t)$, so by the preceding argument we find that the correlation function of $Z_s$ and $Z_t$ is the same as for $x(s)$ and $x(t)$. Hence, for any two loci on a chromosome that does not contain a QTL

$$
\operatorname{cor}(Z_s, Z_t) \approx 1 - 2\theta , \tag{4.7}
$$

where $\theta$ is the recombination fraction between $s$ and $t$. A similar, but more complicated, calculation shows that under local alternatives the same correlation function applies to linked chromosomes.

### 4.2.4 *Sufficiency and the Correlation Structure

In this section we give an indirect argument that proves (4.7) and provides, in our view, more insight into the essence of the problem than the direct proof given above. This argument, however, is not used elsewhere in this book.

The expected value of $Z_\tau$ is given in (4.6) by $\xi$. Its (limiting) standard deviation is equal to one. This is all that is needed in order to determine the statistical properties of a test based on this statistic. In the next two chapters we will deal with the joint statistical properties of a collection of markers. For that we will also need to determine the correlation coefficient between any pair of markers. For unlinked chromosomes, the correlation is given in (4.7). We now compute the same correlation for a linked chromosome. This computation follows directly from (4.6), once one grasps the consequences of using local alternatives and applying the concept of sufficiency in this context.

The fact that we use local alternatives makes not only the variances, but also the covariances, asymptotically stable under the alternative. In our case, when considering local alternatives, the joint limiting distribution of $(Z_t, Z_\tau)$ is bi-normal. The variance of each of the components is equal to one. The covariance between the two components is equal to some $\rho$. (Note that in this case $\rho$ is also the correlation coefficient.) The covariance remains unchanged, regardless of changes in the means, i.e., in the values of the parameters $\alpha_0$ and $\delta_0$ of the local alternatives. The relation between the means of the two components is given in (4.6). We would like to claim that from the notion of statistical sufficiency it must follow that $\rho = 1 - 2\theta$.

The heart of the argument relies on the statement that given the full set of data, $Z_\tau$ is sufficient for determining the status of the QTL. In other words, if one has the genotypic data from the QTL itself, then other markers in the vicinity of the QTL have genotypes determined by recombination with the QTL, and they do not add any information regarding the genetic effect the QTL may have on the trait. This statement holds in particular for the simple regression model we consider. The statistical formulation of sufficiency states that given the value of the sufficient statistic $Z_\tau$, the distribution of the rest of the data, and in particular the value of $Z_t$, is uniquely determined. Hence the conditional expectation of $Z_t$, given $Z_\tau$, does not depend on $\xi$. The formula for the conditional expectation in a bi-normal distribution yields:

$$\mathrm{E}(Z_t \mid Z_\tau) = \mathrm{E}(Z_t) + \rho \times [Z_\tau - \mathrm{E}(Z_\tau)] = [1 - 2\theta - \rho] \times \xi + \rho \times Z_\tau \;,$$

where the second equality follows from (4.6). In order for this conditional expectation not to depend on $\xi$, the terms $\rho$ and $1 - 2\theta$ must coincide.

As a final remark let us add that $1 - 2\theta$ must be the limiting correlation (under local alternatives) for any pair of markers having recombination frequency $\theta$. Indeed, the correlation structure remains unchanged under the null hypothesis of no genetic contribution to the variability in the expression of the trait. In such a case, what we have designated as our QTL is just like any other marker. Hence, its correlation coefficient with some other marker is also the correlation coefficient between two markers. This argument provides an alternative derivation of (4.7)

## 4.3 Intercross and Other Designs

In an intercross design (IC) there are three genotypes at each locus and hence the possibility to estimate $\alpha$ and $\delta$ separately. Recombinant inbred (RI) designs have only two genotypes and hence are similar in several respects to BC designs. In this section we begin with a very brief discussion of RI designs and then proceed to the more complicated case of IC designs, which are studied more thoroughly in the problems at the end of the chapter.

At a locus where the inbred strains are $aa$ or $AA$, RI inbred mice are also either $aa$ or $AA$ homozygotes with probability $1/2$ each. Hence the basic regression model can be written as

$$y = \mu + \alpha x + e \;, \tag{4.8}$$

where $x = 0$ or 2. At a linked marker locus with parental alleles $b$ and $B$, the mouse is again homozygous. Now, however, either pairing of $aa$ or $AA$ with $bb$ or $BB$ is possible. The pairings of $aa$ with $bb$ and of $AA$ with $BB$ are nonrecombinant, because they are found in the inbred parental strains. Each occurs with probability $(1 - \theta_{RI})/2$. The pairings of $aa$ with $BB$ and of $AA$ with $bb$ are recombinant and occur with probability $\theta_{RI}/2$. (Note that these four probabilities add up to one.) The RI recombination frequency can be computed in terms of $\theta$, the recombination frequency for a single meiosis. Since there are many generations, hence many meioses, involved in creating the RI strain, $\theta_{RI}$ is larger than $\theta$. It also depends on the way the RI strain was created. For RI strains created by repeated sib mating we have seen in Chap. 3 that

$$\theta_{RI} = 4\theta/(1 + 6\theta) . \tag{4.9}$$

For RI strains created by repeated selfing, $\theta_{RI} = 2\theta/(1 + 2\theta)$.

Hence in complete analogy with (4.4) we define the statistic

$$Z_t = \frac{(\bar{y}_{BB} - \bar{y}_{bb})[n\hat{p}_{BB}(1 - \hat{p}_{BB})]^{1/2}}{\hat{\sigma}_y} . \tag{4.10}$$

The noncentrality parameter of $Z_\tau$ can be shown to equal $n^{1/2}\alpha/\sigma_y$, while the noncentrality of $Z_t$ is this quantity multiplied by $1 - 2\theta_{RI}$. To the extent that $\sigma_y^2$ is roughly the same for BC and for RI designs, it appears that the RI design will often be more powerful, although the deterioration in the noncentrality due to the recombination frequency between QTL and marker is substantially larger for the RI design. (See Prob. 6.4.) As we shall see in Chap. 6, the larger recombination frequency of a RI design, which works to its disadvantage in testing for a QTL, is actually an advantage in estimating the location of a QTL.

We now turn to an intercross design, for which there are the three genotypes, $aa$, $Aa$, and $AA$ at each locus, which occur with probabilities $1/4$, $1/2$, and $1/4$, respectively. A convenient representation of the basic model is (2.2), which for the present case of $p = 1/2$ is

$$y = m + \alpha(x_F + x_M - 1) + \delta(|x_F - x_M| - 1/2) + e . \tag{4.11}$$

The coefficients of $\alpha$ and of $\delta$ are uncorrelated, so $\sigma_y^2 = \alpha^2/2 + \delta^2/4 + \sigma_e^2$. In this case we can find a statistic with two degrees of freedom to detect a QTL where either $\alpha$ or $\delta$ is different from 0 (or both are). In this case it is convenient to use the regression equation (4.11) to describe the statistic. (See Sect. 4.1.3 for a similar representation of the BC statistic.) A statistic to test whether $\alpha = 0$ is

$$Z_\alpha = \frac{\sum_{i=1}^{n}(y_i - \bar{y})(x_{Fi} + x_{Mi} - 1)}{\sqrt{\sum_{i=1}^{n}(x_{Fi} + x_{Mi} - 1)^2} \cdot \hat{\sigma}_y} ,$$

and the analogous statistic to test $\delta$ is

$$Z_\delta = \frac{\sum_{i=1}^{n}(y_i - \bar{y})(|x_{Fi} - x_{Mi}| - 1/2)}{\sqrt{\sum_{i=1}^{n}(|x_{Fi} - x_{Mi}| - 1/2)^2 \cdot \hat{\sigma}_y}} .$$

A statistic to test for a QTL having either an additive effect $\alpha$ or a dominance effect $\delta$ (or both) is $U = Z_\alpha^2 + Z_\delta^2$, which under the null hypothesis is asymptotically a chi-square random variable with two degrees of freedom. The asymptotic noncentrality parameters of $U$ at the QTL and at linked markers are explored in the problems at the end of the chapter.

## 4.4 Bibliographical Comments

A clear discussion in the framework of a backcross of many of the statistical issues discussed in this and the following chapter is contained in the very influential paper of Lander and Botstein [47].

The idea to use local alternatives to evaluate and compare statistical tests is due to E. J. G. Pitman. A systematic discussion is given, for example, by Cox and Hinkley [15].

The design of Prob. 4.6 has been systematically exploited by D. Burke in a series of experiments exploring QTL for longevity in mice (e.g., refs. [41, 40]).

## Problems

**4.1.** The statistic $U = Z_\alpha^2 + Z_\delta^2$ has been introduced to test for the presence of a QTL in the intercross design. It can be shown that

$$E(Z_\alpha) \approx \frac{n^{1/2}\alpha}{2^{1/2}\sigma_y} , \quad E(Z_\delta) \approx \frac{n^{1/2}\delta}{2\sigma_y} . \tag{4.12}$$

Here, $\sigma_y^2 = \alpha^2/2 + \delta^2/4 + \sigma_e^2$. The standard deviations of both $Z_\alpha$ and $Z_\delta$ are (approximately) equal to one, and they are independent.

(a) Write a function that simulates the statistic. The input to this function should be the relevant parameters and the number of iterations required.
(b) Plot the distribution of the statistic under the null distribution ($\alpha = \delta = 0$) and under various values of the parameters.
(c) Identify the threshold for the test of the null hypothesis that will ensure a significance level of 5%. Compare this threshold determined by simulation with that determined from the chi-square distribution with two degrees of freedom.
(d) Compute the power of the test for various values of the parameters.

**4.2.** Compare the distribution of the statistic $U_{BC} = Z^2$ for the backcross with the distribution of $U_{IC} = Z_\alpha^2 + Z_\delta^2$ for the intercross. You may assume that $\sigma_e^2$ is the same for both designs.

(a) Under the null distribution ($\alpha = \delta = 0$).
(b) For an additive model ($\delta = 0$).
(c) For a dominant model ($\alpha = \delta$).
(d) Why might the assumption that $\sigma_e^2$ is the same for both designs fail to be satisfied?

**4.3.** It can be shown that at a marker having a recombination fraction $\theta$ with the QTL

$$E(Z_\alpha) \approx (1 - 2\theta) \times \frac{n^{1/2}\alpha}{2^{1/2}\sigma_y} , \quad E(Z_\delta) \approx (1 - 2\theta)^2 \times \frac{n^{1/2}\delta}{2\sigma_y} . \qquad (4.13)$$

Plot the noncentrality parameter of the statistic $U$ (in this case the sum of squares of the expectations of $Z_\alpha$ and $Z_\delta$) as a function of $\theta$ for $0 \le \theta < 0.45$.

**4.4.** Consider a trait with two QTL on different chromosomes, with $\alpha_1 = \alpha_2 = 0.5$, $\delta_1 = \delta_2 = 0$, and $\sigma_e^2 = 1.0$. Between a BC design and a RI design based on recurrent sib mating, which leads to a larger noncentrality parameter for a single marker at recombination frequency $\theta = 0.1$ from the first QTL? recombination frequency $\theta = 0.2$? Would your answer change if $\delta_1 = \delta_2 = 0.25$?

**4.5.** (a) If there is only one major QTL, how could you use the phenotypes of the parental generation and of the $F_1$ generation to help you judge whether dominance is important and which parental strain to use for a backcross?
(b) Why might your answer to the preceding question be misleading if there is more than one major QTL?
(c) If the sign of $\delta$ is unknown, one might consider a double backcross design, where some of the $F_1$ generation are backcrossed to one of the parental strains and others are backcrossed to the other parental strain. Suppose that equal numbers of offspring from the two backcrosses are genotyped and two statistics $Z_1$ and $Z_2$ are calculated. How could these be combined to obtain an overall statistic? Would the noncentrality parameter at a QTL of the combined statistic be larger or smaller than the noncentrality parameter of the two-dimensional statistic for the same total number of offspring for an intercross design?

**4.6.** The backcross and intercross starting from two inbred strains is limited to detection of QTL that are already present in one of those strains. A design offering more genetic variability is to start from four inbred strains, with alleles $A_1, a_1, A_2, a_2$. The first cross creates $A_1a_1$ and $A_2a_2$ progeny. An intercross of these progeny creates an $F_2$ population having genotypes $A_1A_2$, $A_1a_2$, $a_1A_2$, and $a_1a_2$. In such a cross, like a backcross, it is not possible to estimate additive and dominance effects separately. To simplify the notation, assume there are only additive effects. Develop a model starting from the regression equation $y = \mu + \alpha_1 x_1 + \alpha_2 x_2 + e$, where $x_i = 1$ or $0$ according as the $F_2$ mouse has allele $A_i$ or $a_i$ for $i = 1$ or $2$. What are the variance components that make up $\sigma_y^2$? Suggest statistics to test the hypothesis of no linkage. Simulate the distribution of your test statistic under the null hypothesis and under the alternative for different values of $\alpha_1, \alpha_2$, and $\theta$. What are advantages and disadvantages of this design compared to two different backcrosses (or intercrosses) starting from the $A_1$, $a_1$ and the $A_2$, $a_2$ inbred strains?

# 5

## Whole Genome Scans: The Significance Level

The task of mapping loci related to a trait is carried out in a number of steps. The aim of the first step is the identification of the chromosome carrying the trait-related polymorphism. Within that chromosome, one finds a wide region which is likely to contain the genetic factor. The second step involves the narrowing down of the region. Finally, once the functional polymorphism has been barricaded within a small enough chromosomal region, the direct examination of polymorphic loci is carried out. An experimental design, which is efficient for one step, may be less efficient for a subsequent step. Therefore, it is useful to apply different experimental resources for each step. Likewise, the statistical tools and the analytic theory may vary from one step to the next. In this and the next chapters we will concentrate on the statistical considerations that are important for the first step: a whole genome scan aimed at the identification of a (relatively wide) chromosomal region containing the functional polymorphism.

In the previous chapter we have dealt with a testing situation where only one marker is examined. In reality, researchers examine in one experiment many molecular genetic markers in an attempt to identify the gene (or genes) associated with variation in expression of the trait. Hence our main focus will shift to the statistical interpretation of data from a collection of markers. The key concepts of a normal approximation in the context of local alternatives and of the role of recombination in the determination of the correlation between two loci will maintain their central role in the current discussion. However, we will also introduce some new concepts, which are relevant in genome scans and in the determination of the region containing the unknown gene.

Genome-wide scanning involves the genotyping of a collection of anonymous markers. These markers are selected across the different chromosomes, so as to cover all the genome. By themselves, they are not suspected of being related to the disease. However, since they cover all the genome, some of them are expected to be close to trait-related genes. Such markers may show significant departure from the null assumption. Our main focus in this chapter will be the issue of statistical significance in a genome-wide scan. In the next chapter we will examine the statistical power – the probability of detecting an association in a genome scan – and the issue of constructing a confidence region – a region likely to contain the gene of interest.

The preliminary question that should be addressed when examining the results from genotyping a collection of markers is this: *Is there some marker that is truly associated with variation in the expression level of the trait?* The natural null assumption in this context is a negative response to that question. Thereby, according to the null assumption, all markers are generated from a distribution where no signal is present. The initial part of this chapter describes the joint distribution of the marker-related test statistics when this null assumption is valid. Not surprisingly, it turns out that the asymptotic distribution is multi-normal, namely that of a Gaussian random process. A simple-minded test rejects the null assumption in favor of the alternative if there exists a marker that is significantly associated with the phenotype. This inferential approach involves conducting a (frequently large) number of tests, since a statistical test of significance is carried out for each one of the genotyped markers. The main new concept that will be discussed in this chapter is the statistical consequence of such multiple testing. In particular, we will examine means for controlling the overall significance level – the probability of rejecting for *some* markers the null hypothesis of no linkage, when the null hypothesis should not have been rejected at all.

*Remark 5.1.* The null hypothesis stated in the preceding paragraph is a convenient mathematical approximation to a more complex set of scientific conditions. This point is discussed in more detail in Sect. 5.2.

## 5.1 The Multi-Marker Process

Whole-genome scans, which involve genotyping a large collection of markers, require a more detailed description of the recombination process. In particular, we need a model for how *crossovers* relate to each other when moving from one position to another along a chromosome.

Crossovers are the genetic mechanism for combining parts of the two homologous chromosomes in the parent during the meiotic process to form the gamete to be transmitted to the child. The result is a mixture of the two homologous chromosomes in each parent. This mixture occurs in segments, so, for example, the gamete transmitted by the mother may consist of a segment inherited from her mother (or father), then a segment inherited from her father (or mother) and so on, up to some random number of segments. The transitions between segments are called crossovers. The simplest mathematical model for the process of crossovers is the Haldane model (after the population geneticist J. B. S. Haldane), in which crossovers occur at the points of a Poisson process. According to this model, the number of crossovers between any two loci on the same chromosome has a poisson distribution with mean value proportional to the *genetic distance* between the markers. Moreover, the number of crossovers occurring over disjoint intervals of a chromosome are independent. (If a crossover in one region promotes or inhibits the occurrence of a crossover in a different region, one speaks of *interference*. The Haldane model is sometimes called the no interference model of crossovers.) Specifically, if the genetic distance between the two loci is $\Delta$ cM (centiMorgan, after the geneticist Thomas Hunt Morgan, in whose laboratory crossovers were first identified and studied), then the expected number of crossovers is $0.01\,\Delta$. The actual number of crossovers is generated according to the poisson distribution that has this mean. Crossovers are in principle unobservable. What can be observed is whether a *recombination* has occurred between two markers, i.e., whether the number of crossovers between the markers is odd. It can be shown (Sect. 5.4) that the recombination fraction equals

$$\theta = [1 - \exp{(-\beta\Delta)}]/2 \,, \tag{5.1}$$

where $\beta = 0.02$ per cM. For small $x$, $\exp(-x) \approx 1 - x$. From this and (5.1) it follows that for small $\Delta$, $\theta \approx 0.01\,\Delta$. For large $\Delta$, $\theta \approx 1/2$. A 20 chromosome mouse genome is about 1600 cM long, which means that about 16 crossovers are expected per meiosis. A human genome is about twice as long. It is important to note that the genetic distance defined here bears no particular relation to distance along a chromosome measured in base pairs. On average the number of base pairs per cM in humans is about $10^6$, but at different places in the genome the expected number of base pairs per crossover may be larger or smaller than this average.

In the preceding paragraph we have regarded the genetic distance as given, and the recombination frequency is defined in terms of the genetic distance by (5.1). In fact, what we can determine experimentally is the recombination frequency between two markers. The genetic distance is a mathematical construct, which is defined by solving (5.1) for $\Delta$ as a function of $\theta$. It is possible to develop other models of recombination, i.e., other functions similar to (5.1). A careful empirical study of the crossover process shows that the Haldane model is overly simplistic, but it offers such substantial computational advantages that it is still widely used in genetic mapping.

Ideally, markers are equally placed across the genome. Thus, for example, a marker may be selected every 30 cM for each of the 20 chromosomes of the mouse genome. Thereby, the collection forms a map of markers with an inter-marker distance of 30 cM. About 60 markers are needed for such a map. A denser map may involve an inter-marker distance of 20, 10, 5, or even 1 cM, and require many more markers. As we will see in the next chapter, the map density is important in the determination of the chances to detect a QTL. Thus, it must be a central parameter in the design of the experiment. Moreover, an inter-marker distance that may be optimal for the backcross design, may be suboptimal for the intercross design or for other types of designs, and vice versa. In a later section of this chapter we will examine the effect of the map density, and the choice of the cross, on the probability of false detection, i.e., the significance level. In this section, however, our aim is to introduce the notation and basic background facts that set the stage for the development of the probabilistic theory. This probabilistic theory produces the formulas for the significance level, along with the formulas for the power and for the construction of confidence intervals for QTL location that will be discussed in the next chapter.

It is convenient to think of the markers as points on the real line. Consider one chromosome. One can set an initial point, say the telomere, and consider the markers consecutively, moving from the telomere to the centromere (which, in the mouse, is on the other end of the chromosome). If the idealized total length of the chromosome is 80 cM, then markers may be placed at the points 0, 10, 20, 30, 40, 50, 60, 70, and 80 cM away from the telomere. At each marker one computes a test statistic of the kind discussed in Chap. 4 to measure the correlation between the genotypic information at the marker and the phenotypic information. The result is a *random process*. In the probability literature, a random processes describes the evolution of a random variable in time. There, with each point in time one associates a random variable. Consequently, if we equate 'time' with 'location on the chromosome', then one can consider the test statistics at a sequence of markers as a discrete time random process, a process for which the random variables are recorded only at discrete points in time. We will use letters like $s$ or $t$ to denote location of markers on the chromosome. For each marker location $t$, $Z_t$ will denote the test statistic computed for that marker.

The distribution of the random variable at a given point in time is called the *marginal distribution* at that point in time. As we saw in the previous chapter, the marginal distribution of the test statistics at a given locus has (approximately) a normal distribution. Under the null hypothesis the mean of that normal distribution is zero and the variance is one. Under the alternative hypothesis the mean is shifted to another value, but the limiting variance remains equal to one. In any case, since it belongs to the normal family of distributions, the marginal distribution is fully determined by its mean and variance.

We also investigated the distribution of two loci jointly, and computed the correlation between them. Specifically, we saw in (4.7) that the correlation of the test statistics between a pair of markers for the backcross design is equal to $1 - 2\theta = \exp\{-0.02\,\Delta\}$, where $\Delta$ is the distance between the two markers, measured in units of cM, and the equation follows from (5.1). The joint distribution of the process $Z$ at a pair of markers is *bi-normal*. This distribution is the extension of the normal distribution of one random variable to the case of a joint distribution of a pair of random variables. The bi-normal distribution is fully determined by the means of the random variable, by their variances, and by the covariance (or correlation) between them.

The notion of a joint distribution of pair of random variables can be taken a step further by considering the joint distribution of a collection of random variables. The proper extension of normal distribution to that case is the *multi-normal* distribution. Again, the multi-normal distribution is determined by the means of the individual variables, their variances, and the covariance coefficients between any pair of variables. It is convenient to write the means in the form of a vector, and to write the variances and covariances in the form of a matrix. The main diagonal of that matrix is formed by the variances, and the covariances are the off-diagonal entries to the matrix. Naturally, the matrix is called the *covariance matrix*. It is important to note that the joint distribution of the test statistics computed over the collection of markers has approximately a multi-normal distribution. Under the null assumption, the vector of means is identically equal to zero. The covariance between any pair of markers on the same chromosome is determined by the genetic distance between the two markers. (If markers are placed on different chromosomes, then the covariance between them is equal to zero. Consequently, the associated normal statistics are independent.) Under the alternative hypothesis the mean vector is no longer equal to the zero vector. However, the covariance structure is unchanged.

The components of a random process have a joint distribution. If the joint distribution of the components is multi-normal, we say that the process is a *Gaussian process*. With each point in time one can associate the mean of the process at that point to form the *mean function*. Similarly, with any pair of points one can associate the covariance between the components associated with these points to form the *covariance function* of the process. The distribution of a Gaussian process is fully determined by its mean function and its covariance function. Specifically, the random process of test statistics over a given chromosome is a Gaussian process. Its mean function when the null hypothesis is in place is $E(Z_t) = 0$, for any $t$ in the set of markers. Its covariance function is given by

$$\text{cov}(t, s) = \text{cov}(Z_t, Z_s) = \exp\{-\beta|t - s|\} \,, \tag{5.2}$$

for any $t$ and $s$ in the set of markers, where $\beta = 0.02$, as above. Note, in particular, that when $s = t$ then $\text{cov}(t, s) = \text{cov}(t, t) = \text{var}(Z_t) = 1$, which is consistent with the definition of the covariance function.

A Gaussian process with a zero mean function and a covariance function of the form $\text{cov}(s, t) = \exp\{-\beta|t-s|\}$ is called an *Ornstein-Uhlenbeck* process, after the physicists who studied it in their research into the mathematics of Brownian motion. The process of test statistics in the backcross design has approximately the Ornstein-Uhlenbeck distribution with $\beta = 0.02$. As we shall find out, the (asymptotic) null distributions of test statistics for other designs also follow the Ornstein-Uhlenbeck distribution, but with a different value of $\beta$. This phenomena will extend beyond experimental genetics. The process of test statistics in some forms of linkage analysis in humans, which will be presented in the next part of this book, is also of the Ornstein-Uhlenbeck form.

The function "OU.sim" simulates the Ornstein-Uhlenbeck process at a collection of markers in a given chromosome. It applies the function "mvrnorm", which generates the multi-normal distribution. The first input argument to this function indicates the number of independent copies to be generated from a multi-normal distribution characterized by the vector of means given in the second argument and the covariance function given as a matrix in the third. The output of the function is a matrix. Each row of the matrix represents an independent copy and each column a single variate of the multivariate distribution.

```
> OU.sim <- function(beta,markers,n.iter=10^4)
+ {
+     V <- outer(markers,markers,function(x,y)
+             exp(-beta*abs(x-y)))
+     mu <- rep(0,length(markers))
+     Z <- mvrnorm(n.iter,mu,V)
+     return(Z)
+ }
```

In the first expression of the R function the covariance matrix is calculated for the twenty-five values $s, t \in \{0, 20, \ldots, 80\}$:

```
> markers <- seq(0,80,by=20)
> beta <- 0.02
> outer(markers,markers,function(x,y) exp(-beta*abs(x-y)))
          [,1]      [,2]      [,3]      [,4]      [,5]
[1,] 1.0000000 0.6703200 0.449329 0.3011942 0.2018965
[2,] 0.6703200 1.0000000 0.670320 0.4493290 0.3011942
[3,] 0.4493290 0.6703200 1.000000 0.6703200 0.4493290
[4,] 0.3011942 0.4493290 0.670320 1.0000000 0.6703200
[5,] 0.2018965 0.3011942 0.449329 0.6703200 1.0000000
```

The function "outer" applies the binary function in its third argument to each of the combinations of components from the first and second argument. If the first two components are vectors, then the outcome is a matrix.

It all seems okay. However, when we try to run the function, we get an error message:

```
> OU.sim(beta,markers,10)
Error in OU.sim(beta,markers,10):
                    couldn't find function "mvrnorm"
```

R tells us that the function "mvrnorm" cannot be found.

The clue for resolving this error message can be found in the help file for the function "mvrnorm". The help file can be accessed via the "Html help" search engine. Examining the help file we can see that at the upper left corner an indication that the function "mvrnorm" belongs to the package "MASS" (named in the curly brackets). Packages are libraries of R functions and objects, written in order to carry a set of tasks in a given context. Many packages are included in the basic R distribution. Many more are contributed packages that can be downloaded from the internet upon demand. Some of these will be introduced as we proceed. The package "MASS", which is associated with the book *Modern Applied Statistics with S-Plus* by Venables and Ripley [87], is part of the basic distribution of R to Windows, but is not automatically uploaded when the session starts. In order to make the functions of the package accessible, the library should be called:

```
> library(MASS)
> OU.sim(beta,markers,10)
           [,1]        [,2]       [,3]        [,4]       [,5]
 [1,] -1.7487580  -1.623945  -1.7348697  -2.2725771  -2.216049
 [2,] -0.6736065  -1.772497  -1.5511478   0.0640753   1.255065
 [3,]  0.0462794   0.632321  -0.1198841   1.6519793   1.199331
 [4,] -0.1431301  -0.960516  -0.7717332  -0.5925920  -0.385762
 [5,] -0.0462375   0.676397   0.1975626   0.1234103   0.231498
 [6,]  0.5996106   1.076913   1.5877974   1.7123749   0.778274
 [7,] -0.7771104  -0.653780  -0.0368332   0.2312790  -0.769185
 [8,]  0.1814315   0.578433  -0.2556236  -0.2689456   0.861487
 [9,]  0.2268141   0.449569  -1.3239697  -0.7302556  -1.020395
[10,] -0.5909323  -0.606835  -0.4042961  -0.5148107  -1.282798
```

A more robust solution to the above mentioned problem is to add the expression "require(MASS)" as the first line of the function. The function "require" is similar to the function "library". It is designed to be used inside other functions since it returns a warning message, rather than an error message, if the library does not exist.

Let us generate and plot some examples of paths of the Ornstein-Uhlenbeck process along a chromosome. The paths are presented in Fig. 5.1:

```
> markers <- seq(0,80,by=10)
> Z <- OU.sim(beta,markers,30)
> plot(range(markers),c(-3,3),type="n",xlab="cM",ylab="Z")
> for (i in 1:30) lines(markers,Z[i,],col=gray((i-1)/30))
> abline(h=c(-1.96,1.96),lty=2)
```

# 5.2 Multiple Testing and the Significance Level

When a single marker was considered, the null hypothesis was rejected if the absolute value of the calculated test statistic reached an extreme level. A cutoff threshold was set, and the significance level corresponded to the (two-tailed) probability of exceeding the threshold for the standard normal distribution, which is approximately the distribution of the test statistic under the null

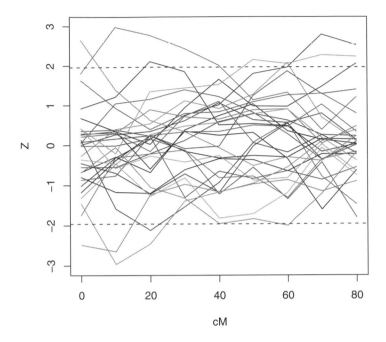

**Fig. 5.1.** Paths of the Ornstein-Uhlenbeck process

hypothesis. In particular, for a significance level of 5%, the threshold was set to be equal to 1.96.

A collection of test statistics are calculated when conducting a genome scan, one for each of the genotyped markers. A statistical test, examining the significance of the correlation with the observed phenotype, is carried out for each of these calculated test statistics. The null hypothesis has a chance of being rejected for each of the tests performed. The genome scan is concluded to have a positive finding if at least one of these tests exceeds the threshold. Otherwise, if none of the tests rejects, one concludes that no significant genetic contribution to the trait has been observed. The issue we would like to explore is the probability of reaching a positive conclusion when, in fact, a negative decision is the appropriate conclusion. In other words, the probability that at least one of the test statistics exceeds the threshold, when the null hypothesis is true. In this section we will explore this issue using simulations. In the next section we will add an analytical component by consideration of asymptotic approximations to that probability.

In testing a single marker we considered two possibilities that would make the null hypothesis true: (i) there was no significant genetic contribution to the trait or more generally (ii) any significant genetic contribution to the trait was simply not linked to the marker being tested. When testing multiple markers distributed throughout the genome, the same conceptual issues become somewhat more subtle. On the one hand, while there may be an overall genetic contribution to the trait, it may be distributed over a large number of genes. In this case the contribution of any individual gene may be comparable to the level of random variation, and hence difficult to detect. On the other hand, since a large value of our statistic at a particular location in the genome leads us to conclude that there is an important genetic contribution to the trait arising from a gene or genes located in that region, even if there are genes that make important contributions to the trait, we can still make a false positive, or type I, error by claiming to have found one in an incorrect region. Hence we want to set a threshold so that it is unlikely we "detect" a genetic contribution to the trait in a region of the genome where none exists or where the effect on the phenotype is negligible. Although we idealize the null hypothesis by the assumption that there is no genetic contribution to the trait, it is important to keep this more realistic set of possibilities in mind.

It is convenient to reformulate the problem using the language of stochastic processes. The test statistics on each chromosome make (approximately) an Ornstein-Uhlenbeck process. The processes for different chromosomes are independent of each other. For testing a single marker, the null hypothesis would be rejected if at least one of the processes attains a value larger than 1.96 or smaller than -1.96, somewhere on the chromosome. In order to evaluate the actual significance level of this threshold when testing multiple markers, we generate 20 independent Ornstein-Uhlenbeck processes, one for each chromosome, a large number of times. For each of these iterations of 20 processes, we record whether the null hypothesis is rejected. The significance level is approximately equal to the proportion, among all iterations, of those iterations for which the null hypothesis is rejected. In order to simplify the programming, we will consider an idealized mouse genome composed of 20 autosomal chromosomes of equal length. The length of each chromosome will be taken to equal 80 cM, which is about the average size of a mouse chromosome.

```
> Z <- NULL
> for (i in 1:20) Z <- cbind(Z,OU.sim(beta,markers))
> dim(Z)
[1] 10000   180
```

The function "cbind" concatenates the columns of the matrices in its argument. (Similarly, "rbind" concatenates rows.)

```
> mean(abs(Z[,1]) >= 1.96)
[1] 0.0461
> mean(apply(abs(Z),1,max) >= 1.96)
[1] 0.993
```

As we can clearly see, the overall significance level is much higher than the nominal significance level of 5% associated with each one of the tests. If we want the probability of declaring a significant finding in the trial, when no such declaration should have been made, to be at most 5%, we must make the significance level of each separate test much smaller. This is a typical property of multiple testing. When our experiment involves scanning over many possibilities, and when an extreme value of the individual test statistics is an indication of departure from the null hypothesis, then a larger threshold should be used compared to the threshold that would have been used had there been only a single test.

Compare the distribution of $|Z_s|$ for a fixed value of $s$ to the distribution of $\max_t |Z_t|$. (See Fig. 5.2.)

```
> plot(c(0,5),c(0,1),type="n",xlab="|Z|",ylab="Density")
> lines(density(Z[,1],from=0))
> lines(density(apply(abs(Z),1,max)),col=gray(0.5))
> legend(0,1,legend=c("|Z_1|","max_t |Z_t|"),lty=c(1,1),
+     col=gray(c(0,0.5)))
```

Although the threshold of 1.96 is suitable for a test based on a single marker, for a test based on the most extreme marker, a threshold in the vicinity of 3.50 is required. A closer look gives the following results.

```
> mean(apply(abs(Z),1,max) >= 3.50)
[1] 0.061
> Z.max <- apply(abs(Z),1,max)
> z.val <- seq(3.52,3.58,by=0.01)
> p <- NULL
> for(z in z.val) p <- c(p, mean(Z.max >= z))
> names(p) <- z.val
> p
   3.52   3.53   3.54   3.55   3.56   3.57   3.58
 0.0569 0.0546 0.0528 0.0514 0.0499 0.0490 0.0465
```

Hence, a threshold of about 3.56 should be used, instead of the threshold 1.96, to ensure and overall significance level of 5% in a genome scan for a backcross design with and inter-marker spacing of 10 cM.

*Remark 5.2.* The LOD score is a different scale of measurement often used, particularly in human genetics, to define test statistics and significance thresholds. If we denote the thresholds given above by $z$, the LOD threshold is given by $LOD = z^2/4.6$. For example, the LOD threshold corresponding to $z = 3.57$ is $LOD = 2.77$. Conversely, in human genetics the LOD threshold of 3 is frequently recommended. This corresponds to a $z$ threshold of $[3 \times 4.6]^{1/2} \approx 3.71$. In this book, we will usually give significance thresholds as a value of $z$ or $z^2$. These can easily be converted to LOD thresholds.

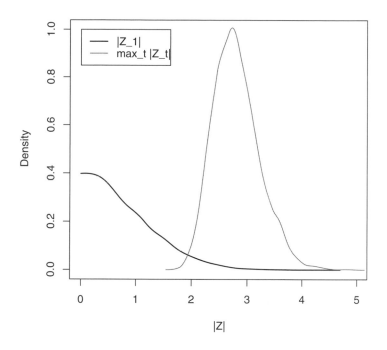

**Fig. 5.2.** The distribution of $|Z_t|$ and of $\max_t |Z_t|$

## 5.3 Mathematical Approximations of the Significance Level

In this section we describe and examine some theoretical results regarding the probability of false detection in a genome scan.indexScan!genome The results will be presented here without proof. Our main goal is to compare the analytical approximations with the numerical results of simulations. The formulas that will be presented can be put in the general context of computing the probability that a Gaussian process exceeds a high threshold. These results are derived in the form of asymptotic approximation to the probability.

We will start, however, with a simple and non-asymptotic upper bound on the probability – the *Bonferroni Inequality*. The asymptotic results presented later can be interpreted as refinements of this simple upper bound. The Bonferroni inequality bounds the probability of a union of events by the sum of the probabilities of the events in the union. Specifically, in our case we consider a collection $\{t_1, t_2, \ldots, t_m\}$ of $m$ markers scattered across the chromosomes. Rejecting the null hypothesis is associated with the event of having an extreme test statistic:

$$\left\{ \max_{1\leq i\leq m} |Z_{t_i}| > z \right\} ,$$

where $z$ is the threshold. This event can be reformulated as:

$$\left\{ \max_{1\leq i\leq m} |Z_{t_i}| > z \right\} = \bigcup_{1\leq i\leq m} \left\{ |Z_{t_i}| \geq z \right\} ,$$

namely that the statistic exceeds the threshold for *some* marker in the range. This leads to the inequality:

$$\Pr\left( \max_{1\leq i\leq m} |Z_{t_i}| \geq z \right) \leq \sum_{i=1}^{m} \Pr(|Z_{t_i}| \geq z) = m \times 2\left[1 - \Phi(z)\right] ,$$

since $2\left[1 - \Phi(z)\right]$ is the two-tailed probability that the absolute value of a standard normal random variable exceeds a threshold $z$. The threshold $z$ at which the Bonferroni upper bound equals 0.05 can serve as a threshold for the test statistics. The threshold for a Bonferroni-corrected significance level of $\alpha$ can be obtained by solving the equation $\alpha = m \times 2\left[1 - \Phi(z)\right]$:

```
> n.tests <- c(1,2,3,5,9,17,81)*20
> bon <- qnorm(1-0.025/n.tests)
> names(bon) <- paste("cM", c(80,40,25,20,10,5,1),sep="")
> bon
    cM80    cM40    cM25    cM20    cM10     cM5     cM1
3.02334 3.22722 3.34148 3.48076 3.63519 3.79598 4.16699
```

Here we computed the Bonferroni-corrected threshold for a range of markers densities, starting from the situation of placing a single marker in each chromosome, and ending in the situation of a 1 cM spacing between markers, which involves the use of 81 markers per chromosome; a total of 1,620 markers.

Let us simulate the actual genome-wide significance levels associated with the Bonferroni-corrected thresholds for each of the proposed marker densities. One strategy for carrying out such simulations is to run them separately for each density by changing each time the definition of "markers". We will use here an alternative approach. We will simulate the process for the densest collection of markers and then compute the appropriate statistics by the selection of subsets of markers:

```
> markers <- 0:80
> Z <- NULL
> for (i in 1:20) Z <- cbind(Z,OU.sim(beta,markers))
> marker.set <- list(cM80=c(41),cM40=c(21,61),
+     cM25=c(16,41,66),cM20=c(1,21,41,61,81),
+     cM10=seq(1,81,by=10),cM5=seq(1,81,by=5))
> marker.set <- lapply(marker.set,
+     function(m) as.vector(outer(m,(0:19)*81,'+')))
```

The function "`lapply`" applies to each of the components of a list the function given in its second argument. In our case, the function expands the indices of markers on the first chromosome to the indices of the corresponding markers in the entire genome. Note that the outcome of the function "`outer`" is a matrix. The function "`as.vector`" converts the matrix into a vector column by column. Having the indices of the sets of markers, we can turn to the computation of the maximal $Z$ score across all markers in the set:

```
> Z.max <- apply(abs(Z),1,max)
> for(i in seq(length(marker.set),1)) Z.max <-
+     cbind(apply(abs(Z[,marker.set[[i]]]),1,max),Z.max)
> colnames(Z.max) <- names(bon)
```

and then the actual significance levels associated with the Bonferroni-corrected thresholds:

```
> p.bon <- NULL
> for(i in 1:length(bon)) p.bon <-
+     c(p.bon, mean(Z.max[,i] >= bon[i]))
> names(p.bon) <- colnames(Z.max)
> p.bon
   cM80   cM40   cM25   cM20   cM10    cM5    cM1
 0.0461 0.0468 0.0468 0.0419 0.0373 0.0286 0.0104
```

In general, a Bonferroni-corrected threshold is conservative in the sense that it leads to an actual significance level smaller than the nominal significance level of 5%. Note that the Bonferroni inequality does not use any information regarding the covariances between the different quantities $Z_s$ and $Z_t$. It is based solely on the marginal distribution of the individual variables. This is a major advantage, since it leads to a very simple formula. However, this simplicity is a double-edged sword because the covariances between the variables may play an important role in the determination of the probability, and ignoring them may lead in some cases to a poor approximation. Indeed, the denser the collection of markers is, the larger the covariance between the markers. And, as we can see numerically, these larger covariances lead to a more conservative inequality.

Let us turn next to the construction of better, albeit more complex, approximations that do take into account the correlation between markers. We will write the formulas for these approximations in the context of a stationary Gaussian process with standardized marginal distributions. A process is called stationary if its distribution does not vary in time. In particular, a gaussian process is stationary if the mean function is constant, and if the correlation between two components is only a function of the time lag between them. Note, in particular, that the Ornstein-Uhlenbeck is a stationary process.

The approximation is determined by several process-specific parameters. One parameter is the total genetic length of that part of the genome spanned by markers, denoted here by $L$. In the context of the mouse genome its value is about 1600 cM, but it may be smaller if there are parts of the genome devoid of markers, for example, near the ends of chromosomes. The second parameter is the number of independent copies of the basic process involved. This number is denoted here by $C$. In our case this number relates to the number of chromosomes, which we take to be 20. (Since the sex chromosomes behave differently from the autosomes, it would also make sense to take $C = 19$. The practical impact of this difference would be negligible.) The effect of the correlation is introduced with the parameter $\beta$, which is the negative of the partial derivative of the correlation function at the origin:

$$\beta = \lim_{s \downarrow 0} \frac{1}{s}[\text{cov}(0,0) - \text{cov}(0,s)] .$$

In particular, for the Ornstein-Uhlenbeck process the covariance is given by $\text{cov}(0,s) = \exp\{-\beta s\}$, for $s \geq 0$. Consequently, the parameter $\beta$ defined here coincides with the rate $\beta$ in the exponent of the covariance function. In the backcross design this parameter takes the value 0.02. The next parameter is the distance between consecutive components of the process. The letter $\Delta$ will be used for this parameter. In genome scans this $\Delta$ corresponds to the spacing between markers. When a marker is placed each 10 cM, the parameter takes the value 10. For a two-sided threshold of $z$ and a genome consisting of $C$ chromosomes of total genetic length $L$, the basic approximation takes the form:

$$\Pr\left(\max_i |Z_{i\Delta}| \geq z\right) \approx 1 - \exp\left\{-2C[1 - \Phi(z)] - 2\beta L z \phi(z)\nu\left(z\{2\beta\Delta\}^{1/2}\right)\right\}.$$
$$(5.3)$$

Three functions appear in (5.3). The first two functions, $\Phi(\cdot)$ and $\phi(\cdot)$, are the cumulative distribution and the density functions, respectively, of the standard normal distribution. The third function $\nu(\cdot)$ appears quite frequently in *renewal theory* – the probability theory that provides many of the tools for the investigation of the probability that a random process exceeds a high threshold. The function $\nu(\cdot)$ can be expressed, in terms of the standard normal cumulative distribution function, as the infinite sum:

$$\nu(y) = 2 y^{-2} \exp\left\{-2\sum_{n=1}^{\infty} n^{-1}\Phi(-yn^{1/2}/2)\right\} .$$

For numerical purposes, it is sufficient to use the approximation:

$$\nu(y) \approx \frac{(2/y)(\Phi(y/2) - 0.5)}{(y/2)\Phi(y/2) + \phi(y/2)} .$$

The function Nu produces the given approximation of the function $\nu(\cdot)$:

```
> Nu <- function(x)
+ {
+     y <- x/2
+     (1/y)*(pnorm(y)-0.5)/(y*pnorm(y) + dnorm(y))
+ }
```

The function "OU.approx" applies approximation (5.3) for the probability that the absolute value of the Ornstein-Uhlenbeck process exceeds a threshold z.

```
> OU.approx <- function(z,beta,Delta,
+     length=1600,chr=20,center=0,test="two-sided")
+ {
+     d <- switch(test,"two-sided"=2,"one-sided"=1)
+     p <- 1-exp(-d*chr*(1-pnorm(z))
+         -d*beta*length*z*dnorm(z)*Nu(z*sqrt(2*beta*Delta)))
+     return(p-center)
+ }
```

Note the argument "test". This parameter determines the type of tail computation. The default is set to a two-sided test, which is the type we will be using in the context of experimental genetics. An alternative is to use a one-sided test, which involves only extreme positive values of the process. The function "switch" is used in order to set the value of "d" to be one or two depending on the value of test. Like in the Bonferroni case, a threshold can be set by solving (5.3) with the target genome-wide significance level at the left-hand side of the equation. A numerical solution can be found by trial and error, or more systematically by application of a root-solver algorithm.

The function "uniroot" searches within an interval for a root of a given function with respect to its first argument. If you inquire the help file of the function, you will note that its last argument takes the form "...". This format allows for the introduction of arguments for functions called by the given function. Hence, in particular, we can set values for the other arguments of the function "OU.approx" as it is being called by the root-finder function:

```
> app <- bon
> app[1] <- uniroot(OU.approx,c(2,4),
+     beta=0.02,Delta=80,length=0,center=0.05)$root
> app[2] <- uniroot(OU.approx,c(2,4),
+     beta=0.02,Delta=40,length=20*(60-20),center=0.05)$root
> app[3] <- uniroot(OU.approx,c(2,4),
+     beta=0.02,Delta=25,length=20*(65-15),center=0.05)$root
> app[4] <- uniroot(OU.approx,c(2,4),
+     beta=0.02,Delta=20,center=0.05)$root
> app[5] <- uniroot(OU.approx,c(2,4),
+     beta=0.02,Delta=10,center=0.05)$root
> app[6] <- uniroot(OU.approx,c(2,4),
+     beta=0.02,Delta=5,center=0.05)$root
> app[7] <- uniroot(OU.approx,c(2,4),
+     beta=0.02,Delta=1,center=0.05)$root
> app
    cM80     cM40     cM25     cM20     cM10      cM5      cM1
 3.01559  3.22481  3.32749  3.45765  3.56272  3.64895  3.78015
```

Observe that for a single marker per chromosome the mathematical approximation and the Bonferroni approximation are simple functions of each other. In the other two examples where markers are not located at the extreme ends of the chromosomes ($\Delta = 40$ and $\Delta = 25$), the effective length of each chromosome is less than 80 cM, and hence the effective length of the genome is less than 1600 cM. In fact, the effective length of each chromosome is the distance between the first and the last markers. These chromosome-specific effective lengths are summed in order to get an effective length for the entire genome, which is the value of $L$ used in the approximation.

The actual significance levels for the approximate thresholds can be obtained from the simulation results which are stored in the matrix "Z.max":

```
> p.app <- NULL
> for(i in 1:length(app)) p.app <-
+        c(p.app,mean(Z.max[,i] >= app[i]))
> rbind(p.bon,p.app)
          cM80    cM40    cM25    cM20    cM10     cM5     cM1
p.bon 0.0461  0.0468  0.0468  0.0419  0.0373  0.0286  0.0104
p.app 0.0478  0.0472  0.0495  0.0464  0.0476  0.0483  0.0470
```

Observe that the actual probabilities for the approximation are not too far from the target significant level of 0.05. Since the correlation between tests is handled properly, we no longer obtain an overly conservative threshold for small values of $\Delta$.

A simpler approximation to the significance level is obtained when one considers the continuous Ornstein-Uhlenbeck process. In this case $\Delta = 0$, thus $\nu = 1$, and the formula becomes:

$$\Pr\left( \max_{0 \leq t \leq L} |Z_t| \geq z \right) \approx 1 - \exp\left\{ - 2\,C[1 - \Phi(z)] - 2\,\beta L z \phi(z) \right\} . \tag{5.4}$$

Similarly to the Bonferroni inequality, this computation also provides an upper bound on the true significance level. However, unlike the Bonferroni upper bound, which is tighter when markers are sparser, this upper bound is tighter when the markers are denser.

## 5.4 Other Methods

There are other methods for approximating the genome-wide significance level. Whereas the simulations discussed above involved an approximating Ornstein-Uhlenbeck process, the most direct method is to simulate the regression statistics themselves. This requires that we simulate the phenotypes according to some hypothetical distribution and the marker genotypes, which under the hypothesis of no linkage are independent of the phenotypes. In principle this should provide a more accurate answer than the simulation of the Ornstein-Uhlenbeck process, since the Ornstein-Uhlenbeck process depends for its justification on the central limit theorem, which may or may not provide an adequate approximation in a particular case. This more direct simulation has two minor disadvantages: (i) it requires more programming and more computing time, and (ii) it requires that we choose a particular distribution for simulating the $y$'s, and to some degree the results will depend on the distribution chosen. (Of course, since the Ornstein-Uhlenbeck process does not involve the specific distribution of the $y$'s, its justification is in effect a mathematical proof that at least in large samples dependence on the distribution of the $y$'s is negligible.)

A conceptually different approach is to regard the phenotypes as given and study the conditional behavior of the process $Z_t$ as a function of the process of marker genotypes. In this case we need make no assumptions about the distribution of the phenotypes. To illustrate this approach we re-define $Z_t$ slightly by

$$Z_t = \frac{\sum_{i=1}^{n}(y_i - \bar{y})[x_i(t) - 1/2]}{n^{1/2}\hat{\sigma}_y/2} .$$

If we regard the $y_i$ as constants, then $Z_t$ is of the form of a weighted linear combination of centered Bernoulli variables divided by the standard deviation of that sum. Consequently for each fixed $t$, $Z_t$ will be approximately standard normal by the central limit theorem, and as a function of $t$ it will behave approximately like a Gaussian process with covariance function the same as the covariance function of $x_i(t)$. Hence without making any assumptions about the distribution of the $y$'s, we see that the process $Z_t$ behaves approximately as an Ornstein-Uhlenbeck process, leading to the same approximations as those given above. Now, strictly speaking, our significance level is conditional on the observed values of the phenotypes. If we want to simulate the exact process in this case, not just the Ornstein-Uhlenbeck approximation, we would choose once and for all some fixed set of phenotypes, perhaps by simulation, although they could also be the phenotypes of the actual study. We would then repeatedly sample processes of genotypes and compute different values of the process $Z_t$ for the fixed phenotypes. The frequency with which $\max_t |Z_t|$ exceeds a threshold $z$ would then be an estimate of the appropriate conditional probability, given the phenotypes; but, as above, the fact that the Ornstein-Uhlenbeck approximation does not involve the phenotypic values is an indication that the conditional probability and the unconditional probability are essentially the same in large samples.

Still another simulation-based method would start from fixed sets of phenotypes and genotypes, which could themselves be simulated or could be the actual phenotypes and genotypes of a particular study. The simulation process would be to randomly reassign the phenotypes to a stochastic process of genotypes. This would make the phenotypes independent of the process of genotypes to which they are assigned. Then the fraction of permutations where the value of $\max_t |Z_t|$ exceeds a threshold would be an estimator of the genome-wide significance level of the threshold. Since evaluating this fraction would be expensive and time consuming computationally, we can simulate a random selection of permutations from which to estimate the genome-wide significance level. While this method is very general with regard to assumptions about the distributions of phenotypes and genotypes, it requires a certain amount of symmetry in order to justify the permutation process. This is usually not a problem in experimental genetics, where one can design the experiment to achieve the desired symmetry, but can be a problem in human genetics, where experimental conditions are not so easily controlled.

## 5.5 P-values

In the preceding discussion, we asked how large a suitable threshold would be in order that the significance level would be a pre-specified small number, customarily taken to be 0.05. A different viewpoint leads to the notion of p-value or *attained significance level*. From this viewpoint one considers the maximum value of the process $|Z_t|$, where $t$ ranges over a particular set of values, say the set of all markers in the experiment. Call this value $z_{max}$. The p-value is then defined to be $\Pr\{\max_t |Z_t| \geq z_{max}\}$, which can be computed by any of the methods discussed above. If the p-value is 0.05, which occurs when $z_{max}$ equals the 0.05 significance threshold, we would be prepared to reject the null hypothesis, but with the realization that a value as large as the observed value might arise by chance 5% of the time even if the null hypothesis is true. But if the p-value is much smaller, say 0.0001, we would feel more confident in rejecting the null hypothesis, since there is a much smaller probability that an outcome as extreme as the one actually observed could occur by chance. Thus p-values are a convenient device for comparing different outcomes and measuring the extent to which we feel comfortable in rejecting a null hypothesis.

For a numerical example we consider the results of Sen and Churchill [70], which involved the phenotype of hypertension in rats. For illustrative purposes we consider the rat and mouse genomes to be the same and assume that markers were evenly spaced at 10 cM. According to the calculations given above a p-value of 0.05 would correspond approximately to the value $z_{max} = 3.56$. In fact there were several places in the genome where $Z_t$ assumed fairly large values, suggesting that several genes might be linked to the phenotype. The observed value of $z_{max}$ on chromosome 4 was about 5.8, for which the p-value is approximately $10^{-6}$, while on chromosome 15 it was 3.7, which has an approximate p-value of 0.03. Thus, while we are inclined to conclude that there may be genes contributing to the phenotype on both chromosomes 4 and 15, we feel substantially more comfortable with this conclusion regarding chromosome 4. This paper involves a number of other interesting problems, some of which are discussed in Chap. 8.

## 5.6 *The Recombination Fraction in the Haldane Model

The probability that a Poisson process with rate parameter $\lambda$ has $n$ jumps in an interval of length $t$ is equal to $\exp(-\lambda t)(\lambda t)^n/n!$, for $n = 0, 1, 2, \ldots$. For a backcross $\lambda = 0.01/\text{cM}$ is the rate of occurrence of crossovers. The recombination parameter $\theta$ is the probability of an odd number of crossovers, which equals the sum of these probabilities over all *odd* values of $n$. To evaluate this sum, recall that $e^x = \sum_{k=0}^{\infty}(x^k/k!)$. Hence, by setting $k = 2m+1$ for odd values of $k$,

$$[e^x - e^{-x}]/2 = \sum_{m=0}^{\infty} \frac{x^{2m+1}}{(2m+1)!} \; .$$

It follows that the recombination fraction for an interval of length $t = \Delta$ is

$$\theta = \exp(-\lambda\Delta) \sum_{m=0}^{\infty} \frac{(\lambda\Delta)^{2m+1}}{(2m+1)!} = \exp(-\lambda\Delta)[\exp(\lambda\Delta) - \exp(-\lambda\Delta)]/2 \; ,$$

which is the same as (5.1) when we put $\beta = 2\lambda$.

To evaluate the covariance function of $Z_t$ for a backcross, write $x(t)$ to denote the number of $A$ alleles at the marker locus $t$. Then, for two markers $t$ and $s$ at a recombination distance $\theta$, we obtain that

$$\mathrm{E}[x(t)x(s)] = \Pr(x_t = 1)\Pr(x(s) = 1|x(t) = 1) = (1/2)(1-\theta) \; .$$

It follows that

$$\begin{aligned}
\mathrm{cov}(x(t), x(s)) &= \mathrm{E}(x(t)x(s)) - \mathrm{E}(x(t))\mathrm{E}(x(s)) \\
&= (1-\theta)/2 - 1/4 = (1 - 2\theta)/4 = \exp(-\beta|t - s|)/4 \; ,
\end{aligned}$$

where the last equality follows from (5.1) with $\Delta = |t-s|$. If we use this result in the approximation (4.4) and recall that $\sigma_y^2 = \mathrm{E}[(y - m)^2]$, we obtain (5.2). Note that this argument already appeared in Sect. 4.2.3, except for the final substitution of (5.1).

## 5.7 Bibliographical Comments

The issue of multiple testing in genome scans and the relevance of the Ornstein-Uhlenbeck process for evaluating (genome-wide) significance levels was recognized by Lander and Botstein [47], who proposed the approximation based on infinitely dense markers. The approximation for equally spaced markers discussed in this chapter is due to Feingold, Brown, and Siegmund [28]. See [73] for additional references. Lander and Kruglyak [48] have been influential in emphasizing the importance of the issue of multiple testing. The permutation based method mentioned in Sect. 5.4 is used in a variety of statistical problems; its implementation and popularity in gene mapping are due to Churchill and Doerge [11].

## Problems

**5.1.** Get a more accurate assessment of the Bonferroni inequality and the approximation formula by: (a) increasing the number of iterations to 100,000; (b) using the R function "uniroot" in order to find the exact location where the curves cross the level 0.05.

**5.2.** (This exercise illustrates the effect of the recombination parameter $\beta$ on the significance thresholds.) In a RI design, unlike the backcross design, both types of homozygote are observable. Suppose we consider a test statistic based on comparing the difference in expression level of the phenotype between $AA$ homozygous and $aa$ homozygous animals. That will lead to a one-dimensional test statistic similar to the one used for the backcross. The main difference between the two cases is that a large number of meiotic events are relevant, compared to only one in the backcross. By using the recombination fraction for the RI found at the end of Chap. 3, it may be shown that the parameter $\beta$ is 0.08.

(a) Redo for this statistic the evaluation of the Bonferroni inequality and the approximation (5.3).
(b) What is the relation between the thresholds for the backcross design and the thresholds you get for the RI design? Explain!

**5.3.** For an intercross we introduced the test statistic $U(t) = Z_\alpha^2(t) + Z_\delta^2(t)$, computed at each marker for the intercross design. $Z_\alpha(\cdot)$ is a realization of an Ornstein-Uhlenbeck process with a covariance structure given by $\operatorname{cor}(Z_\alpha(t), Z_\alpha(s)) = \exp\{-0.02\,|t-s|\}$, and $Z_\delta(\cdot)$ is an independent realization of an Ornstein-Uhlenbeck process with a covariance structure given by $\operatorname{cor}(Z_\delta(t), Z_\delta(s)) = \exp\{-0.04\,|t-s|\}$.

(a) Apply the function "OU.sim" in order to simulate the process $U(\cdot)$ over a collection of markers.
(b) Compare the distribution under the null hypothesis of $U(t)$, the distribution of $Z_\alpha^2(t)$, the distribution of $\max_{t \in T} U(t)$, and the distribution of $\max_{t \in T} Z_\alpha^2(t)$.
(c) Explain any difference you find between the thresholds for a backcross design and for an intercross design.

**5.4.** The formula corresponding to (5.3) for the statistic $U(t) = Z_\alpha^2(t) + Z_\delta^2(t)$ for an intercross is

$$\Pr\big(\max_i U_{i\Delta} \geq u\big) \approx 1 - \exp\big\{-C\exp(-u/2) - \bar{\beta}Lu\exp(-u/2)\nu\big(\{2\,\bar{\beta}\Delta u\}^{1/2}\big)\big\},$$

$$(5.5)$$

where $\bar{\beta} = 0.03$ is the average of the values $\beta_1 = 0.02$ for $Z_\alpha$ and $\beta_2 = 0.04$ for $Z_\delta$. Find thresholds corresponding to markers spaced at $\Delta = 20, 10, 5$, and 1 cM based on this formula and compare them to the Bonferroni bounds and the simulations of Prob. 5.3.

**5.5.** For RI, the approximation given in (5.3) is valid, but the value of $\beta$ is different, to reflect the larger rate of recombination found in RI. In this case the parameter $\beta$ is found by (i) expressing the correlation $1 - 2\theta_{\mathrm{RI}}$ as a function of $\theta$ and hence as a function of $\exp(-0.02\,s)$ for markers at a genetic distance $s$ from each other, then (ii) expanding the exponential for small $s$ using the approximation $\exp(-x) \approx 1 - x$. The appropriate value of $\beta$ is the coefficient of $s$ in this expansion. You must also use for small $x$ the formula $1/(1 - x) = 1 + x + x^2 + \cdots$ and neglect terms like $x^2$, $x^3$, etc., which are smaller than $x$ when $x$ is small. Verify the value of $\beta$ given in Prob. 5.2 for RI obtained by repeated sib mating. What is the value for RI obtained by selfing?

# 6

# Statistical Power and Confidence Regions

The significance level is the determining factor in the specification of the rejection region of a statistical test. Only the distribution under the null assumption of no signal plays a role in setting the level of the threshold, once the test statistic and the general form of test are decided upon. However, after setting that threshold, one can examine other statistical properties of the resulting test. A central property is *statistical power* of the test – the probability to reject the null hypothesis when a signal is present. Since this probability depends on the values of the parameters, one often speaks of the *power function* to emphasize this dependence. For a test at a single marker, this probability is obtained approximately from the normal distribution; it is a function of the noncentrality parameter given by (4.6). In this chapter we will examine the concept of power for a whole-genome scan.

The primary interest is now focused on the case where the null hypothesis is false. Statisticians define the power of an hypothesis test as the probability of concluding correctly the falsity of the null hypothesis. However, the case of a genome scan is more subtle than a simple test of hypothesis. There exists the possibility that due to random fluctuations the significance threshold is exceeded on a chromosome that does not actually contain a QTL. Unless the threshold is also crossed on a chromosome containing a QTL, one would correctly conclude that the simple null hypothesis of no QTL anywhere in the genome is false, but would identify the chromosomal location of the QTL incorrectly. We are particularly interested in the probability of exceeding the threshold on a chromosome containing a QTL, or perhaps even at some marker close to the QTL, say within 20 cM. Although any definition of power in this context is somewhat arbitrary, in this book the power to detect a particular QTL will refer to the probability of correctly identifying the chromosome inhabited by the QTL. This means that when there is more than one QTL, power refers to specific QTL and can vary from one to another, depending on the effect of the QTL on the phenotype. In the case that multiple QTL lie on the same chromosome, one might want to make more subtle distinctions.

While keeping this possibility in mind, we shall for the most part ignore it in our statistical analysis.

At one end of the spectrum, when there is no QTL or only very weak QTL on a given chromosome, the power function is essentially equal to the (chromosome specific) significance level. At the other extreme are parameter values which correspond to a signal that is so large as to make the power approximately equal to one. We will mainly be interested in interim parameter values, values for which the power function is in the range 50%–95%.

The main application of the power function is to help us choose an experimental design – especially the breeding design, marker density, and sample size. The significance level is set to be some fixed value – typically 5% – regardless of the design. The differences between designs will be reflected in their power to detect the signal. Thus, for example, a sample size which is too small, or a collection of markers which is not dense enough, may compromise the chances of successful detection of a QTL. On the other hand, it is neither economically efficient nor ethical to use more animals then needed. Moreover, since genotyping is an expensive component in a genome scan, using more markers than needed is a waste of time and money that can be used for other purposes. Careful planning of an experiment can ensure efficient distribution of resources, without a substantial reduction in power. In the body of the text we focus on issues of sample size determination and the selection of the inter-marker spacing in the context of the backcross design. The power of other breeding designs is left to exercises in the problem set at the end of the chapter.

In the first section we identify the terms that affect the power of detection in a whole-genome scan. In the second section we introduce analytic formulas for the power. These formulas, like the formulas used for the computation of the significance level, are given in the context of the Ornstein-Uhlenbeck process. They will allow us to analyze the power function and to examine the effect of changing the values of various parameters. Consequently, in the third section we apply these formulas to select a good experimental design for the detection of a QTL. In the last two sections we deal with issues related more to estimation. In the fourth section we consider the construction of confidence intervals for the location of a QTL and in the last section the construction of a lower confidence bound for the effect of the QTL.

## 6.1 The Power to Detect a QTL

In this section we identify the parameters that determine the statistical properties of the monitoring process in the presence of a QTL. The examination is carried out in the context of local alternatives. In the case of a single marker, which was analyzed in Chap. 4, this corresponded to considering a shifted normal distribution. Similarly, for the multi-marker process, where the null distribution corresponded to the Gaussian Ornstein-Uhlenbeck process, the computation under the alternative will involve the same process with a shifted mean function. Since we deal with local alternatives, the correlation structure is not affected. The power to detect a signal is the probability that the maximum absolute value of the process, with the shifted mean, exceeds the threshold.

Recall that for the single marker process the expected value under the alternative of the test statistic at a marker equals its expectation at the QTL multiplied by the correlation between the marker and the QTL (see 4.6). For the backcross design and the simple model of QTL, we use the expectation at the QTL itself, which equals $\xi = (\alpha_0 + \delta_0)/(2\sigma_y)$. Here $\alpha_0$ corresponds to the (local) parameter of additive effect, $\delta_0$ is the (local) parameter of dominance effect, and $\sigma_y$ is the standard deviation of the phenotype. In terms of the original parameters of the model, $\alpha_0 = n^{1/2}\alpha$ and $\delta_0 = n^{1/2}\delta$. The correlation between marker and QTL for a backcross design, under the Haldane model of recombination, is equal to $1 - 2\theta = \exp\{-0.02\,|t - \tau|\}$. Here $|t - \tau|$ corresponds to the distance between a QTL located $\tau$ cM from the telomere, and a marker located $t$ cM from the telomere. This information is summarized by the formula

$$\mathrm{E}(Z_t) = \frac{\alpha_0 + \delta_0}{2\,\sigma_y}(1 - 2\,\theta) = \xi \exp(-0.02\,|t - \tau|)\,. \tag{6.1}$$

If the marker and the QTL are not on the same chromosome, then $\theta = 1/2$, the genetic distance from the QTL is defined to be infinite, and the expectation is equal to 0.

Equation (6.1) gives a complete description of the mean function of the multi-marker process under the model of a single QTL. The QTL is located on some chromosome, $\tau$ cM from the telomere. The mean function for the markers on the same chromosome is a function of their distance from the QTL. It decreases exponentially fast, on both sides of the QTL, as the distance from the QTL increases. The mean function of the multi-marker process over the other 19 chromosomes, which do not contain the QTL, is identically equal to 0. The covariance structure of the process under the alternative is identical to the covariance structure under the null assumption. Thus, the standard deviation of the test statistics $Z_t$ is equal to one, regardless of the location of the marker and its distance from the QTL. The correlation between any pair of markers is a function of the genetic distance between them. If the two markers are located on different chromosomes, then the genetic distance between them is infinite, and the two markers are uncorrelated. If the markers are located on the same chromosome, $s$ and $t$ cM from the telomere, then the correlation between them equals $\exp\{-0.02\,|t - s|\}$.

The distribution of the multi-marker process can be generated under the alternative similarly to the way it is generated under the null. The basic random process in both cases is the Ornstein-Uhlenbeck process. This processes describes the stochastic element in the behavior of markers for each chromosome. The deterministic element in the behavior is the mean function. This deterministic element is the difference between the null distribution and the distribution under the alternative. For the latter case a non-zero mean function is added for any chromosome carrying a QTL. Consequently, one can use the R function "OU.sim" that we wrote in the previous chapter in order to simulate the multi-marker process on a given chromosome. The vector of means may be added to the simulated process of marker-specific test statistics. We implement this approach in the function "add.qtl". This function takes as an input the matrix produced by "OU.sim", the location of markers, the coefficient of recombination $\beta$, and two new parameters: "q", the location on the first chromosome of the QTL (measured in cM from the telomere); and "xi", the noncentrality parameter. The mean vector is added and the altered scanning process is returned:

```
> add.qtl <- function(Z,beta,markers,q,xi)
+ {
+     d <- dim(Z)
+     if (length(markers) != d[2])
+         stop("Number of columns of simulated matrix
+         does not match the number of markers")
+     mu <- xi*exp(-beta*abs(markers - q))
+     Z <- sweep(Z,2,mu,"+")
+     return(Z)
+ }
```

The function "stop" may be used in order to stop a function in the case of a fatal error. The argument of the function is printed out if the error occurs. Similarly, a warning may be produced, in the case of a nonfatal errors, with the function "warning".

The function "sweep" returns a matrix obtained from an input matrix by sweeping the elements of a vector. The first argument is the input matrix. The second argument is the margin over which the elements of the vector should be applied. The third argument is the vector, and the fourth argument is the binary function that produces the elements of the output matrix from the application of the binary function to the element of the input matrix and the appropriate element of the vector. More generally, this function may be applied to arrays, which are higher-dimension extension of matrices.

Let us generate some paths of the resulting multi-marker process:

```
> markers <- seq(0,80,by=10)
> beta <- 0.02; q <- 40; xi <- 4
> Z <- NULL
> for (i in 1:20) Z <- cbind(Z,OU.sim(beta,markers,n.iter=5))
Loading required package: MASS
> chr1 <- 1:length(markers)
> Z[,chr1] <- add.qtl(Z,beta,markers,q,xi)
Error in add.qtl(Z, beta, markers, q, xi) :
```

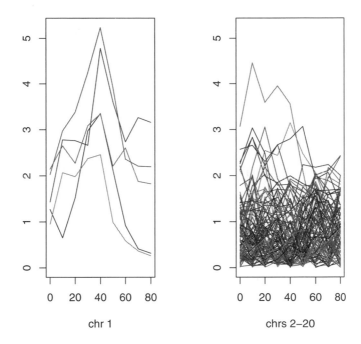

**Fig. 6.1.** Sample paths of the absolute value of the Ornstein-Uhlenbeck process under the alternative.

```
    Number of columns of simulated matrix
    does not match the number of markers
> Z[,chr1] <- add.qtl(Z[,chr1],beta,markers,q,xi)
```

Note an erroneous application of the function "`add.qtl`" resulted in an error message, which helped us to detect the source of the error and debug the mistake. The function "`OU.sim`" did not produce an error message since we added to its definition the expression "`require(MASS)`".

The paths of the scanning process are plotted in Fig. 6.1. The paths for chromosome 1 are shown on the left plot and the paths of chromosomes 2–20 are shown on the right plot. Observe that the values of the test statistics in the middle of chromosome 1 have levels which are consistently large. However, occasional extreme values can occur also in other chromosomes. The code that produces the two plots is:

```
> plot(c(0,80),range(abs(Z)),type="n",xlab="chr 1",ylab="")
> for(i in 1:5) lines(markers,abs(Z[i,chr1]),col=gray(i/7))
> plot(c(0,80),range(abs(Z)),type="n",xlab="chrs 2-20",
```

```
+       ylab="")
> for(j in 2:20)
+ {
+     chr = chr1+(j-1)*length(chr1)
+     for(i in 1:5) lines(markers,abs(Z[i,chr]),col=gray(i/7))
+ }
```

We detect a QTL if the maximum absolute value of the multi-marker process exceeds the significance threshold. Let us examine the distribution of this maximum, both when the signal is absent and when it is present. However, for the sake of determining the significance level we simulate these distributions for a whole-genome scan with markers at 0, 10, 20, ..., 80 cM. The power is considered only in the context of the chromosome that contains a QTL. Specifically, the QTL is located 40 cM from the telomere on chromosome 1. Note that a marker happens to be located right at that spot. The noncentrality parameter at the QTL equals 4 in our simulations.

```
> Z0 <- NULL
> for (i in 1:20) Z0 <- cbind(Z0,OU.sim(beta,markers))
> Z1 <- add.qtl(Z0[,chr1],beta,markers,q,xi)
> d0 <- density(apply(abs(Z0),1,max),from=1,to=7)
> d1 <- density(apply(abs(Z1),1,max),from=1,to=7)
> plot(d0,main="Densities of maximal statistics",
+     xlab="max |Z|")
> lines(d1,lty=2)
> legend(4.7,1,legend=c("Under H0","Under H1"),lty=1:2)
```

Examine the distributions of the test statistic under the two scenarios (Fig. 6.2). Note that although the distribution of the test statistic under the alternative tends to get higher values, still the two distributions cannot be separated perfectly. Any reasonable threshold that eliminates most of the exceedences of the test statistic under the null distribution must eliminate some occurrences of the test statistic under the alternative distribution as well. The traditional way to resolve this difficulty is to set a threshold with an acceptable proportion of the null distribution above it. This proportion is the significance level of the test – the probability of falsely rejecting the null hypothesis – and is typically set to be equal to 5%. The proportion of the distribution under the alternative that is *below* the threshold corresponds to the probability of falsely accepting the null hypothesis, and is called *the probability of a type II error*. Under the alternative the statistical power corresponds in the plot to the portion above the threshold, i.e., one minus the probability of a type II error. The larger this probability the better.

In order to determine the power, one must specify the threshold. In the previous chapter the value of 3.56 was suggested as a threshold for a genome scan in the backcross design with inter-marker spacing of 10 cM. This threshold was derived from an analytical formula for the significance value. Let us see the actual significance level and power via simulations.

**Densities of maximal statistics**

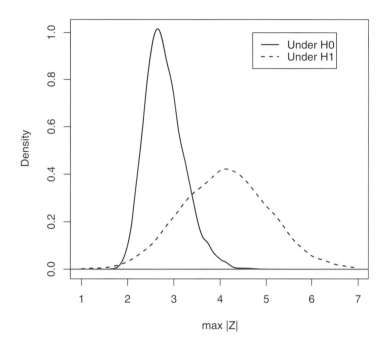

**Fig. 6.2.** The distribution of the test statistic under the null and under the alternative hypotheses.

```
> mean(apply(abs(Z0),1,max) >= 3.56)
[1] 0.0482
> mean(apply(abs(Z1),1,max) >= 3.56)
[1] 0.7044
```

It follows that the power to detect a QTL, located 40 cM from the telomere and with an (asymptotic) noncentrality parameter of 4, is about 70%.

What would be the power if the QTL happened to be between two markers? near the end of the chromosome, rather than near the center? for a smaller value of the noncentrality parameter? a larger value? We can obtain answers to these questions using simulations. For example:

```
> q <- 35; xi <- 4
> Z1 <- add.qtl(Z0[,chr1],beta,markers,q,xi)
> mean(apply(abs(Z1),1,max) >= 3.56)
[1] 0.6482
> q <- 5; xi <- 4
> Z1 <- add.qtl(Z0[,chr1],beta,markers,q,xi)
```

```
> mean(apply(abs(Z1),1,max) >= 3.56)
[1] 0.6455
> q <- 40; xi <- 3
> Z1 <- add.qtl(Z0[,chr1],beta,markers,q,xi)
> mean(apply(abs(Z1),1,max) >= 3.56)
[1] 0.343
> q <- 40; xi <- 5
> Z1 <- add.qtl(Z0[,chr1],beta,markers,q,xi)
> mean(apply(abs(Z1),1,max) >= 3.56)
[1] 0.9363
```

The task of exploring these question can be carried out much more efficiently once we have formulas for the statistical power, similarly to the formulas we have for the significance level. In the next section we describe such formulas.

## 6.2 An Analytic Approximation of the Power

As we saw, the power can be affected quite heavily by the location of the QTL on the chromosome. The probability of detecting a QTL may be substantially reduced if the QTL is located midway between markers, compared to the case where the QTL is located at the same location of a marker in the middle of the chromosome, and even more so if it is near an end of the chromosome. Since the formulas for approximating the power for the case when the QTL is between markers are substantially more complex than the formula when it is located exactly at a marker, we will present here only the formula for the latter case. For the former case, however, we do provide an R function, but not an explicit display of the formula. The interested reader may extract the mathematical expression from the code of the function.

When a marker is located at a QTL, detection occurs on that chromosome if either (i) the test statistic associated with the QTL/marker or (ii) the process associated with the flanking markers exceeds the threshold. The test statistic at the QTL has a normal distribution with mean $\xi$. Thus, the probability of the first case is simply the probability that such normal variable exceeds a threshold $z$. The mathematical derivation of the second case proceeds by conditioning on the value of the test statistic at the QTL, and analyzing the asymptotic conditional distribution of the process at the other markers. The resulting formula for a QTL not near either end of a chromosome is:

$$\Pr\big(\max_i |Z_{i\Delta}| \geq z\big) \approx 1 - \Phi(z - |\xi|) + \phi(z - |\xi|)\big[2\,\nu/|\xi| - \nu^2/(z + |\xi|)\big] \, , \quad (6.2)$$

where $\nu = \nu(z\{2\,\beta\Delta\}^{1/2})$. The first term, $1 - \Phi(z - |\xi|)$, is the probability that the test statistic associated with the QTL/marker exceeds the threshold

$z$. The second term corresponds to the probability of crossing the threshold by one or the other of the two flanking processes when the value of the statistic at the QTL is below the threshold.

*Remark 6.1.* Recall that for $\Delta \approx 0$, i.e., when the distribution of markers on the chromosome is very dense, the correction term $\nu$ is close to one.

*Remark 6.2.* When the QTL is located at the first or last marker on a chromosome, there are flanking markers only to one side. Then the approximation becomes

$$\Pr\big(\max_i |Z_{i\Delta}| \geq z\big) \approx 1 - \Phi(z - |\xi|) + \phi(z - \xi)\nu/|\xi| . \tag{6.3}$$

The function "power.marker" implements Formula (6.2):

```
> power.marker <- function(z,beta,Delta,xi)
+ {
+      nu <- Nu(z*sqrt(2*beta*Delta))
+      return(1-pnorm(z-xi) +
+          dnorm(z-xi)*(2*nu/xi-(nu/(z+xi))^2))
+ }
```

Applying this approximation we get:

```
> z <- 3.56; beta <- 0.02; Delta <- 10;
> xi <- 4
> power.marker(z,beta,Delta,xi)
[1] 0.7194996
```

Compare this to the probability of 0.7044, which was obtained via simulation.

The worst case scenario is to have a QTL midway between markers. The formula corresponding to (6.2) is much more complex since it involves conditioning on the values of the process $Z_{i\Delta}$ at both flanking markers. The expression is omitted, but we use the function power.midway in order to approximate the power in this case.

```
> power.midway <- function(z,beta,Delta,xi)
+ {
+      ul <- 5
+      nu <- Nu(z*sqrt(2*beta*Delta))
+      zz <- z - xi*exp(-beta*Delta/2)
+      cc <- sqrt(1 - exp(-2*beta*Delta))
+      fun1 <- function(x,beta,Delta,zz,cc) dnorm(zz-x)*
+          pnorm((zz-exp(-beta*Delta)*(zz-x))/cc)
+      term1 <- integrate(fun1,0,ul,beta=beta,
+          Delta=Delta,zz=zz,cc=cc)
+      fun2 <- function(x,z,beta,Delta,zz,cc) exp(-z*x)*
```

```
+              dnorm(zz-x)*pnorm((zz-exp(-beta*Delta)*(zz-x))/cc)
+          term2 <- integrate(fun2,0,ul,z=z,beta=beta,
+              Delta=Delta,zz=zz,cc=cc)
+          fun3 <- function(x,z,beta,Delta,zz,cc) dnorm(zz-x)*
+              exp(-z*x-z*(zz-exp(-beta*Delta)*(zz-x))+z^2*cc/2)*
+              pnorm((zz-exp(-beta*Delta)*(zz-x))/cc-z*cc)
+          term3 <- integrate(fun3,0,ul,z=z,beta=beta,
+              Delta=Delta,zz=zz,cc=cc)
+          return(1-term1$value+2*nu*term2$value-nu^2*term3$value)
+ }
```

The analytical expression involves an integral. Numerical integrals of functions with respect to their first argument can be computed with the function "integrate". The output is a list, with the component "value" containing the result of the integration.

In the simulations we obtained a power of 0.6482 when $\xi = 4$, $\Delta = 10$, and the QTL is located halfway between markers. Compare this probability to the analytical approximation:

```
> power.midway(z,beta,Delta,xi)
[1] 0.6498196
```

The *power function* involves the evaluation of the statistical power over the range of parameters under the alternative distribution. In the case of a whole-genome scan using the backcross and a given set of markers, these parameters are the location of the QTL and the strength of the signal, i.e., the noncentrality parameter $\xi$. Let us evaluate the analytical approximations over the range of the power function.

We start with the case of a QTL, which is located next to a marker in the middle of a chromosome. We consider here the case of the backcross design ($\beta = 0.02$), and an inter-marker spacing of 10 cM:

```
> q <- 40
> xi <- seq(0,6,by=0.1)
> n <- length(xi);
> ap.marker <- p.marker <- vector(mode="numeric")
> for (i in 1:n)
+ {
+     Z1 <- add.qtl(Z0[,chr1],beta,markers,q,xi[i])
+     p.marker[i] <- mean(apply(abs(Z1),1,max)>=z)
+     ap.marker[i] <- power.marker(z,beta,Delta,xi[i])
+ }
> plot(c(0,6),c(0,1),type="n",xlab="xi",ylab="Power")
> lines(xi,p.marker)
> lines(xi,ap.marker,lty=2)
```

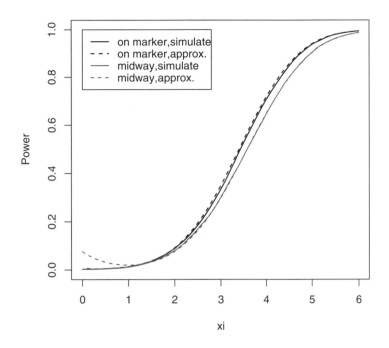

**Fig. 6.3.** The power function when the QTL is next to a marker and when it is midway between markers.

Next, let us consider the case of a QTL midway between markers:

```
> q <- 35
> ap.midway <- p.midway <- vector(mode="numeric")
> for (i in 1:n)
+ {
+     Z1 <- add.qtl(Z0[,chr1],beta,markers,q,xi[i])
+     p.midway[i] <- mean(apply(abs(Z1),1,max)>=z)
+     ap.midway[i] <- power.midway(z,beta,Delta,xi[i])
+ }
> lines(xi,p.midway,col=gray(0.5))
> lines(xi,ap.midway,lty=2,col=gray(0.5))
> legend(0,1,legend=c("on marker,simulate",
+     "on marker,approx.","midway,simulate","midway,approx."),
+     lty=c(1,2,1,2),col=gray(c(0,0,0.5,0.5)))
```

The resulting plot is displayed in Fig. 6.3. Note the reduction in power when the QTL is not perfectly linked to a marker. Observe good agreement between the analytical approximation and the simulated value. This agreement is destroyed when the QTL is between markers for values of $\xi$ less than one. Luckily, the exact evaluation of the power for such low values of the noncentrality parameter is of little practical interest.

## 6.3 Designing an Experiment

Experiments aimed at the dissection of the genetic component of traits in mice require substantial investment. It is unadvisable, therefore, to start such an effort, unless one is likely to obtain a successful outcome. The careful planning of the experiment is key in this regard. It ensures, on the one hand, that sufficient resources are devoted for the task. On the other hand, the optimal distribution of these resources lowers the chance of wasting both time and money.

The role of statistical experimental design is to identify the minimal requirements needed in order to be able to extract scientifically significant signals in the presence of background noise. It helps to use separate terminology in order to distinguish between *statistical significance* and *scientific significance*. Statistical significance is a formal term associated with the properties of the random mechanism underlying the background noise. It contrasts the strength of the observed signal in light of what could have been produced in a scenario where no real signal is present. The statistical significance is computed in the context of a null hypothesis, which assumes the absence of any signal. An observed signal can turn out to be statistically significant even if the underlying true signal is very weak. This can occur if the level of the background noise is low relative the amount of data gathered. The scientific significance, on the other hand, is not determined by statistical considerations. It reflects the specifics of the particular scientific discipline, and is given in terms of the strength of the underlying signal. Thus, in experimental genetics we may aim at detecting QTLs which have a strong enough effect on the phenotypic variance. This corresponds to large enough values of the locus specific heritability. The experiment is designed to discover genetic terms that have an effect above a given threshold.

To be more specific, let us consider an experiment using the backcross design. The strength of the genetic effect is given in terms of the locus specific heritability (see Chap. 2):

$$h^2 = \frac{(\alpha + \delta)^2/4}{\sigma_y^2} \ .$$

In order to design the experiment, we may set a minimal level of this quantity. Based on this level, the specifications of the trials can be determined. In this section we will describe the computations which identify the density of the genotyped markers and the sample size required in order to ensure a reasonably large chance to detect this minimal level of signal. We will carry these computations backwards. First, we will determine, for each inter-marker spacing, the appropriate noncentrality parameter which ensures the minimal statistical power. Second, we will determine the sample size associated with this noncentrality parameter. Finally, we will select the design which minimizes the overall cost.

### Determining the Noncentrality Parameter

Thousands of polymorphic markers, scattered throughout the mouse genome, are available for use. Hundreds of thousands, and even millions more, are expected with the identification of more and more SNP markers. Although not all markers are polymorphic between a given pair of inbred strains, the availability of genetic markers is typically not a limiting factor. Consequently, in principle, we can consider any density of markers. However, in order to simplify the computations, we will analyze here only four different possibilities of inter-marker spacings: $\Delta = 20, 10, 5$, or 1 cM.

In order to have a fair comparison, we will require that all cases possess the same significance level – 5%. Consequently, the thresholds will vary with the spacing. From the computations we made when dealing with the significance level, we found that these thresholds were approximately equal to $z = 3.46, 3.56, 3.65$, and 3.78, respectively.

Let us use the root finder "uniroot" and apply it to the function "power.midway" in order to identify the value of the noncentrality parameter that produces a power of 85% for each one of the designs. Note that the power is computed for a QTL between markers. This makes the conditions less favorable for designs with larger inter-marker spacings:

```
> delta <- c(20,10,5,1)
> z <- c(3.46,3.56,3.65,3.78)
> xi <- vector(mode="numeric")
> ap <- function(xi,z,beta,Delta,p=0)
+          power.midway(z,beta,Delta,xi)-p
> for (i in 1:length(z))
+ {
+     xi[i] <- uniroot(ap,interval=c(4,6),z=z[i],beta=beta,
+         Delta=delta[i],p=0.85)$root
+ }
> names(xi) <- delta
> round(xi,2)
   20   10    5    1
 4.97 4.71 4.64 4.65
```

The last row gives target values for the noncentrality parameter for each of the indicated inter-marker spacings.

## Determining the Sample Size

Next we turn to the determination of the sample sizes. Recall the definition of the noncentrality parameter:

$$\xi = \frac{n^{1/2}(\alpha + \delta)/2}{\sigma_y^2} .$$

Simple algebra can be used in order to express $n$ in terms of the other parameters ($\xi$, $\alpha$, $\delta$, and $\sigma_y$):

$$n = \frac{\xi^2 \sigma_y^2}{(\alpha + \delta)^2/4} .$$

Thus, one can easily compute the required sample size for each combination of parameters. In order to give an example, let $\alpha = \delta = 0.5$, and $\sigma_y^2 = 1.25$. Then

```
> a <- 0.5; d <- 0.5; ss <- 1.25
> n <- xi^2*ss^2/((a+d)^2/4)
> round(n)
 20  10   5   1
154 139 135 135
```

The increase in the required sample size in going from $\Delta = 1$ to $\Delta = 20$ is not as extreme as one might have suspected.

## Selecting the Design

One would like to select the best of the alternative designs. However, a design that involves the use of fewer animals will require more extensive genotyping, and vice versa. A way to evaluate the balance between these competing resources is via analysis of cost effectiveness. For the sake of the comparison we assume here that the expense per animal is $30, including purchase, maintenance, and phenotyping. For the genotyping one must order a set of primers for each marker. The actual genotyping is then carried out for each marker, across all the animals. We take the price of the set of primers to be $70, and the price of a genotype reaction to be $2. This leads to the figures:

```
> Pheno <- 30; Primers <- 70; Geno <- 2;
> cost <- n*Pheno + (1600/delta)*(Primers + n*Geno)
> round(cost)
     20      10       5       1
  34925   59695  112732  549005
```

According to these calculations, if judged only by the power to detect a QTL, the sparsest design is the most efficient. In the following sections we discuss problems of estimating the location and effect of a QTL, for which denser markers promise greater accuracy.

## 6.4  Confidence Sets

Statistical power refers to the probability of detecting a QTL anywhere on a chromosome. It is also important to estimate the genomic location of the detected QTL as precisely as possible. A confidence interval is a range of parameter values that depend on the data and contain the unknown actual value of the parameter with high probability. In this section we will discuss briefly the issue of the construction of confidence interval for the genomic location of the QTL. We will assume that a QTL has been detected on some chromosome. We will also assume that additional markers have been typed in the region suspected to contain the QTL, so the inter-marker spacing is small, say 1 cM.

For markers in the neighborhood of a detected QTL, the values of $|Z_t|$ will be substantially higher than the values for markers in other regions of the genome. Indeed, if we seek to guess the actual location of a QTL, the marker, say $\hat{t}$, where $|Z_t|$ assumes its maximum value is a reasonable choice. Because of random fluctuations, however, this guess will probably not be exactly correct. Consequently, we may want to incorporate the possibility that the QTL is somewhere in the neighborhood of $\hat{t}$. It seems intuitively clear that if $|Z_t|$ is almost as large at nearby markers as at $\hat{t}$, then the QTL might also be close to those markers, while markers where $|Z_t|$ is much smaller are unlikely to be close to the QTL.

To make this idea more precise, we assume that there are several markers in the neighborhood of $\hat{t}$, so to a rough approximation it is reasonable to assume that the QTL coincides with some marker locus. The procedure we propose is to determine a suitable constant $c^2$, so that the QTL might reasonably be located at any marker locus $\tau$ (on the same chromosome as $\hat{t}$) such that

$$\max_t Z_t^2 - Z_\tau^2 < c^2 \ . \tag{6.4}$$

Suppose we can choose $c^2$ to have the property that

$$\Pr_{\tau,\xi}\left( \max_t Z_t^2 - Z_\tau^2 \geq c^2 \right) \leq p \ , \tag{6.5}$$

where $\Pr_{\tau,\xi}$ denotes that the probability is calculated under the condition that the true QTL is at the marker locus $\tau$, and $\xi$ is the noncentrality parameter at that locus. The traditional values for $p$ are 0.1, 0.05, or some other small probability. The set of all marker loci $\tau$ satisfying (6.4) where $c^2$ satisfies (6.5), say $Q$, is called a $(1-p)$-*confidence set* for the QTL. Note that $Q$ is a random set, since it is a function of $\max_t Z_t^2$, $Z_\tau^2$, and $c^2$. It can be shown to have the property that for the true QTL $q$ itself $\Pr_q\{q \in Q\} \geq 1 - p$, hence the name "confidence set".

Unfortunately, there is a technical complication to this approach: the probability on the left-hand side of (6.5) depends both on $\tau$ and on the value of the noncentrality parameter $\xi$, which is unknown. Hence we cannot compute the left-hand side of (6.5), even in principle. However, this probability is fairly constant over a reasonably wide range of plausible values of $\xi$, so knowing $\xi$ exactly is not of critical importance. One can select a representative value for the noncentrality parameter, and use that value in the computation.

Let us demonstrate the approach in an example. Consider an inter-marker spacing of 1 cM. Let us select a critical value $c^2$ that satisfies (6.5) for $p = 0.1$ and $\xi = 6$:

```
> markers <- 0:80
> Z <- OU.sim(beta,markers)
> q <- 40; xi <- 6
> ZZ <- (add.qtl(Z,beta,markers,q,xi))^2
> ZZ.max <- apply(ZZ,1,max)
> d <- dim(ZZ)
> ZZ.dif <- matrix(ZZ.max,nrow=d[1],ncol=d[2])-ZZ
> conf.level <- function(cc,q,ZZ.dif,cl=0)
+       mean(ZZ.dif[,q+1] < cc)-cl
> cc <- uniroot(conf.level, interval=c(0,10),
+            q=q, ZZ.dif=ZZ.dif, cl=0.9)$root
> cc
[1] 4.678781
> 1-conf.level(cc,q,ZZ.dif,cl=0)
[1] 0.1
```

Next we turn to the investigation of the change in the confidence level of the confidence set as a function of the noncentrality parameter. While we are at it, let us compute also the expected lenght of the confidence set:

```
> xi <- 3:8
> cs.length <- cs.level <- vector(mode="numeric")
> for (i in 1:length(xi))
+ {
+       ZZ <- (add.qtl(Z,beta,markers,q,xi[i]))^2
+       ZZ.max <- apply(ZZ,1,max)
+       ZZ.dif <- matrix(ZZ.max,nrow=d[1],ncol=d[2])-ZZ
+       cs.level[i] <- mean(ZZ.dif[,q+1] < cc)
+       cs.length[i] <- mean(apply(ZZ.dif < cc,1,sum))
+ }
> names(cs.level) <- xi
> cs.level
        3       4       5       6       7       8
   0.8593  0.8727  0.8864  0.9000  0.9109  0.9216
> names(cs.length) <- xi
> cs.length
         3        4        5        6        7        8
   31.1014  18.0115  11.2120   7.6695   5.6781   4.3922
```

It can be seen from the results of the simulations that for values of $\xi$ ranging from 3 to 7 the probability that $\max_t Z_t^2 - Z_\tau^2 < c^2$ is not too far from its target value of 0.9. Observe, also, that the expected length of the confidence set is quickly reduced when the noncentrality parameter increases. In essence: the stronger the signal the more accurately it can be located.

Let us propose a theoretical explanation to the phenomena we have just observed. Suppose that markers are equally spaced at inter-marker distance $\Delta$. It can be shown by an argument similar to that used to derive (6.2) that when the QTL $\tau$ coincides with a marker locus not near either end of the chromosome,

$$\Pr_{\tau,\xi}\{\max_i Z_{i\Delta}^2 - Z_\tau^2 \geq c^2\} \approx 2\,\nu\big([2\,\beta\Delta(c^2+\xi^2)]^{1/2}\big)(1+c^2/\xi^2)^{1/2}\exp\left(-c^2/2\right). \tag{6.6}$$

The probability (6.6) is determined primarily by the factor $2\exp(-c^2/2)$ and only to a lesser degree by the factors depending on $\xi$. For a numerical example, suppose $\Delta = 1\,\mathrm{cM}$ and $c^2 = 4.6$, so $2\exp(-c^2/2) = 0.2$. For $\xi$ increasing from 4 to 7, which would be reasonable values for a detectable but not overwhelming QTL, the right-hand side of (6.6) decreases from 0.134 to 0.090, so under these conditions we have approximately a 0.90 confidence set.

We can compare the simulated confidence probabilities to the approximations proposed in (6.6):

```
> Delta <- 1
> cs.level <- rbind(cs.level,1-2*Nu(sqrt(2*beta*Delta*
+                   (cc+xi^2)))*sqrt(1+cc/xi^2)*exp(-cc/2))
> rownames(cs.level) <- c("simulated","analytical")
> round(cs.level,3)
               3     4     5     6     7     8
simulated   0.859 0.873 0.886 0.900 0.911 0.922
analytical  0.848 0.873 0.891 0.904 0.915 0.924
```

Equation (6.1) tells us how fast the process $Z_t$ is expected to drop away from its maximum as we move away from the true QTL. Hence, it also gives us a rough idea of the length of the confidence set: If we assume that $\xi$ approximately equals $\max_t Z_t$ (cf. Prob. 6.6) and $\xi\exp(-0.02\,|\tau - t|)$ equals $Z_t$, then solving the equation $\xi^2 - \xi^2\exp(-0.04\,|\tau - t|) = c^2$, for $2|\tau - t|$, should approximately produce the expected length of the confidence set. Solution to the equation yields the formula $50\log[\xi^2/(\xi^2 - c^2)]$ cM as a rough approximation of the expected length of the confidence set. Compare it to the results of the simulations:

```
> cs.length <- rbind(cs.length,50*log(xi^2/(xi^2-cc))/Delta)
> rownames(cs.length) <- c("simulated","approximation")
> round(cs.length,3)
               3     4     5     6     7     8
```

```
simulated        31.101 18.012 11.212 7.670 5.678 4.392
approximation    36.684 17.295 10.361 6.961 5.018 3.796
```

Note that even under favorable conditions the confidence set is more than 5 cM in width, which is still sizable if measured by the number of genes it can contain. Under less favorable conditions, the interval is much wider. As we saw in Chap. 3, the recombination fraction between two marker loci for a RI design is larger than for a BC or IC design. This means that the rate of decay of the value of $Z_t$ for a RI design from its maximum value near a QTL is more rapid. On the one hand, this has a negative effect on the power to detect linkage, since it means that the noncentrality parameter at markers located at a relatively short distance from a QTL can be substantially smaller than at the QTL itself. On the other hand, it permits the construction of more accurate confidence regions – roughly four times as accurate for a RI strain produced by sib mating and twice as accurate for a RI strain produced by selfing. For the example discussed above with $\xi = 5$, the expected size of the confidence set based on a BC design is about 10 cM. The same argument applied to $\theta_{\mathrm{RI}} = 4\theta/(1 + 6\theta)$ for a RI strain produced by sib mating suggests a confidence set of about 2.5 cM, which would be a substantial improvement in accuracy.

*Remark 6.3.* From the argument given above and from the approximation $\log[\xi^2/(\xi^2 - c^2)] = -\log(1 - c^2/\xi^2) \approx c^2/\xi^2$, we see that the length of the confidence interval is roughly inversely proportional to $\xi^2$. Since $\xi^2$ is itself proportional to the sample size $n$, the length of the confidence interval is inversely proportional to $n$. It is common in statistical problems that confidence intervals are inversely proportional to $n^{1/2}$. The difference reflects special features of gene mapping as a statistical problem. Apart from this theoretical curiosity, we have the more practical implication that the confidence interval can be made half as long by doubling the sample size, whereas one would ordinarily expect that the sample size must be quadrupled in order to halve the confidence interval. Nevertheless, it remains essentially impossible to identify a QTL uniquely by using only methods of gene mapping. We return to this issue briefly in Chap. 8.

## 6.5 Confidence Bounds for the Genetic Effect of a QTL

A confidence region for the genomic location of a QTL gives us an idea of where to concentrate a systematic search for the QTL. A confidence region for the genetic effect of a QTL, as measured, for example, in a backcross by the noncentrality parameter $\xi = \mathrm{E}(Z_\tau) = n^{1/2}(\alpha+\delta)/\sigma_y$, provides information about the importance of the QTL in determining the phenotype, indicates the range of outcomes we might expect to see if we replicated the experiment, and provides a basis for comparing the QTL in different strains of mice. If we could observe $Z_\tau$ itself (and were aware that the marker we are observing is indeed the QTL $\tau$), it would be a simple matter to use the normal distribution to find an approximate confidence interval for $\xi$. However, since we must search for $\tau$, the surrogate observation $\max_t Z_t$ will often not equal $Z_\tau$, especially if the spacing between markers is small, so a large value at one marker may be accompanied by a large value at nearby markers. When $\tau$ itself is a marker or close to a marker, $\max_t Z_t$ may exceed $Z_\tau$, hence giving a biased estimator of $\xi$. (See Prob. 6.6 for an exploration of this issue by simulations.)

To deal with this problem we find in this section a lower confidence bound for $\xi$, which deals with the problem of multiple comparisons and indicates that the genetic effect is at least of some minimal size. The same methods can be used to find an upper confidence bound (hence also a confidence interval), but in view of the applications and the problem of bias, the lower bound seems scientifically more interesting.

Consider $\max_t Z_t$, where the maximum is taken over all chromosomes and all marker loci. In testing the null hypothesis of no genetic effect we ask whether the observed value of $\max_t Z_t$ is incompatible with the null hypothesis in the sense that the probability when the null hypothesis is true of a still larger value is so small that it renders the null hypothesis untenable. A lower confidence bound for $\xi$ is found by asking a very similar question in a slightly different way. We consider a trial value of $\xi$, say $\xi_0$. In the test of hypothesis this was 0. Now, instead of asking whether $\xi_0$ is unreasonable, we ask how large $\xi_0$ must be so that is at least minimally reasonable. This can be formalized by regarding the observed maximum $z_{\max} = \max_t Z_t$ as a constant, then finding the value of $\xi_0$ that satisfies

$$\Pr_{\xi_0}\left(\max_t Z_t \geq z_{\max}\right) = p \tag{6.7}$$

(cf. (6.5)), where $p$ is some (usually small) probability. The value of $\xi_0$, which is a function of $z_{\max}$, say $\xi_0(z_{\max})$, is called a $(1-p)$-*lower confidence bound* for $\xi$, because it can be shown to satisfy

$$\Pr\left(\xi_0\left(\max_t Z_t\right) \leq \xi\right) = 1 - p . \tag{6.8}$$

In particular, if $\max_t Z_t$ were to equal exactly the 0.05 significance threshold, then the 0.95 lower confidence bound would be $\xi_0 = 0$. The case $p = 0.5$ is also of interest. In this case the confidence bound can be regarded as a point estimator of $\xi$, called a *median unbiased estimator* because by (6.8) the median of its cumulative distribution function is $\xi$. As a point estimator it does not have the bias inherent in using $\max_t Z_t$.

To compute the probability (6.7), we assume that there is only one QTL contributing to the trait. The complementary event, $\{\max_t Z_t < z_{\max}\}$, consists of the intersection of the events that the indicated inequality holds for all $t$ on the chromosome where the QTL is located and on all other (unlinked) chromosomes. These two events are independent, so the probability of their intersection is the product of their probabilities. Moreover, the first probability is just one minus the power, which is given approximately in (6.2) in the special case that the QTL coincides with a marker locus. Let $Q_1(z, \xi_0)$ denote the power. The second probability is just one minus the type I error probability associated with a genome containing one less chromosome (namely the linked chromosome), which is given approximately by (5.3). If we denote the type I error probability by $Q_0(z, C, L)$, then the probability on the left-hand side of (6.7) is

$$Q_1(z_{\max}, \xi_0) + Q_0(z_{\max}, C, L) - Q_1(z_{\max}, \xi_0)Q_0(z_{\max}, C, L) \, ,$$

which can be used, as indicated above, to give a confidence bound for $\xi$.

For a numerical example, suppose that $z_{\max} = \max_t Z_t = 3.9$ with a marker spacing of $\Delta = 1$ cM. Recall that the 0.05 significance level with this marker spacing was about 3.78, so the assumed value of $z$ is only marginally significant. If this were the outcome of directly observing $Z_\tau$, which is normally distributed with mean $\xi$ and variance one, a point estimator of $\xi$ would be 3.9 itself while a 0.95-lower confidence bound would be $3.9 - 1.65 = 2.25$. The method described above to account for the genome scan gives a 0.95-lower confidence bound of 1.35 if the QTL is assumed to coincide with a marker and we use the approximations (6.2) and (5.3) for $Q_1$ and $Q_2$. If we assume that the QTL is midway between markers, we require a more complicated calculation of $Q_1(z, \xi)$ (using, for example, the program given at the end of Sect. 6.2). The result would be a lower confidence bound of 1.45. Assuming the QTL is located exactly at a marker gives the most conservative result (smallest lower confidence bound).

The 0.5-confidence bound, i.e., the median unbiased estimator for the hypothetical data of the preceding paragraph when the QTL coincides with a marker locus is 3.57, somewhat less than the naïve biased estimate of 3.9.

Consider now the example from Sen and Churchill [70] mentioned in the preceding chapter, where $\Delta \approx 10$ cM. For the value $z_{\max} = 5.8$ on chromosome 4, the 95% lower confidence bound would be approximately 4.10. For the value $z_{\max} = 3.7$ on chromosome 15 it would be approximately 1.08. In the first case the confidence bound is very close to the naïve lower confidence bound of $z_{\max} - 1.645 \approx 4.15$; but in the second case the problem of bias is more serious and hence the difference from the naïve confidence bound is substantial.

*Remark 6.4.* The approximation for $Q_1(z, \xi)$ given in (6.2), which involves division by $\xi$, is a poor approximation for very small values of $\xi$. Since the lower confidence bound involves evaluation of $Q_1(z, \xi)$, for small $\xi$, before putting too much faith in the approximation, it is advisable to check its accuracy by simulation. (Consider Prob. 6.7 and see also Fig. 6.3.)

## 6.6 Bibliographical Remarks

The analytic approximations to the power are found in Feingold, Brown and Siegmund [28]. The discussion of confidence regions is taken from [22] and [74].

# Problems

**6.1.** The two parameters that determine the power for a given cross and a given inter-marker spacing are the proximity of the QTL to the markers and the noncentrality parameter ($\xi$).

(a) Use the programs in the text to simulate the power function, for the backcross design with 10 cM, 5 cM, and 1 cM inter-marker spacings, over a grid of values of the noncentrality parameter.
(b) Plot the power function and interpret the resulting plots.

**6.2.** Given a budget of $50,000 and other costs as described in the text, what would be the smallest effect that can still be detected with a power of 85% for the backcross design?

**6.3.** Write a function that simulates the test statistic for the intercross when the QTL is present. Use this function in order to examine the power function for detecting a QTL using this design. Consider both the cases that the trait is additive ($\delta = 0$) and the trait is either dominant or recessive ($\delta = \pm\alpha$). (Remember that the test statistic for an IC has two degrees of freedom and hence requires a different threshold from a BC – Probs. 4.3 and 5.3.) Assuming $\sigma_e^2$ is the same for an IC and for a BC, discuss the comparative advantages and disadvantages of a BC and an IC design.

**6.4.** In this problem we use the notation of Prob. 5.4. A mathematical approximation for the power of an intercross design for a QTL at 0 recombination distance from the nearest marker is given by

$$\Pr\left(\max_i U_{i\Delta} \geq u\right) \approx$$

$$1 - \Phi\left(u^{1/2} - \xi\right) + \phi\left(u^{1/2} - \xi\right)\left[\frac{1}{2\xi} + \left(\frac{u^{1/2}}{\xi}\right)^{1/2}\left\{\frac{2\nu}{\xi} - \frac{\nu^2}{u^{1/2} + \xi}\right\}\right]. (6.9)$$

In this equation $\xi = (\xi_1^2 + \xi_2^2)^{1/2}$, where $\xi_1 = (n/2)^{1/2}\alpha/\sigma_y$ and $\xi_2 = n^{1/2}\delta^2/2\sigma_y$ are noncentrality parameters for the additive and dominance effects, respectively, and $\nu = \nu((2\bar{\beta}\Delta u)^{1/2})$, where $\bar{\beta} = (\xi_1^2\beta_1 + \xi_2^2\beta_2)/\xi^2$. Compare the numerical results you obtained by simulation in the preceding problem with those obtained from this approximation.

**6.5.** Use the approximate power formula given in (6.3) to compare the power of a BC and a RI design based on repeated sib mating. Assume that the trait is additive and that $\sigma_e^2$ is the same for both designs. (This latter assumption may not be satisfied if there is more than one major QTL. See Prob. 4.4.) Recall that the appropriate value of $\beta$ for the RI design is 0.08 (Prob. 5.5), so the RI design will have both a higher threshold and a steeper drop-off in power when markers are relatively widely spaced and the QTL is located between markers.

**6.6.** A parameter of interest is the locus-specific heritability, defined in a back-cross by $h^2 = (\alpha + \delta)^2/4\,\sigma_y^2$. This parameter gives the proportion of the total phenotypic variance that is attributable to the QTL. Suppose we would like to use the data from a genome scan to get some idea of the magnitude of $h$. We found in Chap. 4 that $\mathrm{E}(Z_\tau) \approx n^{1/2}h$. Hence if we had a single marker at the QTL itself, we could use our test statistic to obtain an unbiased estimate of $h$. In a genome scan we could use the statistic $\max_t Z_t/n^{1/2}$, as a surrogate for $Z_\tau/n^{1/2}$. Simulate the expected value of this statistic for different values of $h$ ranging from 3.0 to 6.0 and $\Delta$ ranging from 1 to 10 cM, and for different positions of $\tau$ with respect to the nearest flanking markers. Does it seem reasonable from your simulations to conclude that $\max_t Z_t/n^{1/2}$ provides an acceptable estimator of $h$? Discuss.

**6.7.** Calculate lower confidence bounds for values of $\max_t Z_t$ ranging from 3.85 to 6.0. How does the "bias adjustment" for multiple comparisons change as $\max_t Z_t$ increases? Use simulations to approximate $Q_1(z, \xi)$ for some of the smaller values of $\xi$.

**6.8.** A strategy to minimize the cost of genotyping, while maximizing the amount of marker information in the neighborhood of a QTL, is to select markers to genotype in two stages. In the first stage one uses fairly widely spaced markers, say at 20 or 40 cM intervals. Then in regions where there is some evidence of linkage, say a $Z$ value in the range of 1 to 2, one adds more markers. For a specific example, suppose that markers are originally genotyped at 20 cM intervals. If $Z_t > z_1$, where $z_1$ is a parameter to be chosen, then in each of the 20 cM intervals flanking $t$, additional markers are placed at 5 cM intervals. Linkage is detected if for the final collection of markers, $\max_t Z_t > z_2$. Simulate this experiment with different values of $z_1$, and determine $z_2$ so that the significance level is 0.05. Find the power of this procedure for several values of the noncentrality parameter. How do the significance threshold and power compare with the case of a single stage of 5 cM genome scan? On average, how many fewer markers are genotyped per mouse? Repeat the same experiment with the modified first stage rule: if for two consecutive markers $(Z_t + Z_{t+\Delta})/2 > z_3$, then add new markers between the loci $t$ and $t + \Delta$. Does one of these first stage procedures seem much better than the other?

# 7

## Missing Data and Interval Mapping

In the preceding chapters we have assumed that all marker genotypes are known exactly. In practice there are often missing genotypes; and even when the genotypes of actual markers are not missing, one can regard the positions between actual markers as potential markers, the genotypes of which are missing. One can then make an attempt to reconstruct the missing information by statistical means. For important reasons that will become apparent in Chap. 9, problems of missing information are even more acute in outbred populations, and in particular humans. In this chapter we introduce, in the relatively simple situation of crosses between inbred strains, some basic techniques that are particularly useful for dealing with different kinds of missing information that arise in problems of gene mapping. In later chapters we will discuss more general approaches that can handle more complex situations.

Regarding positions between markers as missing and applying statistical tools in order to reconstruct genotypes of these imaginary markers from the observed genotypes is called *interval mapping* in the genetic literature. A large part of this chapter discusses the statistical properties of interval mapping. In particular, we present approximations that extend the approximations presented in Chaps. 5 and 6 to the case of interval mapping. We also use this opportunity to outline, in the context of interval mapping, some of the considerations that are applied in the development of the mathematical theory for gene mapping. Instead of mathematical proofs, we investigate, using simulations, the validity of some critical steps in the approximations. These investigations, which are discussed in the star-marked Sect. 7.2.1, involve more time-consuming simulations. They are not essential for the understanding of the statistical properties of interval mapping and may be skipped by readers who have less interest in this more technical issue.

## 7.1 Missing Marker Genotypes

In a previous chapter we introduced the notion of genetic markers, and discussed means for measuring the genotype of a given animal at a given marker. In general, the protocols for genotyping are quite established, in particular for SSR markers, and one can expect to obtain reliable outcomes for most of the animals and for most of the markers. However, no measurement technique is perfect and failed measurements are unavoidable. Indeed, every now and then one may find that the quality of DNA extracted from a given animal is poor or that a marker consistently produces unreliable genotypes. In the former case one may consider a second attempt of extracting DNA or one may drop the animal from the sample altogether. In the later case one may prefer to replace the failed marker by another marker in the same chromosomal region. The focus of this section, however, will be on the intermediate situation in which the genotyping of a marker is generally producing good results; but for a small fraction of the animals, due mainly to technical reasons, the marker is not working. Repeating the process of genotyping may be undesirable, since this will substantially increase the workload and the cost. Alternatively, one may apply some of the statistical remedies that have been developed for dealing with missing data.

Consider, for example, the case where a marker genotype is missing for a given animal, but we know the genotypes of flanking markers. If those flanking markers are sufficiently close, we have some information about the missing marker by virtue of the probabilities of recombination. To be specific, consider a backcross and adopt the notation of Sect. 4.2.3, which takes $x(t)$ to denote the indicator of heterozygosity at the locus $t$. Suppose that the three markers are $t_1 < t_2 < t_3$, and that $x(t_1) = x(t_3) = 1$, while $x(t_2)$ is missing. If $\theta_{12}$ denotes the recombination fraction between $t_1$ and $t_2$, $\theta_{23}$ denotes the recombination fraction between $t_2$ and $t_3$, and $\theta_{13}$ the recombination fraction between $t_1$ and $t_3$, then the conditional probability that $x(t_2) = 1$ is

$$\frac{(1 - \theta_{12})(1 - \theta_{23})}{1 - \theta_{13}} = \frac{(1 - \theta_{12})(1 - \theta_{23})}{(1 - \theta_{12})(1 - \theta_{23}) + \theta_{12}\theta_{23}}.$$

The numerator is the probability that there is no recombination in the interval from $t_1$ to $t_2$ and likewise no recombination in the interval from $t_2$ to $t_3$, which when multiplied by $1/2$ gives the probability that $x(t_1) = x(t_2) = x(t_3) = 1$. The denominator is the sum of this term and the smaller – usually much smaller – probability that there are recombinations in both these intervals, which when multiplied by $1/2$ gives the probability that $x(t_1) = x(t_2) = 1$. We can perform similar calculations for other values of $x(t_1)$ and $x(t_3)$. Thus, although we do not know the value of $x(t_2)$ exactly, we do know its probability distribution conditional on the observed data at the flanking markers.

The recombination fractions may be expressed as a function of the genetic distance between markers. It follows that

$$\hat{x}(t_2) = \Pr\big(x(t_2) = 1 | x(t_1), x(t_3), t_1, t_2, t_3\big) = \pi_{12}\pi_{23}/\pi_{13} \,, \qquad (7.1)$$

where

$$\pi_{12} = 1/2 + \big(x(t_1) - 1/2\big)\exp\{-0.02\,|t_2 - t_1|\},$$
$$\pi_{23} = 1/2 + \big(x(t_3) - 1/2\big)\exp\{-0.02\,|t_3 - t_2|\},$$
$$\pi_{13} = 1/2 + \big(1/2 - |x(t_1) - x(t_3)|\big)\exp\{-0.02\,|t_3 - t_1|\}$$

(see Prob. 7.1). Observe that $\pi_{12}$ may be used in the absence of an informative marker to the right of $t_2$ and $\pi_{23}$ may be used in the absence of an informative marker to the left of $t_2$.

In our regression statistic we can replace missing values of $x$ by these conditional expectations $\hat{x}$. If the number of missing values is relatively small, and the distance between markers not too large, the modified statistic can be expected to perform about the same as the statistic based on the assumption of complete information.

In order to explore the relative efficiency of the proposed procedure let us run a small simulation experiment. In the simulation we will generate a backcross experiment which is genotyped over a collection of markers in a given chromosome. Some of the genotypes will be randomly deleted. The interpolation procedure will be applied to the remaining genotypes in order to fill-in the missing genotypes. Three types of test statistics will be considered. The first is the test statistic computed for the complete collection of genotypes, prior to the random deletion. This statistic will serve as the gold standard against which the other two statistics compare. These other two will use only the undeleted genotypes. One will attempt to complete the missing genotypes with the aid of (7.1). The second will compute a regression statistic for a marker based solely on the undeleted genotypes of that marker.

We start by extending the functions "meiosis" in order to allow a general collection of markers:

```
> meiosis.chr <- function(GF,GM,markers)
+ {
+       rec.frac <- (1-exp(-0.02*diff(markers)))/2
+       n <- nrow(GF)
+       GS <- GF
+       from.GM <- rbinom(n,1,0.5)
+       GS[from.GM==1,1] <- GM[from.GM==1,1]
+       for (i in 1:length(rec.frac))
+       {
+           rec <- rbinom(n,1,rec.frac[i])
+           from.GM <- from.GM*(1-rec) + (1-from.GM)*rec
+           GS[from.GM==1,i+1] <- GM[from.GM==1,i+1]
+       }
+       return(GS)
+ }
```

Observe that for this version of the function "GF" and "GM" are matrices with a column dimension equal to the length of the collection of markers. We have used the function "diff" which, when applied to a vector, produces consecutive differences of the elements of the vector. Consider an example:

```
> n <- 10
> markers <- seq(0,80,by=20)
> gam1 <- matrix(1,n,length(markers))
> gam2 <- matrix(0,n,length(markers))
> x <- meiosis.chr(gam1,gam2,markers)
> x
        [,1] [,2] [,3] [,4] [,5]
 [1,]     0    1    1    1    1
 [2,]     0    0    0    1    1
 [3,]     0    0    0    1    1
 [4,]     0    0    0    0    0
 [5,]     0    0    0    0    0
 [6,]     0    0    0    1    1
 [7,]     1    1    1    0    0
 [8,]     1    1    0    0    0
 [9,]     0    0    0    0    0
[10,]     0    0    0    1    0
```

Five markers at 20 cM intervals are genotyped for 10 mice over a chromosome that is not linked to the phenotype. The matrix "x" corresponds to the full genotypic information, with one denoting heterozygosity and zero denoting homozygosity.

Next we erase 20% of the genotypes at random:

```
> p.miss <- 0.2
> index.miss <- sample(1:length(x),p.miss*length(x))
> obs <- x
> obs[index.miss] <- NA
> obs
        [,1] [,2] [,3] [,4] [,5]
 [1,]     0    1    1    1    1
 [2,]    NA    0    0    1   NA
 [3,]     0   NA   NA   NA    1
 [4,]     0    0    0    0    0
 [5,]     0    0    0    0    0
 [6,]     0    0    0    1    1
 [7,]     1    1   NA    0    0
 [8,]     1   NA   NA    0    0
 [9,]     0   NA    0    0    0
[10,]    NA    0    0    1    0
```

The function "`sample`" randomly selects as many values from its first argument as indicated by its second argument. The default, when the second argument not is provided, is uniform sampling with no replacement. In this example we use the function in order to sample indices of genotypes to be deleted. Observe that for the sake of deleting values we treat "x" and its copy "obs" as vectors. Yet, the resulting object "obs" is still a matrix. This matrix "obs" (and not "x") represents the actual observations the statistician gets to observe.

Next we prepare a function that takes as input a vector of observed genotypes and a locus. The output of the function is the expected genotype at the locus:

```
> expected.genotype <- function(geno,markers,locus)
+ {
+      index <- 1:length(markers)
+      x.hat <- NA
+      l.i <- index[!is.na(geno)&(markers < locus)]
+      if(length(l.i)>0)
+      {
+          l.i <- max(l.i)
+          l.m <- markers[l.i]
+          l.g <- geno[l.i]
+          p12 <- 0.5+((l.g==1)-0.5)*exp(-0.02*abs(locus-l.m))
+          x.hat <- p12
+      } else l.i <- NA
+      r.i <- index[!is.na(geno)&(markers > locus)]
+      if(length(r.i)>0)
+      {
+          r.i <- min(r.i)
+          r.m <- markers[r.i]
+          r.g <- geno[r.i]
+          p23 <- 0.5+((r.g==1)-0.5)*exp(-0.02*abs(locus-r.m))
+          x.hat <- p23
+      } else r.i <- NA
+      if (!is.na(l.i) & !is.na(r.i))
+      {
+          p13 <- 0.5+((l.g==r.g)-0.5)*exp(-0.02*abs(r.m-l.m))
+          x.hat <- p12*p23/p13
+      }
+      return(x.hat)
+ }
```

The function works by locating, first to the left and then to the right, the nearest informative marker. The probabilities p12, p23, and p13 are computed using (7.1) based on these distances and on the genotypes and the nearest informative markers. One can check that the definition of the probability p13 given here and the definition in (7.1) coincide. The situation is different when the missing information involves the first or last marker on the chromosome, as the reader can verify.

Let us apply the function "expected.genotype" in order to reconstruct the missing genotypes in "obs". For each marker we identify the animals that have a missing genotype at that marker. The function reconstructs, for each animal, the genotype at that marker. The reconstructed genotype matrix is called "x.hat":

```
> x.hat <- obs
> n.mark <- length(markers)
> for(i in 1:n.mark)
+ {
+     no.geno <- is.na(obs[,i])
+     obs.no.geno <- matrix(obs[no.geno,],ncol=n.mark)
+     x.hat[no.geno,i] <- apply(obs.no.geno,1,
+       expected.genotype,markers=markers,locus=markers[i])
+ }
> round(x.hat,3)
         [,1]  [,2]  [,3]  [,4]  [,5]
 [1,]  0.000 1.000 1.000 1.000 1.000
 [2,]  0.165 0.000 0.000 1.000 0.835
 [3,]  0.000 0.269 0.500 0.731 1.000
 [4,]  0.000 0.000 0.000 0.000 0.000
 [5,]  0.000 0.000 0.000 0.000 0.000
 [6,]  0.000 0.000 0.000 1.000 1.000
 [7,]  1.000 1.000 0.500 0.000 0.000
 [8,]  1.000 0.658 0.342 0.000 0.000
 [9,]  0.000 0.037 0.000 0.000 0.000
[10,]  0.165 0.000 0.000 1.000 0.000
```

Compare the matrix "x.hat" with the the two matrices "obs" and "x". Do the estimated probabilities make sense?

Next we look at more realistic simulations. Our goal is to understand the effect of the marker density and the proportion of failed genotypes on the relative statistical power. Specifically, we consider here three inter-marker spacings and three proportions:

```
> Delta <- c(40, 20,10)
> p.miss <- c(0.2,0.1,0.05)
> n <- 200
> qtl <- 40
> mu <- 5; alpha <- 0.3; sig <- 0.5
```

We imagine an experiment with $n = 200$ backcross mice. We assume that a QTL having an additive effect $\alpha = 0.3$ is located 40 cM from the telomere and that the residual variance is $\sigma_e^2 = 0.25$. It follows that the asymptotic noncentrality of the test statistic at the QTL itself is $\xi = n^{1/2}\alpha/2\,\sigma_y = 4.08$. Observe that one of the markers, for each of the three spacings considered here, is completely linked to the QTL. (In Chap. 6 we saw that if there were no missing information, sparser genotyping would be advantageous when a marker is completely linked to a QTL. However, our primary goal is to compare the relative loss in power for the two competing procedures when there is missing information.)

For each marker spacing and for each iteration we simulate a backcross experiment. The simulation involves the formation of a matrix which contains the genotypes at the marker and at the QTL. The genotypes at the QTL are used in order to simulate the vector y of phenotypes and the other genotypes form the matrix "x" of markers genotypes. These marker genotypes are modified, according to each of the proportions of failed genotypes, in order to form the matrix "x.obs". The matrix "x.hat" of reconstructed genotypes is created with the aid of the function "expected.genotype". Finally, the three test statistics are computed for each of the markers and their maximal absolute values are stored in three arrays "Z.full", "Z.hat", and "Z.obs":

```
> n.iter <- 5000
> Z.obs <- array(dim=c(n.iter,length(Delta),length(p.miss)))
> Z.full <- Z.hat <- Z.obs
> for(l in 1:length(Delta))
+ {
+     markers <- seq(0,80,by=Delta[l])
+     n.mark <- length(markers)
+     gam1 <- matrix(1,nrow=n,ncol=n.mark+1)
+     gam2 <- matrix(0,nrow=n,ncol=n.mark+1)
+     loci <- sort(c(markers,qtl))
+     qtl.index <- which.max(qtl <= loci)
+     for (i in 1:n.iter)
+     {
+         x <- meiosis.chr(gam1,gam2,loci)
+         y <- rnorm(n,mean=mu+alpha*x[,qtl.index],sd=sig)
+         x <-x[,-qtl.index]
+         for (p in 1:length(p.miss))
+         {
+             obs <- x
+             obs[sample(1:length(x),
+                 p.miss[p]*length(x))] <- NA
+             x.hat <- obs
+             for(m in 1:n.mark)
+             {
+                 no.geno <- is.na(obs[,m])
+                 obs.no.geno <- matrix(obs[no.geno,],
+                     ncol=n.mark)
```

```
+                        x.hat[no.geno,m] <- apply(obs.no.geno,1,
+                            expected.genotype,markers=markers,
+                            locus=markers[m])
+                    }
+                    Z.full[i,l,p] <- max(abs(sqrt(n)*cor(y,x)))
+                    n.hat <- apply(!is.na(x.hat),2,sum)
+                    Z.hat[i,l,p] <- max(abs(sqrt(n.hat)*cor(y,x.hat,
+                                        use="pairwise.complete.obs")))
+                    n.obs <- apply(!is.na(obs),2,sum)
+                    Z.obs[i,l,p] <- max(abs(sqrt(n.obs)*cor(y,obs,
+                                        use="pairwise.complete.obs")))
+                }
+            }
+ }
```

Observe that the test statistics are based on the correlation coefficients between the vector y and each of the columns of the matrices "x", "x.hat", and "x.obs". The correlations are normalized by multiplying them by the square root of the sample size. The option "use="pairwise.complete.obs"" is selected for the function "cor" in order to handle missing observations. Note that if all the markers on the chromosome are missing, the function "expected.genotype" will produce "NA" as an output.

After the simulation of the test statistics we can compute the power:

```
> z <- c(3.34,3.46,3.56)
> p.full <- matrix(ncol=3,nrow=3)
> colnames(p.full) <- paste("p=",p.miss,sep="")
> rownames(p.full) <- paste("Delta=",c(40,20,10),sep="")
> p.obs <- p.hat <- p.full
> for (l in 1:3)
+ {
+     for (p in 1:3)
+     {
+         p.full[l,p] <- mean(Z.full[,l,p] >= z[l])
+         p.hat[l,p] <- mean(Z.hat[,l,p] >= z[l])
+         p.obs[l,p] <- mean(Z.obs[,l,p] >= z[l])
+     }
+ }
```

We are using the significance thresholds for a genome scan which were found in Chap. 5 to be appropriate for the Ornstein-Uhlenbeck process. This process is the limit distribution of test statistics for the case of full information. However, it need not be the limit in the case of missing information or when genotypes are reconstructed. Here we ignore this issue, but we discuss similar questions in the next two sections, which deal with interval mapping.

The power, under ideal conditions of complete genotypic information, is given below. Not surprisingly, the power does not depend on the proportion of missing genotypes since the same unaltered matrix of genotypes "x" is used for all the proportions:

```
> round(p.full,2)
          p=0.2 p=0.1 p=0.05
Delta=40  0.79  0.79   0.79
Delta=20  0.77  0.77   0.77
Delta=10  0.75  0.75   0.75
```

Observe that there is a 5% relative loss of power for using a marker every 10 cM compared to using a marker every 40 cM. The order would have been reversed had we considered a QTL located midway between markers.

The relative efficiency of the procedure, which we define here to be the ratio of the power with reconstructed genotypes divided by the ideal power with complete genotype information, is given next. The actual power with reconstructed genotypes can be obtained by multiplying the efficiency by the ideal power given above.

```
> round(p.hat/p.full,2)
          p=0.2 p=0.1 p=0.05
Delta=40  0.86  0.94   0.97
Delta=20  0.92  0.97   0.98
Delta=10  0.96  0.98   0.99
```

Note that only when a large proportion of the genotypes are missing and the markers are sparse does one lose a noticeable part of the power.

Finally, we can investigate the *relative reduction* in power that results from not attempting to estimate the missing genotypes:

```
> round((p.hat-p.obs)/p.full,2)
          p=0.2 p=0.1 p=0.05
Delta=40  0.06  0.03   0.01
Delta=20  0.10  0.06   0.02
Delta=10  0.12  0.05   0.02
```

The reduction is more substantial when the collection of markers is relatively dense and the proportion of missing genotypes relatively large. This suggests that it may be beneficial to apply the more sophisticated statistical approach when the proportion of missing genotypes is large, say 10% or more.

## 7.2 Interval Mapping

Interval mapping is a method for enhancing the statistical power in gene mapping by a more complete exploitation of the statistical characteristics of the problem. In its most widely used formulation, interval mapping examines a candidate QTL, which usually is located in an interval between markers, by attempting to reconstruct, for each animal, the genotypes at the candidate locus from genotypes at nearby markers. The method presented in the previous section can be used for that reconstruction. Then the correlation between the reconstructed genotypes and the phenotypes can be tested by an appropriate statistic, with a large enough absolute value of that statistic indicating a significant divergence from the null hypothesis. Since the actual location of a QTL is unknown, the reconstruction of the genotype and the computation of the test statistic are carried out for each locus, resulting in a process across the entire genome. The significance level corresponds to the null probability that the process exceeds a threshold somewhere on the genome, and statistical power corresponds to exceeding the threshold in the vicinity of a genuine QTL.

The goal of Sects. 7.2.1 and 7.2.2 is to prepare the ground for the investigation of the statistical properties of interval mapping, which will be conducted in 7.2.3. That investigation will involve a comparison between interval mapping and the marker-by-marker approach that was developed in Chaps. 5 and 6. We will also describe in the 7.2.3 some analytic approximations to the significance level and the statistical power. In 7.2.1 we examine the approximation of the process of interval mapping statistics by an appropriate Gaussian process. This Gaussian process is much more amenable to analytic assessments and to simulation, as we show in 7.2.2. The material in 7.2.1 is somewhat technical and can be omitted by the reader willing to take on faith the accuracy of the approximating normal processes.

### 7.2.1 *Approximating the Process of Interval Mapping

Recall the Ornstein-Uhlenbeck process of Chap. 5. It is claimed that the distribution of the process of the marker-by-marker test statistics converges, as the sample size goes to infinity, to this Gaussian process. Relying on that assertion, we investigated the statistical properties of genome scans by simulating from the limiting process. Also, the appropriateness of the analytical approximations was tested against these simulations. A point we did not check is the level of discrepancy between the large-sample Gaussian approximations and the actual processes. Indeed, if the approximation of an actual trial by the limiting Gaussian process is not accurate enough, then the fact that one can have very good approximations to the distribution of the limit process is of questionable practical value. This problem is particularly acute in the determination of the significance threshold, which involves assessing the probability of rare events and may require large sample sizes before the central limit theorem applies. In this section we carry out an investigation of the effect of a finite sample size on the significance threshold for a sample of $n = 200$ and for various inter-marker spacings. The investigation of the appropriateness of the Gaussian approximation for the significance level for other sample sizes and for the statistical power is left as an exercise.

The function "expected.genotype", which was presented in the previous section, is too slow for conducting large scale simulations. In order to proceed with the investigation we should have a much faster version:

```
> inferred.genotypes <- function(x,n.grid,Delta)
+ {
+       d <- Delta/(n.grid+1)
+       n.mark <- ncol(x)
+       gr <- seq(d,Delta-d,length=n.grid)
+       p12 <- 0.5+outer(exp(-0.02*gr),t(x[,1:(n.mark-1)]-0.5))
+       p23 <- 0.5+outer(exp(-0.02*(Delta-gr)),
+               t(x[,2:n.mark]-0.5))
+       p13 <- 0.5+outer(exp(-0.02*Delta*rep(1,n.grid)),
+               t(x[,1:(n.mark-1)]==x[,2:n.mark])-0.5)
+       x.grid <- matrix(p12*p23/p13,nrow=nrow(x),byrow=TRUE)
+       return(x.grid)
+ }
```

The function "inferred.genotypes" takes as an input a matrix of genotypes over equally spaced markers. It returns a matrix of inferred genotypes over a grid between the markers. The other arguments of the function are "n.grid", which indicates the number of grid points between any pair of markers, and "Delta", the genetic distance between the actual markers.

With this function we can run the simulation:

```
> n = 200
> sigma = 0.5
> Delta = c(40,20,10,5)
> n.iter = 50000
> z.200 = matrix(nrow=2,ncol=length(Delta))
> colnames(z.200) = paste("Delta=",Delta,sep="")
> rownames(z.200) =c("Marker","Interval")
```

We consider four different inter-marker spacings. The thresholds resulting from the simulations will be saved in the matrix "z.200". The number of iteration is relatively large, n.iter = 50,000, since we want to assess very small tail probabilities. The simulations will be conducted for a single representative chromosome. In order to obtain an overall significance level of 5% for the entire 20 chromosomes, the probability of crossing the threshold over that one chromosome should roughly equal $0.05/20 = 0.0025$. Using 50,000 iterations should most likely put us within 10% of the true value. This is border line accuracy. However, using a much larger number of iterations will make the time consuming simulations we present next even more so. (A more efficient alternative would be to use the statistical technique of importance sampling for improving the accuracy of the simulations. Since this complicates the analysis and the computing, we do not pursue that route here.)

```
> for(l in 1:length(Delta))
+ {
+     n.grid <- Delta[l]-1
+     markers <- seq(0,80,by=Delta[l])
+     gam1 <- matrix(0,n,length(markers))
+     gam2 <- matrix(1,n,length(markers))
+     Z.mark <- Z.inter <- vector(length=n.iter)
+     for(i in 1:n.iter)
+     {
+         x <- meiosis.chr(gam1,gam2,markers)
+         x.sd <- sqrt(diag(var(x)))
+         x.grid <- inferred.genotypes(x,n.grid,Delta[l])
+         x.grid.sd <-  sqrt(diag(var(x.grid)))
+         y <- rnorm(n,sd=sigma)
+         sig.hat <- sd(y)
+         Z <- sqrt(n)*cov(x,y)/(x.sd*sig.hat)
+         Z.mark[i] <- max(abs(Z))
+         Z.grid <- sqrt(n)*cov(x.grid,y)/(x.grid.sd*sig.hat)
+         Z.inter[i] <- max(Z.mark[i],abs(Z.grid))
+     }
+     z.200["Marker",l] <-
+         sort(Z.mark)[round(n.iter*(1-0.05/20))]
+     z.200["Interval",l] <-
+         sort(Z.inter)[round(n.iter*(1-0.05/20))]
+ }
```

Observe that we are using the sample correlation coefficient (times the square root of the sample size) as the test statistic. We compute the statistic, and the resulting thresholds, both for the marker-by-marker process and for interval mapping over a 1 cM grid.

Consider the resulting thresholds:

```
> round(z.200,3)
         Delta=40 Delta=20 Delta=10 Delta=5
Marker      3.300    3.428    3.557   3.603
Interval    3.398    3.484    3.608   3.615
```

Comparing the thresholds for the marker-by-marker process with the thresholds for interval mapping, we see that slightly higher thresholds should in general be used for interval mapping. This observation applies in particular for larger inter-marker spacings.

## 7.2.2 Normal Approximation of the Process for Interval Mapping

The Ornstein-Uhlenbeck process is used to model the marker-by-marker process of test statistics when the sample size is large. It has a Gaussian distribution, as the limit of the test statistics should have, and its covariance function coincides with the covariance function of the test statistics themselves. In the same spirit, we would like to identify a Gaussian process that approximates the process of interval mapping.

In principle, an interval mapping statistic can be computed for any locus on the genome. It may be observed, in particular, that for the locus of a marker, since no reconstruction is needed, the interval mapping statistic is exactly the same statistic we have defined in Chap. 4. Consequently, the limiting interval mapping process coincides with the Ornstein-Uhlenbeck process at the locations of markers. Moreover, since reconstruction of the genotype at a location between markers is carried out by interpolation from the genotypes of the flanking markers, it can be shown that the limit process at a location between markers is obtained by an appropriate interpolation of the test statistics at the flanking markers.

To be more specific, let $t$ be a locus located between the two markers $t_i$ and $t_{i+1}$. Assume that the genetic distance between these two markers is $\Delta$ cM. Denote by $Z = (Z_{t_i}, Z_{t_{i+1}})'$ the vector of the test statistics at the two flanking markers and let

$$W = \frac{1}{1 - e^{-2\beta\Delta}} \begin{pmatrix} 1 & -e^{-\beta\Delta} \\ -e^{-\beta\Delta} & 1 \end{pmatrix} \quad \text{and} \quad \sigma_t = \begin{pmatrix} e^{-\beta(t-t_1)} \\ e^{-\beta(t_2-t)} \end{pmatrix}$$

be, respectively, the inverse of the correlation matrix of the vector $Z$ and the vector of correlations between the Ornstein-Uhlenbeck process at the markers and the true, unobserved value of the process to be constructed at $t$. It can be shown that the interpolated process at the location $t$ results from finding the conditional expectation of $Z_t$ given the observed data $Z$ to obtain:

$$\hat{Z}_t = E[Z_t|Z] = \sigma_t' W Z/(\sigma_t' W \sigma_t)^{1/2} . \tag{7.2}$$

Observe that $\hat{Z}_t$ defines a continuous process over the entire length of the chromosome. This process is Gaussian and is described by a smooth curve between markers. We will be using this process in order to investigate the statistical properties of interval mapping and to develop analytic approximations. However, before rushing ahead, let us take a moment and check that using this approximating process produces probabilistic statements that are similar to those produced by the actual process of test statistics. In particular, let us check that the thresholds obtained for this process resemble thresholds we obtained by simulating a process for 200 animals.

A convenient feature of the limit process is the fact that the extreme values of the expression in (7.2) can be found analytically. Indeed, it can be shown that if $\hat{t} = t_i + \Delta/2 + (2\beta)^{-1} \log\{Z_{t_{i+1}}/Z_{t_i}\}$ is located in the interior of the interval between the markers then it is the extremal point. Otherwise, the extreme occurs at one of the flanking markers. We exploit this fact in the function "interval.mapping", which takes as an input a matrix "z", the inter-markers spacing "Delta" and the recombination rate "beta" and returns as output a vector of the per row maximal absolute values of the interval mapping process:

```
> interval.mapping <- function(Z,Delta,beta)
+ {
+     n.mark <- ncol(Z)
+     r <- exp(-beta*Delta)
+     W <- matrix(c(1,-r,-r,1),2,2)/(1-r^2)
+     Z0 <- Z[,-n.mark]; Z1 <- Z[,-1]
+     t.hat <- log(pmax(Z1/Z0,0))/(2*beta)
+     t.hat <- pmin(t.hat,Delta/2)
+     t.hat <- pmax(t.hat,-Delta/2)
+     t.hat <- Delta/2+ t.hat
+     sig0 <- exp(-beta*t.hat)
+     sig1 <- exp(-beta*(Delta-t.hat))
+     term1 <- sig0*(W[1,1]*Z0+W[1,2]*Z1)
+     term2 <- sig1*(W[2,1]*Z0+W[2,2]*Z1)
+     term3 <- sig0^2*W[1,1]+2*sig0*sig1*W[1,2]+sig1^2*W[2,2]
+     Z.hat <- apply(abs((term1+term2)/sqrt(term3)),1,max)
+     return(Z.hat)
+ }
```

The functions "pmin" and "pmax" produce the point-wise minimum and maximum of their inputs, respectively. If these inputs are in the form of a vector and a scaler, as in the current case, the output is a vector of the same length, which is truncated by the scaler.

We use this function in order to compute a new matrix of threshold values "z.gauss", which are based on the limit Gaussian processes:

```
> z.gauss <- z.200
> for(l in 1:length(Delta))
+ {
+     markers <- seq(0,80,by=Delta[l])
+     Z <- OU.sim(0.02,markers,n.iter=n.iter)
+     Z.mark <- apply(abs(Z),1,max)
+     Z.inter <- interval.mapping(Z,Delta[l],beta=0.02)
+     z.gauss["Marker",l] <- sort(Z.mark)[n.iter*(1-0.05/20)]
+     z.gauss["Interval",l] <-
+         sort(Z.inter)[n.iter*(1-0.05/20)]
+ }
Loading required package: MASS
```

Examining the thresholds obtained by a simulation from the Gaussian process we see the basic phenomena we saw before, namely that interval mapping calls for slightly higher thresholds. The thresholds converge as markers become more dense.

In order to compare the Gaussian approximation with the finite-sample behavior let us look at the *difference* in thresholds:

```
> round(z.gauss-z.200,3)
          Delta=40 Delta=20 Delta=10 Delta=5
Marker       0.063    0.071    0.013   0.008
Interval     0.036    0.083   -0.010   0.024
```

Observe that the Gaussian approximation produces thresholds that are typically slightly conservative. Overall, the approximation is within one or two percent of the simulated value. However, based on the simulations we have conducted it is difficult to pinpoint the exact source of the discrepancy. First, there is the conservativeness that results from considering a Gaussian process as opposed to the finite sample process. Second, there is the conservativeness that stems from the consideration of a 1 cM grid in the finite sample simulations as opposed to the continuous maximization applied for the limit process. Finally, for both thresholds, simulation variability plays some role. All in all, we can conclude that, at least for a sample size of 200, the limit process produces good approximations.

### 7.2.3 The Statistical Properties of Interval Mapping

In this section we investigate the statistical properties of interval mapping in the context of a backcross experiment. In the previous section we described a Gaussian process that mimics the process of interval mapping for large samples. This Gaussian process is obtained by a smooth interpolation of the marker-by-marker process of test statistics. This marker-by-marker process is asymptotically the discrete skeleton of an Ornstein-Uhlenbeck process. Asymptotically for large sample sizes, the interval mapping process consists of smooth interpolations of the discrete skeleton of the Ornstein-Uhlenbeck process.

It is illuminating to examine the smooth paths of the interpolated process and compare it to the very irregular paths of the underlying Ornstein-Uhlenbeck process. Three simulated paths of both processes are presented in Fig. 7.1. These paths were simulated for a scenario in which a QTL, with a noncentrality parameter of size 5, is located between two markers at 50 cM from the telomere. Both the Ornstein-Uhlenbeck and the interval mapping processes coincide at the actual markers. However, their behavior between markers is very different. The paths for interval mapping connect markers with smooth curves, whereas the paths of the Ornstein-Uhlenbeck process are very ragged. The ragged process corresponds to the ideal situation in which we have a very dense collection of markers with test statistics computed for each. Interval mapping corresponds to an attempt to reconstruct the ideal situation with the available information. It should be clear from this picture that attempting to reconstruct genotypes at unmeasured markers may produce results which are quite different from the actual situation. Obviously, the

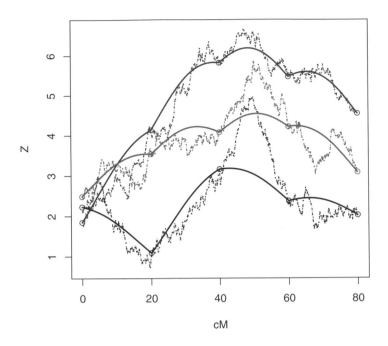

**Fig. 7.1.** Paths of the Ornstein-Uhlenbeck and the interval mapping processes.

statistical properties of interval mapping are not the same as the statistical properties of the underlying process. We will return to this point below.

Before doing so, let us present the code that produced Fig. 7.1. This program will not be used again, so is not necessary for understanding the rest of this section.

```
> markers <- seq(0,80,by=20)
> interval.grid <- seq(0,80,by=0.1)
> n.iter <- 3
> Z.den <- OU.sim(0.02,interval.grid,n.iter)
> Z.den <- add.qtl(Z.den,0.02,interval.grid,q=50,xi=5)
> mark.index <- seq(1,length(interval.grid),length=5)
> Z <- Z.den[,mark.index]
> V <- outer(markers,markers,
+   function(x,y) exp(-0.02*abs(x-y)))
> W <- solve(V)
> sig <- outer(markers,interval.grid,
+   function(x,y) exp(-0.02*abs(x-y)))
```

```
> std.int <- apply(sig,2,function(x) sqrt(x %*% W %*% x))
> Z.int <- t(apply((Z%*%W%*%sig),1,function(x) x/std.int))
> plot(range(interval.grid),range(c(Z.int,Z.den)),
+     type="n",xlab="cM",ylab="Z")
> for (i in 1:n.iter)
+ {
+       gr <- gray(0.5*(i-1)/n.iter)
+       lines(interval.grid,Z.int[i,],col=gr)
+       lines(interval.grid,Z.den[i,],col=gr,lty=3)
+       points(markers,Z[i,],col=gr)
+ }
```

In general, "solve" is used for solving a system of linear equations. In the specific implementation presented here, it produces the inverse of a matrix. We use it in order to compute the inverse of the correlation matrix. The interval mapping process is obtained by regressing the Ornstein-Uhlenbeck process on the marker by marker skeleton. The regression is performed by taking the matrix that results from the matrix product "Z%*%W%*%sig" and dividing it row by row by the elements of "std.sig". (Alternatively, we could have used the function "sweep".) Observe that the inverse of the covariance matrix for the markers is used in the regression formula.

The object "Z.int" stores the paths of the interval mapping process. When forming it, we use the function "t" that transposes a matrix. The reason we do so is to have "Z.int" in the same format as "Z", with rows containing independent copies of the scanning process.

Our main goal in this section is to investigate the statistical properties of interval mapping and to compare them to the statistical properties of the marker-by-marker approach from previous chapters. The investigation is carried out for the limiting Gaussian processes, which was shown in Sect. 7.2.1 to provide a reasonable approximation to the true, but much more complicated process of test statistics. We use the function "OU.sim" in order to simulate the null distribution of the marker-by-marker process. The function "add.qtl" is used for adding the effect of a QTL. Both these functions were used in previous chapters. The new function "interval.mapping", which is used for the computation of the maximal absolute interval mapping statistic, was presented in Sect. 7.2.2. It takes as an input a matrix of a marker process, a scaler "Delta" which represents the inter-marker spacing, and the rate of recombination "beta" (= 0.02 for the backcross). The output is a vector of length equal to the row dimension of the input matrix. Each component of the output vector contains the maximal absolute value of the interval mapping process, which is formed from the corresponding row of the matrix.

Let us consider in the simulation four different inter-marker spacings $\Delta = 40, 20, 10$, and $5$. First, we should obtain the appropriate thresholds for these spacings, both for the interval mapping method and for the marker-by-marker process. While we could use the thresholds that were computed in the past, an advantage of the Gaussian approximation is that the thresholds can be recomputed with almost no additional effort. Note, however, that because of sampling variability the thresholds will differ slightly from those obtained previously. We base our simulations for the thresholds on $10^6$ iterations in order to ensure accuracy. Recall that the default number of iterations in the function "OU.sim" is $10^4$. We repeat the simulations 100 times and average the results. Attempting to simulate directly $10^6$ repetitions would have exhausted the memory resources of our machine:

```
> Delta <- c(40,20,10,5)
> z <- matrix(nrow=2,ncol=length(Delta))
> colnames(z) <- paste("Delta=",Delta,sep="")
> rownames(z) <-c("Marker","Interval")
> n.rep <- 100
> for(l in 1:length(Delta))
+ {
+      markers <- seq(0,80,by=Delta[l])
+      z.mark <- z.int <- 0
+      for(rep in 1:n.rep)
+      {
+          Z <- OU.sim(0.02,markers)
+          Z.mark <- apply(abs(Z),1,max)
+          Z.int <- interval.mapping(Z,Delta[l],beta=0.02)
+          z.mark <- z.mark +
+              sort(Z.mark)[length(Z.mark)*(1-0.05/20)]
+          z.int <- z.int +
+              sort(Z.int)[length(Z.int)*(1-0.05/20)]
+      }
+      z["Marker",l] <- z.mark/n.rep
+      z["Interval",l] <- z.int/n.rep
+ }
```

Next we simulate the power curves for a QTL, which is located midway between the two central markers, and for a noncentrality parameter in the range between 3.5 and 6.5. To ensure accuracy we use $10^5$ iterations in the simulations. In order to save memory, we perform the simulations in 10 batches:

```
> ncp <- seq(3.5,6.5,by=0.1)
> qtl <- rep(40,length(Delta))+Delta/2
> Marker.power <- matrix(nrow=length(ncp),ncol=length(Delta))
> Interval.power <- Marker.power
> n.rep <- 10
> for (l in 1:length(Delta))
+ {
+      markers <- seq(0,80,by=Delta[l])
+      m.p <- i.p <- rep(0,length(ncp))
+      for(rep in 1:n.rep)
```

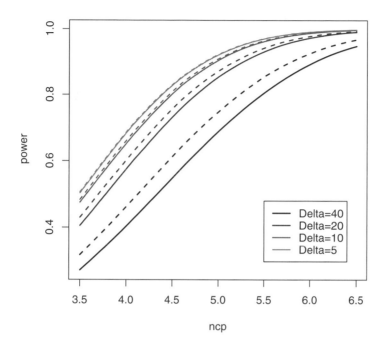

**Fig. 7.2.** Power functions for the marker process (*full line*) and interval mapping (*broken line*).

```
+      {
+          Z <- OU.sim(0.02,markers)
+          for(p in 1:length(ncp))
+          {
+              Z1 <- add.qtl(Z,0.02,markers,qtl[l],ncp[p])
+              Z.mark <- apply(abs(Z1),1,max)
+              Z.int <- interval.mapping(Z1,Delta[l],beta=0.02)
+              m.p[p] <- m.p[p]+mean(Z.mark >= z["Marker",l])
+              i.p[p] <- i.p[p]+mean(Z.int >= z["Interval",l])
+          }
+      }
+      Marker.power[,l] <- m.p/n.rep
+      Interval.power[,l] <- i.p/n.rep
+ }
```

Finally, the curves are plotted:

```
> plot(range(ncp),range(c(Marker.power,Interval.power)),
+      type="n",xlab="ncp",ylab="power")
> gr <- gray(0.75*(0:(length(Delta)-1))/length(Delta))
> for (l in 1:length(Delta))
+ {
+      lines(ncp,Marker.power[,l],col=gr[l])
+      lines(ncp,Interval.power[,l],col=gr[l],lty=2)
+ }
> legend(5.5,0.48,legend=paste("Delta=",Delta,sep=""),
+      lty=rep(1,length(Delta)),col=gr)
```

The resulting power curves are given in Fig. 7.2. A noticeable advantage in favor of interval mapping can be observed when $\Delta = 40$ and a smaller advantage when $\Delta = 20$. The advantage essentially disappears for smaller inter-marker distances. Another point worth making is that the gain in power using interval mapping for $\Delta = 40$ is much less than the gain in power obtained by decreasing the inter-marker distances. Hence, only a small fraction of the information lost is regained using genotype reconstruction.

We now turn to the construction of analytic approximations for the significance level and of the statistical power of interval mapping. Consider first the significance level. Apart from the Bonferroni inequality, two approximations for the probability of crossing a high threshold were presented in Chap. 5. In (5.3) an expression was given for the process observed only at markers. The simpler expression in (5.4) refers to the maximum of the ragged process under the assumption of infinitely dense markers. Neither expression seems appropriate for interval mapping. The former approximation will produce anti-conservative thresholds, since obviously $\max_t \hat{Z}_t \geq \max_m Z_{t_m}$. The latter approximation will tend to produce overly conservative thresholds, since the ragged process can more easily produce extreme values. Indeed, for conditions of a backcross experiment the threshold for the ragged process of Fig. 7.1 is approximately equal to 3.90. Compare this to the 3.51 threshold, which is appropriate for the process obtained by a smooth interpolation between markers which are placed 20 cM apart. Using the higher threshold would result in a reduction in the power which is far more than the gain that interval mapping can produce.

An alternative approximation of the significance level is the following:

$$\Pr\Big( \max_{1 \leq t \leq L} |\hat{Z}_t| \geq z \Big) \approx$$

$$1 - \exp \Big\{ - 2\, C[1 - \Phi(z)] - 2\, \beta L z \phi(z) \nu\big(z\{2\,\beta\Delta\}^{1/2}\big) \tag{7.3}$$

$$-2 \frac{L}{\Delta} \Big[ \frac{2\,\phi(z)}{\sqrt{2\pi}} \tan^{-1} \Big(\frac{1-r}{1+r}\Big) - \Pr\big(Z_1 \leq z, Z_2 > z\big) \Big] \Big\}. \tag{7.4}$$

Here $\nu$ is the function given in Chap. 5, $r = \exp(-\beta\Delta)$ is the correlation between the coordinates of the vector $Z$ at consecutive markers, and $(Z_1, Z_2)$ are two standard normal random variables having correlation equal to $r$. Observe that the expression in (7.3) is identical to the expression given for the process of markers in (5.3). The expression in (7.4) is a modification that results from the smooth interpolation between markers. The theoretical justification for this modification involves a classical formula of Rice for the expected number of upcrossings of a smooth Gaussian process. A detailed treatment is beyond the scope of this book. See [76].

We now turn to the investigation of the accuracy of the proposed approximation. First we write a function that implements the approximation.

```
> interval.sig <- function(z,beta,Delta,Length=1600,
+       chr=20,center=0)
+ {
+       require(mvtnorm)
+       r <- exp(-beta*Delta)
+       R <- matrix(c(1,r,r,1),2,2)
+       p <- 1-exp(-2*chr*(1-pnorm(z))
+           -2*beta*Length*z*dnorm(z)*Nu(z*sqrt(2*beta*Delta))
+           -2*(Length/Delta)*((dnorm(z)/sqrt(2*pi))*
+               2*(atan(sqrt((1-r)/(1+r)))))
+           -(as.vector(pmvnorm(lower=c(-Inf,z),
+               upper=c(z,Inf),corr=R)))))
+       return(p-center)
+ }
```

Note that the term in (7.4) requires the evaluation of the joint distribution function of a two-dimensional Gaussian vector. The package "mvtnorm" contains functions for the numerical evaluation of the multi-dimensional distribution function of both the normal and the $t$ distributions. Unfortunately, this contributed package is not part of the basic distribution of R for Windows. However, it can be easily installed by selecting "Packages" from the upper menu bar of the R Console window and then the option "Install package(s)..." from the menu that opens. A window with a list of the available contributed packages on CRAN is then opened (provided the computer is connected to the internet and a mirror of CRAN has been set). Selecting "mvtnorm" and clicking "OK" will initiate the process of installing the required package.

Let us examine the analytic approximation of the significance level for the interval mapping threshold that we found via simulation:

```
> Delta <- c(40,20,10,5)
> sig.approx <- vector(length=length(Delta))
> for(l in 1:length(Delta)) sig.approx[l] <-
+       interval.sig(z["Interval",l],0.02,Delta[l])
```

```
Loading required package: mvtnorm
> round(rbind(z["Interval",],sig.approx),3)
          Delta=40 Delta=20 Delta=10 Delta=5
             3.422    3.506    3.601   3.653
sig.approx   0.053    0.054    0.052   0.055
```

We can see that the analytic approximation is slightly conservative, giving probabilities within about 5%–10% of the simulated value. Significance thresholds obtained from the analytic approximation would be about 0.01–0.03 larger than those obtained by simulation.

For an approximation of the power let us consider the marker process about the QTL: $\ldots, Z_{t_{i-1}}, Z_{t_i}, Z_{t_{i+1}}, Z_{t_{i+2}}, \ldots$, where $t_i$ and $t_{i+1}$ are the two markers flanking the QTL. The probability is computed under the alternative. Therefore, the test statistics have non-zero expected values. At a marker, the expected value depends on the noncentrality parameter of the Ornstein-Uhlenbeck process at the QTL, where it (usually) is not observed, and on the correlation between the values of that process at the QTL and at the marker, where it is observed. In the formula for the statistical power we approximate the contribution of interval mapping between markers by the standardized average of the two flanking test statistics:

$$\hat{Z}_j = \frac{Z_{t_j} + Z_{t_{j+1}}}{\sqrt{2(1 + e^{-\beta \Delta})}} \ .$$

This can be shown by a large amount of tedious algebra to be a rough approximation to the true value, when the statistics at the flanking markers are about equal and hence the maximizing value of $t$ is close to the midpoint of the interval. As usual $\beta = 0.02$ is the rate of recombination in the backcross design and $\Delta$ is the genetic distance between the two markers. The approximation takes the form

$$\Pr\left(\max_{1 \leq t \leq T} |\hat{Z}_t| \geq z\right) \approx$$
$$1 - \Pr\left(Z_{t_j} < z, \hat{Z}_j < z, Z_{t_{j+1}} < z, \ i - m \leq j \leq i + m\right) . \qquad (7.5)$$

The left-hand side of (7.5) is a probability of the maximum over a continuum of points of the process created by interval mapping, whereas the right-hand side involves only a finite-dimensional multivariate normal distribution. The dimension of the multivariate distribution depends on $m$. For example, if $m = 0$, then we consider only the two flanking markers and their standardized sum, which involve a three-dimensional multi-normal distribution. If $m = 1$, then two more markers and two more sums are considered, making a total of seven variables.

Implementation is carried out with the function "interval.power". The input to the function is the threshold z, the rate beta, and the spacing $\Delta$. The vector "markers" gives the location of the markers to include in the computation, and "qtl" gives the location of the QTL with respect to the markers. The noncentrality parameter "xi" can enter as a vector, in which case the output is the vector of approximated probabilities, one for each value of "xi". The function is:

```
> interval.power <- function(z,beta,markers,qtl,xi)
+ {
+       require(mvtnorm)
+       V <- outer(markers,markers,function(x,y)
+            exp(-beta*abs(x-y)))
+       sig <- sqrt(sum(V[1:2,1:2]))
+       n.mark <- length(markers)
+       R <- matrix(nrow=2*n.mark-1,ncol=2*n.mark-1)
+       R[1:n.mark,1:n.mark] <- V
+       for(i in 1:(n.mark-1)) for(j in (1:n.mark-1))
+            R[n.mark+i,n.mark+j] <- sum(V[i+0:1,j+0:1])/sig^2
+       for(i in 1:n.mark) for (j in 1:(n.mark-1))
+            R[i,n.mark+j] <- sum(V[i,j+0:1])/sig
+       for(j in 1:n.mark) for (i in 1:(n.mark-1))
+            R[n.mark+i,j] <- R[j,n.mark+i]
+       Mu <- exp(-beta*abs(markers-qtl))
+       Mu <- c(Mu, (Mu[-1]+Mu[-n.mark])/sig)
+       p <- vector(length=length(xi))
+       for (i in 1:length(p)) p[i] <-
+         as.vector(pmvnorm(upper=z,mean=xi[i]*Mu,corr=R))
+       return(1-p)
+ }
```

Most of the code is devoted to the construction of the correlation matrix for the multivariate normal distribution and for the computation of the correlation component of the mean. In the last loop the function "pmvnorm" is called in order to compute the probability of the appropriate event.

In order to test the accuracy of the approximation, we consider using four markers ($m = 1$ in (7.5)). We recompute the power curves using the approximation and then plot them alongside the simulated power curves:

```
> power.approx <- matrix(nrow=length(ncp),ncol=length(Delta))
> n.mark <-4
> for(l in 1:length(Delta))
+ {
+       markers <- (1:n.mark)*Delta[l]
+       qtl <- (n.mark+1)*Delta[l]/2
+       power.approx[,l] <- interval.power(z["Interval",l],
+            0.02,markers,qtl,ncp)
+ }
> plot(range(ncp),range(c(power.approx,Interval.power)),
+       type="n",xlab="ncp",ylab="power")
> gr <- gray(0.75*(0:(length(Delta)-1))/length(Delta))
```

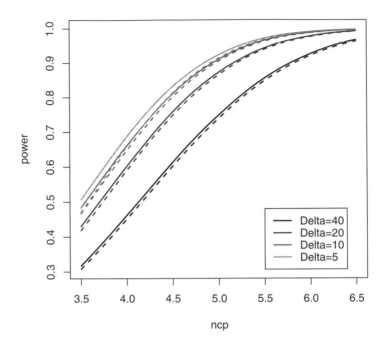

**Fig. 7.3.** Analytic approximation of the statistical power in interval mapping (simulation = *solid line*, analytic approximations = *broken line*).

```
> for (l in 1:length(Delta))
+ {
+      lines(ncp,Interval.power[,l],col=gr[l])
+      lines(ncp,power.approx[,l],col=gr[l],lty=2)
+ }
> legend(5.5,0.48,legend=paste("Delta=",Delta,sep=""),
+      lty=rep(1,length(Delta)),col=gr)
```

The plot is presented in Fig. 7.3. Note the close proximity of the *broken line*, which represents the approximation, to the *solid line*, which represents the simulations. Only for $\Delta = 5$ can one note a substantial discrepancy. To some extent this discrepancy can be reduced by increasing the number of markers considered.

# 7.3 Bibliographical Comments

Interval mapping was proposed by Lander and Botstein in the 1980s. The standard reference is Lander and Botstein [47]. This was not their first paper on the subject, but it stands out as a landmark for the lucidity of the exposition and the variety and richness of the ideas. In that paper the authors used a likelihood analysis, based on the assumptions that the residuals in the regression model for the QTL are normally distributed, to implement interval mapping. For a more detailed exposition see [19].

In this chapter we have followed the suggestion of Haley and Knott [36], who noted that a regression approach simplifies the computational burden substantially. It may be shown that the regression statistic is the score statistic under the normal model, and under an approximating Gaussian model the score and likelihood ratio statistics are the same. For these reasons the two methods can be expected to lead to similar results, at least when the noncentrality parameter is not too large. In some cases the regression and likelihood methods can lead to somewhat different results. The regression method is robust in the sense that its properties in large samples follow from the central limit theorem and do not require additional assumptions. The likelihood ratio statistic is calculated under the assumption that the residuals are normally distributed, and its properties may change if that assumption is not at least approximately true.

Judging by the power to detect linkage, we find only a modest value for interval mapping, and then only when markers are widely separated. This judgment is consistent with simulations described in [17] and [22], and stands in contrast to Lander and Botstein's enthusiasm for and the wide acceptance of interval mapping.

For additional theoretical discussions of interval mapping see [62], [63], and [76].

# Problems

**7.1.** Prove (7.1).

**7.2.** Use simulations in order to investigate the appropriateness of the Gaussian approximation for the significance level for different sample sizes. Do the simulation both for interval mapping and for the marker-by-marker statistics.

**7.3.** Use simulations in order to investigate the appropriateness of the Gaussian approximation for the power. Do the simulation both for interval mapping and for the marker-by-marker statistics.

# 8

## Advanced Topics

In this chapter we introduce briefly a number of more advanced topics, some of which are of current research interest. Some of the topics involve multidimensional parameters and are closely related to Probs. 5.4 and 6.4, which are concerned with detection of linkage in an intercross when the model permits both an additive and a dominance effect. (See also Prob. 4.3.) In these cases the new problems can be solved by conceptually straightforward extensions of results derived previously. In other cases the analysis becomes substantially more difficult and the results less conclusive, so the discussion here is at best an introduction.

## 8.1 Multivariate Phenotypes, Gene-Environment Interaction, and Longitudinal Studies

In an intercross, if we allow for both additive and dominance effects, our model involves two parameters; and we saw in Chap. 4 that the statistic to test for *either* an additive *or* a dominance effect is two dimensional. Multidimensional statistics can arise in a variety of other ways. Two examples are discussed here: (i) multidimensional phenotypes and (ii) gene covariate interactions.

### 8.1.1 Multivariate Phenotypes

Some traits are associated with a complex of phenotypes, which may be genetically related. For example, blood pressure is measure by systolic and diastolic pressure; cholesterol can be high density or low density, and diseases associated with cholesterol may also be affected by other lipids.

Assume then that we have a backcross in which we want to study two phenotypes $y_1$ and $y_2$, both of which satisfy the model (4.1), where $x_1$ and $x_2$ are the genotypes at the QTL $\tau_1$ and $\tau_2$, respectively. These may be the same, linked, or unlinked QTL. In this section we consider methods to study both phenotypes simultaneously.

In addition to the assumption that $y_1$ and $y_2$ both satisfy a model of the form of (4.1), assume also that $e_1$ and $e_2$ have correlation $r$. Let $\rho = \text{cor}(y_1, y_2)$. (See Problem 1 for the evaluation of $\rho$ in terms of $r$ and the genetic effects at $\tau_1$ and $\tau_2$.) Suppose we form a bivariate statistic, $(Z_1(t), Z_2(t))'$, expressed in the regression form of Sect. 4.2.3 for the two phenotypes $y_1$ and $y_2$ relative to the same marker data $\{x(t)\}$. As stochastic processes the $Z_i(t)$ are both asymptotically Gaussian processes, as discussed in Chaps. 5 and 6, and $\text{cov}[Z_1(t), Z_2(t)] = \rho$. One can use (4.4) to show along the lines of the calculations in Sect. 4.2.4 that if the QTL $\tau_1$ and $\tau_2$ are linked with recombination fraction $\theta$, then $E[(Z_1(\tau_1), Z_2(\tau_1))] = (n^{1/2}/2)((\alpha_1 + \delta_1)/\sigma_{y1}, (1 - 2\theta)(\alpha_2 + \delta_2)/\sigma_{y2})$, with a similar formula holding at $\tau_2$.

In our discussion of the intercross design in Chap. 4, we used the two degree of freedom statistic $U = Z_1^2 + Z_2^2$ to combine information from the one-dimensional statistics that estimate the additive effect and the dominance effect. The present case is similar, but slightly more complicated because now $Z_1$ and $Z_2$ are correlated. Let $Z(t) = (Z_1(t), Z_2(t))'$, and let $\Sigma$ be the $2 \times 2$ matrix with 1's on the main diagonal and $\rho = \text{cov}[Z_1(t), Z_2(t)]$ on the off diagonal. Multivariate statistical analysis suggests that in this case an appropriate statistic is $U(t) = Z'(t)\Sigma^{-1}Z(t)$, where the inverse matrix $\Sigma^{-1}$ is $1/(1 - \rho^2)$ times the matrix that has 1's on the main diagonal and $-\rho$ on the off diagonal. Since $\rho$ is typically unknown, we estimate it by the sample correlation coefficient between $y_1$ and $y_2$,

$$\hat{\rho} = \frac{\sum_{i=1}^{n}(y_{1i} - \bar{y}_1)(y_{2i} - \bar{y}_2)}{[\sum_{i=1}^{n}(y_{1i} - \bar{y}_1)^2 \sum_{i=1}^{n}(y_{2i} - \bar{y}_2)^2]^{1/2}} ,$$

and use this estimator in the definition of $U(t)$.

The approximation for the genome-wide significance level given by (5.5) is again valid, with $\bar{\beta} = 0.02$. An approximation for the power when $\tau_1 = \tau_2$ (so the two QTL are actually the same locus) is given in Prob. 6.4, where now $\xi = [\mu'\Sigma^{-1}\mu]^{1/2}$, where $\mu = (\mu_1, \mu_2)' = EZ'(\tau)$. The noncentrality $\xi$ can be re-written as

$$\xi = \{\mu_1^2 + [\mu_2 - \rho\mu_1]^2/(1 - \rho^2)\}^{1/2}. \tag{8.1}$$

Observe that if we use only the process $Z_1(t)$ associated with the first phenotype, the noncentrality is $\mu_1$. If either $\mu_2 \neq 0$ or $\rho \neq 0$, the noncentrality parameter of the bivariate statistic is larger than $\mu_1$. From (8.1) we see that the bivariate statistic has its largest noncentrality parameter if the sign of $\rho\mu_1$ is opposite to the sign of $\mu_2$, although one might question the biologically plausibility of such a relation. Even if $\mu_2 = 0$, the two-dimensional statistic can have a slightly larger noncentrality, provided the two phenotypes are correlated.

To analyze more carefully the power of the two-dimensional statistic, we must also consider its significance threshold, which will be larger than that of a one-dimensional statistic. For an inter-marker distance of 10 cM, the threshold for the significance level 0.05 given by (5.5) is approximately $u = 15.84$.

Equivalently, if we re-write the rejection region as $\max_t U^{1/2}(t) > z$, then $z = u^{1/2} = 3.98$. For a single phenotype the threshold found in Chap. 5 would be approximately $z = 3.57$. It follows that the noncentrality parameter $\xi$ must be larger when we consider two phenotypes simultaneously or we can actually *lose* power to detect the QTL. For example, suppose that QTL associated with the second phenotype are actually unlinked to the major QTL at $\tau_1$ associated with the first phenotype and that the phenotypes themselves have negligible correlation. At a QTL for the first phenotypes with noncentrality parameter $\mu_1 = 4.0$, the power of the one-dimensional statistic $Z_1$ is about 0.7, while that of the two-dimensional statistic is only about 0.6. Now suppose that $\mu_2 = 2.0$, a value for which the statistic $\max_t Z_2(t)$, when used alone with the threshold of 3.57, would have power of less than 0.1, so this QTL would most likely go undetected. Again assume the correlation $\rho$ is negligible. Now the power of the two-dimensional statistic would be about 0.77, somewhat larger than the power of $Z_1$ alone.

One can also consider three- and higher-dimensional phenotypes, although the increase in significance threshold raises questions about the efficiency of this approach. If the different phenotypes are linked to the same QTL, we can expect to gain power to detect the QTL and increase our understanding of the biological relatedness of the traits. Unfortunately we can rarely be confident that this will be the case.

### 8.1.2 Gene-Covariate Interaction

Phenotypes are a complex function of genes and environment. An attractive feature of experimental genetics is that we can often control the environment, so differences in phenotype can be attributed to differences in genotype. A disadvantage is that what we learn in one environment may not apply to another environment. To get some idea of the complexities involved without substantial new statistical problems, we discuss here the simplest possible case of two equally likely environments. To do so we consider a backcross with the model (4.1) expanded to

$$y = \mu + \beta w + (\alpha + \gamma w)x + e(w) . \tag{8.2}$$

In this equation $w$ designates a covariate, say a particular environment or the sex of the organism. It can affect the phenotype by itself through the term $\beta w$, or by a multiplicative genetic interaction through the term $\gamma w x$. Insofar as we are interested primarily in genetic effects the parameter $\beta$ is a nuisance parameter like $\mu$. The parameters $\alpha$ and $\gamma$ are the parameters of interest. As before, we assume that $e(w)$ is uncorrelated with $x$ and has variance $\sigma^2_{e(w)}$.

We have for simplicity taken the parameter $\delta$ in (4.1) equal to 0. As long as we are concerned with a backcross to one parental line, this involves no loss of generality because we cannot estimate $\alpha$ and $\delta$ separately, only their sum. For an intercross, we might want to include the dominance deviation $\delta$

in the model – as well as a dominance deviation for the term interacting with the covariate.

Suppose that $w$ can take only two values, say 0 and 1. This model is then closely related to the multivariate model. We can let $y_1$ denote the phenotype when $w = 1$ and $y_2$ the phenotype when $w = 0$, so, for example in the notation of the preceding section $\alpha_1 = \alpha + \gamma$ and $\alpha_2 = \alpha$. Now $y_1$ and $y_2$ refer to the same phenotype for different individuals, whereas previously they referred to different phenotypes for the same individual. Nevertheless, the results of the preceding section can be applied essentially without change. In principle, the same method will also work when the covariate $w$ takes more than two different values, but it becomes increasingly cumbersome. In experimental genetics, although not in human genetics, we can control the environments reasonably well, and hence control to some extent the number of dimensions to the model.

### 8.1.3 Longitudinal Studies

In some cases we may be interested in a phenotype that changes as a function of age, and we want to explore the effect of genetics on these changes. We can again use the model (8.2) with $w$ denoting age (or some function of age). Alternatively, we can consider this another variation of the multivariate model if the number of different values of $w$ is reasonably small or if we restrict our consideration to a small number of different values.

If the number of ages at which we make phenotypic measurements is not small, say more than 5, or if the values of $w$ at which measurements are made are not the same for all organisms, one may want to consider a structured model like (8.2). Depending on what we want to assume about $e(w)$, the analysis of (8.2) can be relatively straightforward or quite complicated. The simplest assumption would be that the residuals are independent with a fixed variance $\sigma_e^2$ for all individuals and all ages. However, in earlier chapters we have assumed that even when we explicitly model only a single QTL, the residual $e$ implicitly contains the effects of other QTL (provided they are unlinked to the modeled QTL, so that $e$ is uncorrelated with the effect of the modeled QTL). If we take that approach here, then we should presumably assume that $\mathrm{var}[e(w)]$ is a quadratic function of $w$, since the variance of the effect of the modeled QTL is. One may also want to consider the possibility that for the same individual the residual phenotypic measurements at different ages are correlated, i.e., $\mathrm{cov}[e(w), e(w')] = r(w, w') \neq 0$. Even without this last complication, a proper analysis would be substantially more complex than anything we have already considered, and we would want to ask how much we gain in comparison to simpler analyses.

## 8.2 Multiple Gene Models and Gene-Gene Interaction

While we expect that there may be several genes contributing to a quantitative trait, the model introduced in Chap. 2 focused on a single QTL. (See, however, Prob. 2.5.) Consider the basic model as it is written in (2.3), and suppose now that there is a second QTL. We attach subscripts to the parameters associated with each QTL, so, for example, $p_i$ denotes the frequency of $A_i$ alleles associated with the $i$th QTL, $\tau_i$. Assuming for simplicity that there is no dominance, we have the obvious extension of (2.3) that allows for two QTL:

$$y = m + \alpha_1[x(\tau_1) - 2p_1] + \alpha_2[x(\tau_2) - 2p_2] + e .$$

The two QTL may also interact in their contribution to the trait, i.e., they may have an effect that is not simply the sum of their individual effects. The simplest way to incorporate such an interaction is to consider the model

$$y = m + \alpha_1[x(\tau_1) - 2p_1] + \alpha_2[x(\tau_2) - 2p_2] + \gamma[x(\tau_1) - 2p_1][x(\tau_2) - 2p_2] + e , \quad (8.3)$$

where $\gamma$ is called the additive-additive interaction effect, and we assume that $e$ is uncorrelated with the QTL effects. The model can be made much more complicated to account for dominance, separately at each QTL, and dominance in the interaction, but for our purposes this simpler model will suffice.

Even without dominance this model is complicated if the QTL are linked, so we assume for our introductory discussion that they lie on different chromosomes, hence segregate independently. Note that with this assumption, the model (8.3) is consistent with the model (2.3) in the sense that the residual $e$ in (2.3) can be assumed to contain the last three terms in the model (8.3). Our basic assumption about the variable $e$ in (2.3) was that it is uncorrelated with the genetic effects of the modeled QTL. That is true of each of the last three terms in (8.3) when the two QTL segregate independently, so it is also true of their sum.

It is not necessary to use a model that explicitly contains multiple QTL. If the statistic $Z_t$ of a genome scan exceeds the significance threshold at markers on different chromosomes (or even at widely spaced, hence loosely linked markers on the same chromosome), we would interpret that to mean there are multiple QTL contributing to the trait. The focus in this section is to ask whether we can achieve more power for QTL detection or better understanding of the effect of genetics on the phenotype by explicitly including more than one QTL in the model.

### 8.2.1 Strategies for Detecting Multiple Genes

Assume now that we are concerned with an intercross, which is the special case of (8.3) with $p_1 = p_2 = 1/2$. The phenotypic variance has the decomposition $\sigma_y^2 = \sigma_{A_1}^2 + \sigma_{A_2}^2 + \sigma_{A_1A_2}^2 + \sigma_e^2$, where $\sigma_{A_1}^2 = \alpha_1^2/2$, $\sigma_{A_2}^2 = \alpha_2^2/2$ and $\sigma_{A_1A_2}^2 =$

$\gamma^2/4$. This last quantity is called the additive-additive interaction variance component.

There are roughly two ways that we can try to use the information from the multiple regression model. The first involves simultaneously estimating the three genetic parameters of the model and testing simultaneously the null hypothesis that $\alpha_1 = \alpha_2 = \gamma = 0$. For testing a single pair of markers, multiple regression analysis provides an ideal tool. However, the multiple comparison problem becomes very complicated, because a genome scan would involve all possible pairs of (unlinked) marker loci. A second approach is sequential. We use a simple genome scan to detect at least one of the two loci. If this scan finds one locus, we consider a new regression model formed by estimating the location and effect of the detected locus, subtracting it from the model, then performing a second genome scan with the new model.

## 8.2.2 Multiple Regression and Model Selection

To use multiple regression to map two QTL simultaneously, we can calculate three statistics of the type discussed in Chap. 4: for unlinked markers $s$ and $t$, we form regression statistics of the phenotype $y$ on $x(s) - 1$, on $x(t) - 1$, and on the product $[x(s) - 1][x(t) - 1]$. Standardizing these to have unit variance under the hypothesis of no linkage, we obtain a three-dimensional statistic $Z(s, t) = (Z_1(s), Z_2(t), Z_3(s, t))'$. For fixed markers that are unlinked to the QTL, the coordinates of this vector would be approximately independent standard normal random variables. If $s$ is linked to $\tau_1$ with recombination fraction $\theta_1$ and $t$ is linked to $\tau_2$ with recombination fraction $\theta_2$, the expected value of the vector $Z(s, t)$ would be

$$n^{1/2}\left((\alpha_1/2^{1/2})(1 - \theta_1), (\alpha_2/2^{1/2})(1 - \theta_2), (\gamma/2)(1 - \theta_1)(1 - \theta_2)\right).$$

To test the hypothesis of no linkage against an alternative of two linked genes one can use $U_2(s, t) = Z_1^2(s) + Z_2^2(t)$ or $U_3(s, t) = Z_1^2(s) + Z_2^2(t) + Z_3^2(s, t)$, according as one wants the alternative hypothesis to include the possibility of interaction or not. The statistic $U_2$ would have a chi-square distribution with two degrees of freedom at unlinked marker loci, while the second would have a chi-square distribution with three degrees of freedom. To develop a genome scan we would consider maximizing $U_j(s, t)$ over different $s$ and $t$. Approximations similar to those given in Chap. 5 ((5.3) and (5.5)) for a genome of total length $L$ and markers spaced at inter-marker distance $\Delta$ are

$$\Pr\left(\max_{s<t} U_2(s, t) \ge u\right) \approx 1 - \exp\left\{-16^{-1}(\beta L)^2 u^2 \exp(-u/2)\nu^2[(\beta\Delta u/2^{1/2})^{1/2}]\right\}$$

and

$$\Pr\left(\max_{s<t} U_3(s, t) \ge u\right) \approx 1 - \exp\left\{-c^2(32\pi)^{-1/2}\left(\beta L\nu\left(\sqrt{2\beta\Delta u c}\right)\right)^2 u^{5/2} e^{-u/2}\right\}$$

$$(8.4)$$

where $c^2 = 8/5$. Because of the maximization over pairs of markers to reflect the possibility of two QTL, the problem of multiple comparisons requires that the threshold to maintain an overall type I error rate of 0.05 must be considerably higher than it was before. For example, for $\Delta = 10$, and an idealized mouse genome of total length 1600 cM, the 0.05 threshold for the two degree of freedom statistic is 22.7, while that of the three degree of freedom statistic is $u = 26.7$ compared with $z^2 = 3.57^2 = 12.7$ for the simple genome scan of Chap. 5. Note the quite substantial increase in threshold attributable to searching for a second QTL, but the much smaller increase for adding the interaction term.

An illuminating numerical example is provided by the data described by Sen and Churchill [70], which involved a backcross of 250 rats with markers placed about every 10 cM. A rat genome is similar to a mouse genome. For purposes of this example we assume they are the same – 20 chromosomes of 80 cM each. Among other results (which we do not discuss here), Sen and Churchill report apparent linkage to a gene on chromosome 6 having locus-specific heritability $h^2 = 0.03$, a gene on chromosome 15 having heritability 0.055, and a heritability of 0.06 for the interaction between these two QTLs. By the argument of Prob. 6.6, this can be interpreted as indicating that the maximum value of $Z^2(s)$ on chromosome 6 is approximately $nh^2 = 7.5$, the maximum value on chromosome 15 is 13.8, and the value of the interaction component, $Z_3^2(s,t)$ is 15.0. For a simple genome scan, we find that the p-value associated with the putative QTL on chromosome 15 is 0.03, but on chromosome 6 it is only 0.5. We would conclude that there is some evidence for a QTL on chromosome 15 but very little for a QTL on chromosome 6. However, the sum of all three terms is 36.3, and the p-value given by (8.4) is 0.0006, which provides comparatively strong evidence of interacting QTL on chromosomes 6 and 15. By way of comparison the p-value for $\max_{s<t} U_2(s,t)$ is 0.09, i.e., because of the larger number of pairs of markers associated with the two-dimensional statistic, the very weak suggestion of a main effect on chromosome 6 combined with the marginally significant effect on chromosome 15 leads to an increase in the p-value compared to that for an effect on chromosome 15 alone.

For this example it seems useful to consider the more complex multiple regression statistics, although new problems arise concerning the proper interpretation of a statistically significant outcome. The statistic $U_3$ is made up of the sum of three different putative contributions to the phenotype. These could be equally important, or one could be much more important than the other two, or two could be more important than the third. One can get an informal idea about the relative importance of these three contributions by looking at the individual coordinates of the vector $Z_{s,t}$, which was the starting point for the example of the preceding paragraph. Another informal guideline is provided by comparison of the three p-values computed above. For the particular case of our example one might conclude that there is a QTL on chromosome 15 that has a main effect of its own and also interacts with a

QTL on chromosome 6 that has no effect by itself. However, experience shows that interactions without main effects are rare, so one is inclined to think that there may also be a main effect on chromosome 6 in spite of the weakness of the evidence.

A more formal statistical procedure for making this kind of inference is called "model selection". Is the phenotype better explained by a model with a single QTL, by a model with two QTL that do not interact, by a model with two QTL that do interact? Is a model with dominance better than a purely additive model? Is there evidence for more than two QTL, etc.? Such questions do not have simple answers and are the subject of current research to explore a number of possible approaches. See the references suggested in the bibliographical notes.

### 8.2.3 Sequential Detection

While the approach of the preceding paragraphs is built around multiple regression of the phenotype on a number of putative QTLs, a complementary approach can be useful if there is at least one QTL that can be easily detected by an initial genome scan. To be specific, suppose that $\alpha_1$ is quite large, so the associated QTL is easily detected. We can then consider a modified model that takes the detected QTL into account and then repeat the genome scan to see if this makes it easier to detect at least one more QTL. In practice, we must estimate $\alpha_1$ and the QTL location, say $\tau_1$, but to simplify the problem as much as possible, assume that $\alpha_1$ and $\tau_1$ are known and that $\tau_1$ coincides with the location of a marker. We can then define a new phenotype $\tilde{y} = y - \alpha_1[x(\tau_1) - 1]$, which will satisfy a reduced model

$$\tilde{y} = m + \alpha_2[x(\tau_2) - 1] + \gamma[x(\tau_1) - 1][x(\tau_2) - 1] + e.$$

We can now consider new regression statistics by regression of $\tilde{y}$ on $x(t) - 1$ and on the product $[x(\tau_1)) - 1][x(t) - 1]$. Let $Z_1(t)$ and $Z_2(t)$ denote the standardized statistics. To search for a second QTL one could use either the one-dimensional statistic $\max_t Z_1(t)$ or the two-dimensional statistic $\max_t[Z_1^2(t) + Z_2^2(t)]$. (For the two-dimensional statistic one can set a significance threshold by using (5.5) with $\bar{\beta} = 0.02$.)

Observe that for an intercross $\sigma_{\tilde{y}}^2 = \sigma_y^2 - \alpha_1^2/2 = \sigma_y^2(1 - h_1^2)$, where $h_1^2$ is the locus-specific heritability for the QTL at $\tau_1$. It follows from the regression equation for $\tilde{y}$ by arguments similar to those in Chap. 4 that the noncentrality parameter of $Z_1(\tau_2)$ is $n^{1/2}\alpha_2/(2^{1/2}\sigma_{\tilde{y}})$, which would be reduced by a factor of $(1 - 2\theta)$ at a marker $t$ at a recombination distance $\theta$ from $\tau_2$. This is larger than the noncentrality parameter at $\tau_2$ (or at a linked marker) for the initial genome scan, because the effective phenotypic variance has been reduced by subtracting out the relatively large effect of the QTL at $\tau_1$. By a similar argument the noncentrality parameter of $Z_2(\tau_2)$ can be shown to equal $n^{1/2}\gamma/(2\sigma_{\tilde{y}})$. Hence the power of the two-dimensional statistic can be approximated by using (6.9) with $\xi = n^{1/2}(\alpha_2^2/2 + \gamma^2/4)^{1/2}/\sigma_{\tilde{y}}$. The two-dimensional statistic will presumably have more power than the one-dimensional statistic if the gene-gene interaction plays an important role.

For a simple numerical example, suppose that the heritability at the first locus is 0.2. Then under the idealized conditions of the preceding analysis, the effect of considering the residual $y - \alpha_1[x(\tau_1) - 1/2]$ is to reduce the noise $\sigma_y^2$ by the factor $1 - 0.2 = 0.8$. This would increase the noncentrality of the conditional statistic by $1/0.8^{1/2} \approx 1.1$ and would make the (one-dimensional) test statistic at a second locus about 10% larger. Although this might in principle change a result that is statistically not significant into one that is statistically significant, the impact is actually quite modest. It would not, for example, reveal a signal at a locus where there had previously been no evidence for one. If $\alpha_1$ were larger, the increase in noncentrality would also be larger, but simple numerical examples suggest it is unlikely that this increase would make an important difference. Hence it is most likely that this method would be useful when there is gene-gene interaction, so the second coordinate of the two-dimensional statistic would be the source of an increase in power.

Note, however, that this analysis depends in two respects on our ability to identify $\tau_1$ accurately: (i) in the definition of $\hat{y}$ and (ii) in the definition of $Z_2(t)$. We must estimate the value of $\tau_1$, and our estimate may not be especially accurate unless the noncentrality parameter at $\tau_1$ is large. In addition, the estimate of $\alpha_1$ used to define $\tilde{y}$ may be positively biased (cf. Chap. 6), which would tend to make the situation look more favorable than it actually is. Consider the Sen-Churchill data described in the preceding section with $\tau_1$ denoting the putative QTL on chromosome 15. Since the signal at $\tau_1$ is relatively modest, it is unlikely that sequential search would be especially useful.

Another example is provided by the Sen and Churchill data, which also identified a QTL on chromosome 4 with the much larger heritability of $h_1^2 = 0.138$. In this case one would expect the noncentrality at a second QTL to increase by the factor $1/(1 - 0.138)^{1/2} \approx 1.08$. However, there was no evidence of an interaction between the QTL on chromosome 4 and any of the other putative QTLs identified in this study. As a result the benefit of a sequential approach would be quite modest at best. See Prob. 8.6 for a computational experiment.

## 8.3 Selective Genotyping

Because of variability in both genotype and phenotype, different individuals provide different amounts of information toward the detection of linkage. A suggestion designed to save some of the costs of genotyping is to identify and genotype only those individuals who provide large amounts of information. Although this procedure can result in a larger noncentrality parameter *per genotyped individual*, it also results in a smaller effective sample size, hence a smaller total noncentrality parameter than genotyping all available individuals.

To see how this might work, consider an intercross and the test statistic for an additive effect. For simplicity assume that the overall mean $\mu = 0$ and the phenotypic variance $\sigma_y^2 = 1$. Then the noncentrality parameter at a QTL $\tau$ derived in Chap. 4 would be $(n/2)^{1/2}\alpha$. Consider the strategy of genotyping only individuals having either a large or a small phenotype, say determined by the requirement that $|y| > c$, for some threshold $c$ to be determined. If we make the additional assumption that the residuals are normally distributed, it can be shown that the noncentrality parameter under local alternatives per *genotyped* individual is

$$\alpha\{[1 + c\phi(c)/(1 - \Phi(c))]/2\}^{1/2}.$$

This is greater than $\alpha/2^{1/2}$, and simple numerical examples show that it increases approximately linearly with $c$. However, only one out of every $2[1 - \Phi(c)]$ phenotyped individuals is used, so the effective sample size is $2n[1 - \Phi(c)]$. Hence the noncentrality parameter per *phenotyped* individual is

$$\alpha\{2[1 - \Phi(c) + c\phi(c)]/2\}^{1/2},$$

which is maximized at $c = 0$. Nevertheless, if one can assess the relative costs of phenotyping (which must include the costs of maintaining the animals) and genotyping, there is the possibility of savings.

A related idea would be in a first stage to genotype those individuals whose phenotypes indicate they are especially informative and then in a second stage genotype the remainder, but only in those regions of the genome that on the basis of the first stage appear likely to contain QTL.

## 8.4 Advanced Intercross Lines and Fine Mapping

The accuracy with which we can estimate the location of a QTL, say by a confidence interval as discussed in Chap. 6, depends on the rate of recombinations, hence on the number of meioses in a breeding experiment. One can create additional meioses in various ways. For example, we found that recombinant inbred lines have a higher recombination rate than either backcrosses or intercrosses, and hence can be expected to yield shorter confidence intervals.

In this section we mention another breeding design that has been suggested as a method to obtain more precise estimates of QTL location. This is the *advanced intercross line*, which is created by crossing the offspring of an intercross for several, say $n$, additional generations.

To see the effect of repeated intercrossing, consider two loci separated by the recombination fraction $r_2 = \theta$ in the intercross (i.e., second) generation.

We can use an argument like those in Chap. 3 to determine $r_n$, the recombination fraction in the $n$th generation.

Consider an $n + 1$st generation gamete and the two $n$th generation homologous chromosomes that created this gamete during meiosis. Suppose, to be definite, that the gamete has the allele $A$ at locus 1. The gamete is non-recombinant with probability $1 - r_{n+1}$. This can happen in two ways. In the first, the chromosome contributing the $A$ allele does not recombine during meiosis and was initially non-recombinant. This happens with probability $(1 - \theta)(1 - r_n)$. In the second, the chromosome that contributes the $A$ allele does recombine during meiosis, and the homologous chromosome has a $B$ allele at the second locus. This happens with probability $\theta/2$. Putting these two possibilities together, we obtain

$$1 - r_{n+1} = (1 - r_n)(1 - \theta) + \theta/2 .$$

It can be shown that the solution of these equations is

$$r_n = [1 - (1 - \theta)^{n-2}(1 - 2\theta)]/2 .$$

For small $\theta$ we can simplify this relation to obtain the approximation $r_n \approx n\theta/2$. The same calculations as in Chap. 4 show that the correlation in the genotype between two markers is $1 - 2r_n$, or approximately $n\theta$. From the reasoning in Chap. 6 about the expected length of confidence intervals, we conclude that after $n$ generations we will have reduced the expected length of a confidence interval by a factor of approximately $2/n$. Hence if an intercross produces a confidence interval for QTL location of about 5 cM, two more generations of intercrossing would be expected to produce a confidence interval of about 2.5 cM, while eight more generations would produce a confidence interval of about 1 cM.

Unfortunately even 1 cM, which is a small genetic distance, can harbor a large number of genes. When one considers the amount of breeding and animal care that an advanced intercross line requires, it seems unlikely that one will want to use it to identify a single gene, but it might be possible to achieve economies of scale if one were able to develop one such line to identify several different genes that segregate in the same cross.

## 8.5 Bibliographical Comments

Model selection for gene mapping has been discussed in a number of recent papers, but as in statistics generally the problem does not seem to admit completely satisfactory solutions. Broman and Speed [9] compare several competing procedures. They favor a modification of the Bayes Information Criterion (BIC) originally suggested by Schwartz [81] for a different class of problems, but they consider only single gene (additive) effects. Bogdan et al. [8] propose a different modification of BIC. While they also consider interactions, their procedure seems to be best suited to cases where the interacting genes do not themselves have single gene effects. Sen and Churchill [70] and Yi *et al.* [92] use Monte Carlo methods to implement an exact Bayesian criterion that selects the model with the highest posterior probability. Siegmund [75] suggests a still different version of the BIC and an alternative method based on comparison of p-values. A Monte Carlo comparison of a number of methods is contained in the Ph. D. thesis of Jianxin Shi [72].

The strategy of sequential detection of Sect. 8.2.3 has been discussed somewhat more completely in the literature of human genetics. Although the idea there is similar, the details of implementation are quite different. See Chap. 11 and references given there. For a recent discussion of longitudinal traits see [91]. Advanced intercross lines were first suggested by Darvasi and Soller [18].

# Problems

**8.1.** Show that under the assumptions of our model for a bivariate phenotype that $\rho = \mathrm{cor}(y_1, y_2) = [\alpha_1 \alpha_2 (1 - 2\theta)/4 + r\sigma_{e_1}\sigma_{e_2}]/\sigma_{y_1}\sigma_{y_2}$.

**8.2.** For the model of bivariate phenotypes, suppose that QTL contributing to $y_1$ and to $y_2$ are linked, but are not necessarily the same QTL. Discuss possible advantages and possible disadvantages of trying to use the bivariate process $(Z_1, Z_2)'$ to develop confidence regions.

**8.3.** Suppose that $y$ satisfies $y = \mu + bx(\tau_1) + cx(\tau_1)x(\tau_2) + e$, where $\tau_1$ and $\tau_2$ are unlinked loci. Write this model in a form where it is convenient to compute the additive genetic variances and the additive-additive interaction variance for a BC design and for an IC design. Find formulas for these variance components. Suppose $b = 0$. For which design (BC or IC) is the interaction variance component a larger part of the total genetic variance? Suppose $\sigma_e^2 = 1$, $b = 0.7$, and the sample size is $n = 150$ for an intercross and 300 for a backcross. For the idealized sequential approach described in Sect. 8.2.3, find the power of the one- and two-dimensional statistics for various values of $c$. (Remember to use thresholds that are appropriate for each statistic.)

**8.4.** Consider the breeding design introduced in Prob. 4.6 and a model with two unlinked QTL that can interact additively. What are variance components for this model that could not be estimated from separate backcrosses (or intercrosses) starting from the $A_i \times a_i$ initial crosses for $i = 1, 2$?

**8.5.** Generalize (8.3) to allow for dominance in the main effects and in the interactions. Find expressions for the eight possible variance components.

**8.6.** In this problem we study the sequential strategy proposed in Sect. 8.2.3. For an intercross, write a program to simulate the detection of a QTL on a chromosome of length 80 cM, where $\alpha_1 = 0.707$, $\sigma_y^2 = 1$, and $n = 200$. Your program should keep track of the locus where the scan statistic is maximized and the maximum value. To provide favorable conditions for locating the QTL and estimating $\alpha_1$ reasonably well, place markers every 4 cM, and consider the cases where the QTL lies on a marker or midway between two markers, near the middle of the chromosome. What is the probability of detecting a second QTL, unlinked to the first, that has $\alpha_2 = 0.35$ (a) with a simple genome scan, or (b) after implementing the conditional search strategy described in the text. For (b) consider also the possibility that the second QTL interacts (additively) with the first and has $\gamma = \alpha_2$. Repeat the experiment with $\alpha_1 = 1$, $n = 100$, and $\alpha_2 = 0.5$. How does the simulated power for the sequential strategy compare with the analysis presented in the text, where it was assumed that the location of the first QTL and its effect were known without error? Which case is more favorable to a simple genome scan? to the sequential strategy?

# Part III

# Human Genetics

# 9

## Mapping Qualitative Traits in Humans Using Affected Sib Pairs

In humans we cannot create inbred lines, backcrosses, etc. Consequently, it is more difficult to study directly the correlation of phenotypes and genetic markers. We can proceed indirectly by noting that relatives frequently have more similar phenotypes than non-relatives, presumably because they have more similar genotypes. For studying human diseases, particularly convenient units are *affected sib pairs* (ASP), which are the subject of this chapter. We delay until Chap. 11 a discussion of the substantially more complex problem of pedigrees involving variable numbers and relationships of affecteds.

Since humans are members of populations, not subject to breeding experiments, we shall want to use some of the material on population genetics from Chap. 3, notably that concerned with the ideas of random mating/Hardy-Weinberg equilibrium and identity by descent (IBD).

If two relatives are affected with the same disease, which is caused to some extent by the individual's genotype and is relatively rare in the population, it seems plausible to hypothesize that they have the disease because both inherited one or more disease-predisposing alleles from a common ancestor.

Recall that two relatives are said to have inherited an allele identical by descent (IBD) at a given locus, if they have inherited the same allele from a common ancestor. At any genetic locus, two siblings inherit their paternal allele IBD with probability $1/2$ and independently inherit their maternal allele IBD with probability $1/2$. Thus they inherit 0, 1, or 2 alleles IBD with probabilities $1/4$, $1/2$, $1/4$, respectively; and on average they inherit one allele IBD. (This argument presupposes that the parents are not inbred, so they do not already contain alleles inherited IBD from a remote ancestor.) If we have a sample of, say, $n$ sib pairs, at a randomly selected genetic marker, there will be about $n$ alleles inherited IBD. If the sib pairs share a given phenotype, e.g., the same disease, then we expect that at a marker tightly linked to a gene or genes contributing to the phenotype there will be more than $n$ alleles IBD. In the following sections we develop a genetic model for a qualitative trait and discuss genome scans to detect genes contributing to the trait. For simplicity we refer to the trait as a disease and individuals having the disease as affected. Analysis of QTL in humans is discussed in Chap. 11.

We use a simple genetic model in order to describe a potential connection between the disease status of the siblings and the distribution of the number of alleles shared IBD. This connection is then used in order to derive the conditional distribution of the number of alleles shared IBD, given that both siblings are affected. This leads in turn to a relation that connects the frequency in the population of the susceptibility alleles and their contribution to the risk of getting the disease to the distribution of the test statistics. These issues are discussed in the following two sections.

The third section deals with the asymptotic distribution of the test statistic, and the properties of the associated test, when large samples are used in order to detect a risk factor that has a relatively small effect on the probability of being affected.

The IBD status for a pair at a given locus is inferred from the genotypic information at hand, which may include the genotypes of the siblings and their parents or the genotypes of the siblings alone. In the preceding discussion we make the assumption that the IBD status can be perfectly reconstructed based on the genotypic information. In practice, this is seldom the case and an estimate of the IBD number has to replace the unknown true value. In the fourth section we will present statistical tools for the estimation of the IBD state from the genotypic information and assess the effect of partial information on the statistical properties of the scanning procedure.

## 9.1 Genetic Models

We assume initially a single susceptibility locus. The polymorphism at that locus consists of two alleles – a susceptibility allele $D$, and wild-type allele $d$. The genetic model provides the ingredients which are needed in order to compute the conditional distribution of the IBD status, given the phenotypes of the siblings. It consists of two components: a model connecting phenotypes to genotypes at the susceptibility locus and a population genetic model, describing the population joint distribution of genotypes at the trait loci for the parents of the siblings. Although we discuss in detail the case of a bi-allelic disease locus, essentially all the results described are valid for loci having an arbitrary number of alleles.

### A Model for the Trait

Here we consider a single autosomal trait locus with allele $D$, associated with the disease, and a wild-type allele $d$. In the model we allow for sporadic cases and partial penetrance. Specifically, define the three *penetrance* probabilities:

$$g_0 = \Pr(\text{Affected} \,|\, dd)\,,$$
$$g_1 = \Pr(\text{Affected} \,|\, Dd)\,,$$
$$g_2 = \Pr(\text{Affected} \,|\, DD)\,.$$

In order to emphasize certain similarities with the models used for experimental genetics, it is often convenient to re-write the penetrances in the form

$$g_1 = g_0 + \alpha + \delta; \quad g_2 = g_0 + 2\alpha\,.$$

We assume that $\alpha > 0$ to be consistent with $g_2 > g_0$. With this notation $y$, which refers here to the *probability* that an individual is affected as a function of genotype, has the form of equation (2.2) with $e = 0$. Now $x_{\mathrm{M}}$ (resp. $x_{\mathrm{F}}$) equals the number of $D$ alleles, 0 or 1, inherited from the mother (resp. father). Note, however, that while in Chap. 2 $y$ is an observed quantitative phenotype that is allowed to take on any value, here the phenotype is 0 (no disease) or 1 (disease), while $y$ is the *unobserved* (conditional on the genotype) probability that the phenotype is 1. Hence $y$ itself must be between 0 and 1.

The *additive model*, which we emphasize in what follows, relates the three penetrance parameters to each other by requiring that $g_1 = (g_0 + g_2)/2$, or equivalently that $\delta = 0$. Thus, one needs to specify only $g_0$ and $\alpha$. Moreover, as we shall see later, the quantity of interest will depend only on the ratio $R_2 = g_2/g_0 = 1 + 2\alpha/g_0$ – the relative risk of a *risk-allele* homozygote with respect to a *wild-type* homozygote. One can envision other special cases of this model, even in the simple case of a single trait locus. Important examples are the *recessive model* which assumes $g_0 = g_1 < g_2$, equivalently $\delta = -\alpha$ or the *dominant model*, which assumes $g_0 < g_1 = g_2$, or $\delta = \alpha$. The statistic we consider below, which counts the total number of alleles shared IBD, is most appropriate for the additive model and to a good approximation for a dominant model as well.

We complete the description of the relation between the genotypes and the probability that the siblings are affected by adding the assumption that within a pedigree, the phenotypes of relatives are conditionally independent given their genotypes. As a result, we find that

$$\Pr(\text{Both affected}\,|\,G_1, G_2) = \Pr(\text{Affected}\,|\,G_1) \times \Pr(\text{Affected}\,|\,G_2) = y_1 y_2\,, \tag{9.1}$$

where $G_1$ and $G_2$ are the genotypes of the first and the second sibling respectively. An important consequence of this assumption is the exclusion of environmental effects on susceptibility to the disease. (See Prob. 9.5 for possible generalizations.)

## A Population Genetic Model

The second component in the genetic model is a population genetic model that describes the frequencies of the pedigree founders' genotypes. For the case of a sib-pair, there are two founders – the mother and the father, whom we assume mate at random in an infinitely large population and are themselves the product of random matings. Hence their genotypes are in Hardy-Weinberg equilibrium, i.e., the two alleles at a given locus are randomly sampled from the population pool. If the population frequency of the allele $D$ is denoted by $p$ (and the frequency of the allele $d$ is $1 - p$), then the probability of the genotype $DD$ is $p^2$. Likewise, the probability of the genotype $dd$ is $(1 - p)^2$, and the probability of the genotype $Dd$ is $2p(1 - p)$. Random mating also implies independence between the parents' genotypes. For example, the probability that both parents' genotypes are $DD$ is $p^4$. In a similar fashion, one can compute the probability of all other combinations of parents' genotypes as a function of a single parameter $p$, the frequency of the allele $D$ in the genetic pool. It also follows that each child individually has a genotype that satisfies Hardy-Weinberg frequencies. However, the genotypes of two children are dependent.

From (2.2), we obtain (2.3), which for convenience we repeat here (with $e = 0$):

$$y = m + \{\alpha + (1 - 2p)\delta\}[(x_{\mathrm{M}} - p) + (x_{\mathrm{F}} - p)] - \{2\delta\}[(x_{\mathrm{M}} - p)(x_{\mathrm{F}} - p)] .$$

Combining this expression with the assumption of Hardy-Weinberg equilibrium, we also obtain the variance decomposition (2.5):

$$\sigma_y^2 = \sigma_{\mathrm{A}}^2 + \sigma_{\mathrm{D}}^2 ,$$

where $\sigma_{\mathrm{A}}^2 = 2p(1-p)[\alpha + (1 - 2p)\delta]^2$ and $\sigma_{\mathrm{D}}^2 = 4p^2(1 - p)^2\delta^2$. For a dominant trait $(\delta = \alpha)$ $\sigma_{\mathrm{A}}^2 = 8p(1 - p)^3\alpha^2$, while for a recessive trait $(\delta = -\alpha)$ $\sigma_{\mathrm{A}}^2 = 8p^3(1 - p)\alpha^2$. In the usual case that $p$ is substantially less than $1/2$, the additive variance is much larger than the dominance variance for a dominant trait, smaller for a recessive trait. The simplest case is an additive trait, for which $\delta = 0$, hence $\sigma_{\mathrm{D}}^2 = 0$.

By taking expectations in (9.1) and using the representation of $y_i$ given above to obtain an expression for the product $y_1 y_2$, we can calculate the probability $\Pr(A)$ that two sibs are both affected:

$$\Pr(A) = \mathrm{E}(y_1 y_2) = m^2 + \mathrm{cov}(y_1, y_1) = m^2 + \sigma_{\mathrm{A}}^2/2 + \sigma_{\mathrm{D}}^2/4 . \tag{9.2}$$

To see how (9.2) is derived, let $x_{\mathrm{M}i}$ $(x_{\mathrm{F}i})$ denote the number of $D$ alleles inherited by the $i$th sib from their mother (father). First recall that $\mathrm{E}[(x_{\mathrm{M}i} - p)^2] = p(1-p)$. Now consider the product $(x_{\mathrm{M}1} - p)(x_{\mathrm{M}2} - p)$. If $x_{\mathrm{M}1}$ and $x_{\mathrm{M}2}$ are IBD, then the product equals $(x_{\mathrm{M}1} - p)^2$, so in this case the expected product is just $p(1-p)$, as before. If $x_{\mathrm{M}1}$ and $x_{\mathrm{M}2}$ are not IBD, then by the Hardy-Weinberg assumption, they are independent and the expected product is the product of the expectations, which equals 0. Since $x_{\mathrm{M}1}$ and $x_{\mathrm{M}2}$ are IBD with probability $1/2$, we find that $\mathrm{E}(x_{\mathrm{M}1} - p)(x_{\mathrm{M}2} - p) = p(1-p)/2 + 0/2 = p(1-p)/2$. Similarly $\mathrm{E}[(x_{\mathrm{M}1} - p)(x_{\mathrm{M}2} - p)(x_{\mathrm{F}1} - p)(x_{\mathrm{F}2} - p)] = [p(1 - p)]^2/4$, since alleles inherited from the father and from the mother are independent. Also, terms like $E[(x_{M1} - p)(x_{M2} - p)(x_{F1} - p)] = 0$, since one factor is independent of the other two. Collecting together the various products gives (9.2).

## 9.2 IBD Probabilities at the Candidate Trait Locus

Given a a pair of affected sibs, let $J_M$, resp. $J_F$, be 1 or 0 according as the alleles at the trait locus from the mother, resp. from the father, are inherited IBD or not. Let $J = J_M + J_F$ denote the total number of alleles inherited IBD at a trait locus. Note that $J_M$ and $J_F$ are independent random variables taking values 0 and 1 with probability $1/2$ each. The argument given above can be expressed conditionally, as $E[(x_{M1} - p)(x_{M2} - p)|J_M] = p(1 - p)J_M$. Other terms can be evaluated similarly, leading to $E(y_1 y_2 | J_M, J_F) = m^2 + J\sigma_A^2/2 + J_M J_F \sigma_D^2$, which in turn implies

$$E(y_1 y_2 | J) = m^2 + J\sigma_A^2/2 + I_{\{J=2\}}\sigma_D^2 , \qquad (9.3)$$

where $I_{\{J=2\}}$ is the indicator of the event that the IBD count is two. Let $Q_2 = \Pr(A) = E(y_1 y_2)$ be the probability given in (9.2) that both sibs are affected. Then by Bayes' formula:

$$\Pr(J = j | A) = \Pr(J = j)\frac{\Pr(A | J = j)}{\Pr(A)} = \Pr(J = j)\frac{E(y_1 y_2 | J = j)}{E(y_1 y_2)} ,$$

Substituting (9.2) and (9.3), we find after some algebraic simplification that $\pi_j = \Pr(J = j | A)$ is given by

$$\pi_0 = [1 - (\check{\alpha} - \check{\delta}/2)/Q_2]/4 ,$$
$$\pi_1 = [1 - \check{\delta}/2Q_2]/2 , \qquad (9.4)$$
$$\pi_2 = [1 + (\check{\alpha} + \check{\delta}/2)/Q_2]/4 ,$$

where $\check{\alpha} = (\sigma_A^2 + \sigma_D^2)/2$ and $\check{\delta} = \sigma_D^2/2$. We have used the notation $\check{\alpha}$, $\check{\delta}$ because these quantities play roles in human genetics, here and in Chap. 11, similar to $\alpha$, $\delta$ in the analysis of an intercross (cf. (9.6)). Note, however, that $0 \le \check{\delta} \le \check{\alpha}$, although there is no similar restriction on $\alpha$ and $\delta$. In the special case of an additive model, $\sigma_D^2 = 0$, the equations simplify accordingly. While the terms in (9.4) are very simple, tedious calculation is required to evaluate them in terms of the allele frequency $p$ and penetrances. Special cases are explored in the problems at the end of the chapter. A case of particular interest is the additive case, where $g_1 = g_0 + \alpha$, $g_2 = g_0 + 2\alpha$, so $\sigma_D^2 = 0$. Let $R_2 = y_2/y_0 = 1 + 2\alpha/y_0$ denote the ratio of the penetrance of a $DD$-homozygote to that of a $dd$-homozygote. By solving for $\alpha = g_0(R_2 - 1)/2$, we find that $m = g_0 + 2p\alpha = g_0[1 + p(R_2 - 1)]$ and $\sigma_A^2 = g_0^2 p(1 - p)(R_2 - 1)^2/2$. Thus $\sigma_A^2/2Q_2$ and hence the IBD probabilities in (9.4) depend only on $p$ and $R_2$

The case $R_2 = 1$, which is equivalent under the additive model to the case $g_0 = g_1 = g_2$, corresponds to no relation between the disease and the investigated gene. Indeed, when $R_2 = 1$, $\sigma_A^2 = \sigma_D^2 = 0$, so (9.4) gives the null distribution of the IBD status: $\pi_0 = \pi_2 = 1/4$, $\pi_1 = 1/2$. This distribution is the $B(2, 1/2)$ distribution. The expected number of alleles IBD in this case is 1. However, when $R_2 > 1$, the relation between the probabilities is $\pi_0 < \pi_2$. Also, since $\pi_1 = 1/2$, $\pi_2 = 1/2 - \pi_0$. The expected number of alleles IBD becomes $1/2 + 2\pi_2 = 1 + (\pi_2 - \pi_0) > 1$. Thus the equations (9.4) give quantitative meaning to the intuitive idea expressed in the introduction to this chapter that two siblings affected with the same disease are likely to have inherited the same disease predisposing allele from a parent.

The function "DistIBD" computes the IBD probabilities as a function of the allele frequency p and the penetrance probabilities g0, g1 , and g2:

```
> DistIBD <- function(p,g0,g1,g2)
+ {
+       alpha <- (g2-g0)/2
+       delta <- g1 - g0 - alpha
+       m <- g0 + 2*p*alpha + 2*p*(1-p)*delta
+       a <- alpha + (1-2*p)*delta
+       d <- delta
+       sig.A <- 2*p*(1-p)*a^2
+       sig.D <- 4*p^2*(1-p)^2*d^2
+       Q <- m^2 + sig.A/2 + sig.D/4
+       pi.0 <- (1- (sig.A + sig.D/2)/(2*Q))/4
+       pi.1 <- (1-sig.D/(4*Q))/2
+       pi.2 <- (1 + (sig.A + 3*sig.D/2)/(2*Q))/4
+       return(data.frame(pi.0=pi.0,pi.1=pi.1,pi.2=pi.2))
+ }
```

Let us explore the effect of the parameters on the distribution of IBD for $g_0 = 0.05$ and $p = 0.1$, for an additive model with different values of $\alpha$:

```
> alpha <- seq(0,0.4,by=0.1)
> IBD.prob <- DistIBD(0.1,0.05,0.05+alpha,0.05+2*alpha)
> IBD.e <- IBD.prob$pi.1+2*IBD.prob$pi.2
> IBD.sd <- sqrt(IBD.prob$pi.1+4*IBD.prob$pi.2 - IBD.e^2)
> round(cbind(alpha,IBD.prob,IBD.e,IBD.sd),3)
  alpha  pi.0 pi.1  pi.2 IBD.e IBD.sd
1   0.0 0.250  0.5 0.250 1.000  0.707
2   0.1 0.211  0.5 0.289 1.078  0.703
3   0.2 0.173  0.5 0.327 1.154  0.690
4   0.3 0.150  0.5 0.350 1.200  0.678
5   0.4 0.135  0.5 0.365 1.230  0.669
```

Of no surprise is the fact that the expectation increases with $\alpha$. Note that the standard deviation remains more or less constant.

A more systematic exploration will also consider the effect of the allele frequency, which we vary from 0.1 to 0.5:

```
> alpha <- seq(0,0.4,by=0.02)
> g0 <- 0.05
> R2 <- 1+2*alpha/g0
> p <- seq(0.1,0.5,by=0.1)
> rep.a <- rep(alpha,length(p))
> rep.p <- rep(p,rep(length(alpha),length(p)))
> IBD.prob <- DistIBD(rep.p,g0,g0+rep.a,g0+2*rep.a)
> IBD.e <- IBD.prob$pi.1+2*IBD.prob$pi.2
> ncp <- (IBD.e-1)^2/0.5
> plot(range(R2),range(ncp),type="n",xlab="R2",ylab="ncp/n")
> for(i in 1:length(p)) lines(R2,ncp[rep.p==p[i]],lty=i)
> legend(1,max(ncp),legend=paste("p = ",p),lty=1:length(p))
```

The increase of the squared standardized difference of the expectation of IBD as a function of the parameter of genetic relative risk $R_2$ and for various values of allele frequency $p$ is presented in Fig. 9.1. As we will see below, this parameter is the square of the noncentrality parameter of the test statistic and is the basis for the evaluation of the statistical power when scanning for disease predisposing genes. Roughly speaking, rarer disease susceptibility alleles with larger genetic relative risk are easier to detect.

## 9.3 A Test for Linkage at a Single Marker Based on a Normal Approximation

In the previous section we considered the distribution of a single quantity – the IBD status of a given pedigree – both under the null assumption $H_0 : \pi_2 = 1/4, \pi_1 = 1/2, \pi_0 = 1/4$ and under the additive alternative $H_1 : \pi_2 - \pi_0 > 0$. In this section we investigate the properties of a test statistic calculated from a sample of such observations. These properties form the basis for the justification of the use of the total number of alleles shared IBD as an appropriate test statistic under the additive model. We begin with the test for linkage of a single marker and move from there to consideration of tests using a genome scan.

*Remark 9.1.* It is not immediately clear that one can actually determine the number of alleles shared IBD. In fact, if one parent is homozygous at a marker, the sibs must inherit the same allele, and one cannot say with certainty whether it is inherited IBD. This problem makes for another level of difficulty in human genetics, compared to experimental genetics. For the moment we assume that this problem does not exist and return to it in Sect. 9.6.

In the case of siblings, the number of alleles IBD at a given marker locus can be 0, 1, or 2. Let $N_0$, $N_1$, and $N_2$ be the total number of pedigrees

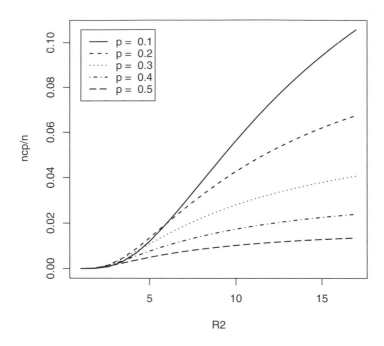

**Fig. 9.1.** The squared noncentrality parameter (per unit sample size) of the IBD statistic.

that share 0, 1, or 2 alleles IBD. The joint distribution of these counts is Multinomial$(n, \pi)$, with $\pi = (\pi_0, \pi_1, \pi_2)$. Under the null distribution $\pi = (1/4, 1/2, 1/4)$. For an additive model (i.e., $\sigma_D^2 = 0$) the alternative distribution at the trait locus takes the form $\pi = (1/4 - \sigma_A^2/(8Q), 1/2, 1/4 + \sigma_A^2/(8Q))$. The total IBD, standardized to have mean 0 and variance 1 under the hypothesis of no linkage, is a reasonable statistic for testing $H_0 : \sigma_A^2 = 0$ versus the alternative $H_1 : \sigma_A^2 > 0$.

The total number of alleles shared IBD is $N_1 + 2N_2$. Since the expected number of alleles shared IBD is $n$ and the variance is $n/2$ (both computed under the null distribution), and since $n = N_0 + N_1 + N_2$, the standardized statistic is

$$Z = \frac{2N_2 + N_1 - n}{(n/2)^{1/2}} = \frac{N_2 - N_0}{(n/2)^{1/2}} . \tag{9.5}$$

Observe that the hypothesis tested is one sided, hence the null is rejected for large positive values of the test statistic. The threshold for significance is determined by the null distribution of the test statistic. As the sample size increases $(n \to \infty)$ the distribution of this test statistic resembles more and more that of the standard normal distribution.

The mean of the test statistic under the alternative hypothesis for a marker perfectly linked to the trait locus $\tau$ is given by

$$\xi = \mathrm{E}(Z_\tau) = (2n)^{1/2}(\pi_2 - \pi_0) = (n/2)^{1/2}\check{\alpha}/Q_2 . \qquad (9.6)$$

The variance of $Z$ is $1 - (\check{\alpha}/Q_2)^2/2 \approx 1$, for local alternatives. The noncentrality parameter at a linked marker is calculated below.

## 9.4 Genome Scans

For a genome scan, we use $\max_t Z_t$, where we now introduce the subscript $t$ to denote marker location. The significance level and power can be found exactly as in Chaps. 5 and 6, provided we use an approximation suitable for a one-sided test and the appropriate value of $\beta$. It turns out that $\beta$ is 0.04, exactly twice what it was for a backcross. The reason is that along each chromosome (one maternally inherited and the other paternally inherited), the two siblings involve two meiotic events. In contrast a backcross involved only one. A more detailed mathematical analysis follows.

To study the properties of $Z_t$ and hence to approximate the significance level and power of a genome scan using the results of Chaps. 5 and 6, it is helpful to use the representation of the numerator $2N_{2t} + N_{1t} - n = \sum_{i=1}^{n}[J_i(t) - 1]$. This representation shows that the correlation function and mean value of the standardized statistic $Z_t$ can be obtained directly from the correlation function and mean value of each term in the numerator, namely $J(t)$, the number of alleles shared IBD by a sib pair at the marker $t$.

We first consider the case of markers that are unlinked to the trait locus. Under local alternatives, the same results for the covariance function hold at linked markers. Let $J(t)$ be written as $J_{\mathrm{M}}(t) + J_{\mathrm{F}}(t)$ where $J_{\mathrm{M}}(t)$ is the number of alleles, 0 or 1, inherited by the siblings IBD from their mother at locus $t$ and $J_{\mathrm{F}}(t)$ is the number inherited from their father. Let $s$ be a locus at recombination distance $\theta$ from $t$. If $J_{\mathrm{M}}(s) = 1$, then $J_{\mathrm{M}}(t) = 1$ if and only if both sibs have recombinations between $t$ and $s$ on their maternally inherited chromosome or neither sib does. The probability of no recombinations in the two maternal meioses is $(1 - \theta)^2$, while the probability of two recombinations is $\theta^2$. Thus $\Pr(J_{\mathrm{M}}(t) = 1|J_{\mathrm{M}}(s) = 1) = \theta^2 + (1 - \theta)^2$. For future notational convenience, let $\varphi$ be defined by $1 - \varphi = \theta^2 + (1 - \theta)^2$. Similar reasoning applies to $J_{\mathrm{F}}$, so $\Pr(J(t) = 2|J(s) = 2) = (1 - \varphi)^2$. By similar arguments one sees that $\Pr(J(t) = 1|J(s) = 2) = 2\varphi(1 - \varphi)$, $\Pr(J(t) = 1|J(s) = 1) = \varphi^2 + (1 - \varphi)^2$, $\Pr(J(t) = 2|J(s) = 1) = \varphi(1 - \varphi)$, etc., so we obtain a $3 \times 3$ matrix of transition probabilities from state $J(s) = i$ to state $J(t) = j$, for $i, j = 0, 1, 2$. Some calculation with these probabilities leads to

$$\mathrm{E}[J(t) - 1|J(s)] = (1 - 2\varphi)[J(s) - 1] . \qquad (9.7)$$

Multiplying by $J(s) - 1$ and taking expectations, we obtain the important relation

$$\mathrm{cov}[J(t), J(s)] = (1 - 2\,\varphi)/2 = 2\,\theta(1 - \theta) = \exp(-0.04\,|s - t|)/2\,,$$

where the third equality in the preceding expression follows from the equation $\theta = [1 - \exp(-0.02\,|t - s|)]/2$, for the recombination fraction $\theta$ in terms of genetic distance $|t - s|$ in cM. We conclude by observing that the preceding conditional probabilities and the resulting covariance are exactly the same as for $x(t)$, the number of $A$ alleles in an intercross design, except that $\theta$ has been replaced by $\varphi$, which has the effect of turning the parameter 0.02 into 0.04 in the exponent of the correlation coefficient.

As an illustration, let us determine the thresholds for a genome scan with various inter-marker spacings. Note that the genetic length of the human genome is very roughly about twice that of a mouse, or about 3,200 cM. Moreover, the genetic material in humans is distributed among 23 pairs of chromosomes (22 pairs of autosomes and a pair of sex chromosomes). We use these values in the approximation (5.3), but we divide the expression in the exponent by 2, since we are now interested in a one-sided test, and we set $\beta = 0.04$, to obtain:

```
> Delta <- c(35,20,10,5,1)
> z <- vector(length=length(Delta))
> names(z) <- paste("Delta=",Delta,sep="")
> for (i in 1:length(Delta)) z[i] <-
+     uniroot(OU.approx,c(3,4),beta=0.04,Delta=Delta[i],
+     length=3200,chr=23,center=0.05,test="one-sided")$root
> round(z,3)
Delta=35 Delta=20 Delta=10  Delta=5  Delta=1
   3.337    3.459    3.601    3.721    3.906
```

The noncentrality parameter for a marker at no recombination distance from the trait locus itself was given in the preceding section. To evaluate power in a genomic scan, we must also know the effect of recombination on the noncentrality parameter. Observe that the reasoning behind (9.7), which depends only on the recombination fraction between the loci $s$ and $t$ continues to apply if we set $s = \tau$, the trait locus. By taking expectations, we conclude that at a marker $t$ linked to the trait locus $\tau$,

$$\mathrm{E}(Z_t) = \mathrm{E}(Z_\tau)(1 - 2\varphi) = \xi \exp(-0.04|t - \tau|)\,.$$

The rate of decay of the noncentrality parameter is twice what it was for a backcross or an intercross. This means that as one increases the inter-marker distance, there is a greater loss of power for detecting a trait locus midway between markers in sib pairs than in a backcross or an intercross.

We now explore numerically the power to detect a trait locus in a genome scan as a function of the noncentrality parameter at the trait locus. We consider separately the case where the trait is perfectly linked to a marker and

the case where the trait locus is midway between two markers. The functions
"power.marker" and "power.midway", which were developed in Chap. 6, are
used in order to obtain analytic approximations:

```
> xi <- seq(3,6,by=0.1)
> power.mark <- matrix(nrow=length(xi),ncol=length(Delta))
> colnames(power.mark) <- names(z)
> power.between <- power.mark
> for (j in 1:length(Delta))
+ {
+     power.mark [,j] <- power.marker(z[j],0.04,Delta[j],xi)
+     for (i in 1:length(xi)) power.between[i,j] <-
+         power.midway(z[j],0.04,Delta[j],xi[i])
+ }
```

Let us plot the power functions for the two cases:

```
> old.par = par(mfrow=c(1,2))
> plot(range(xi),c(0,1),type="n",xlab="xi",ylab="power",
+      main="Trait locus on a marker")
> for(j in 1:length(Delta)) lines(xi,power.mark[,j],lty=j)
> legend(3.5,0.3,bty="n",legend=names(z),lty=1:length(z))
> plot(range(xi),c(0,1),type="n",xlab="xi",ylab="power",
+      main="Trait locus between markers")
> for(j in 1:length(Delta)) lines(xi,power.between[,j],lty=j)
> par(old.par)
```

The output is given in Fig. 9.2. Observe the relatively small decrease in power
as the markers density increases in the case that the trait locus is perfectly
linked to a marker compared to a more substantial increase in the power when
the locus is between markers. The difference is more pronounced than for a
backcross or intercross. The reason is the greater recombination parameter
($\beta = 0.04$ instead of 0.02) in sib pairs, because two parental meioses are
involved. The result is a more rapid decay of the linkage signal as one moves
away from a marker, so that markers in sib pairs about 10 cM apart lead to
about the same loss of power to detect a gene at the midpoint as markers
about 20 cM apart in a backcross. This difference is even larger when we
consider in the following section markers that are only partly informative. As
a result, for a genome scan based on affected sib-pairs, it appears advisable to
use a substantially denser collection of markers than we found necessary for a
backcross or intercross in experimental genetics. The results of the following
section will reinforce this conclusion.

An interesting example is provided by John et al. [43], who reanalyzed
with SNPs an earlier study of rheumatoid arthritis that had been conducted
with microsatellite markers at an approximately 10 cM inter-marker spacing.
The earlier analysis gave the value $\max_t Z_t = 4.22$ on chromosome 6 (in the
Human Leucocyte Antigen, or HLA, region, which harbors a large number of

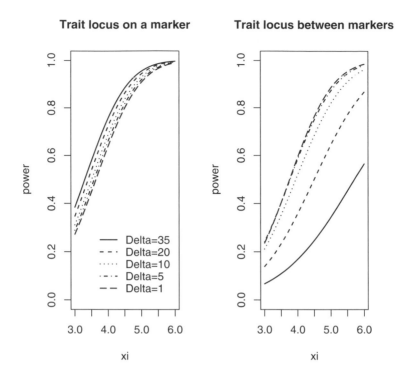

**Fig. 9.2.** The power for detecting a trait locus in a genome scan

genes associated with the function and diseases of the immune system). The p-value based on the approximation in Chap. 5 would be 0.004.

SNPs at an inter-marker distance of about 0.3 cM led to $\max_t Z_t = 3.97$ (at very close to the same genomic location), with a p-value of just slightly more than 0.05. Since markers placed this close together might fail to satisfy the important assumption of linkage equilibrium, the authors selected one (particularly informative) marker per cM. This reduced the value of $\max_t Z_t$ to 3.54, which has a p-value of 0.18.

Although these results suggest that the SNPs overall were not quite as efficient as SSRs, the authors indicated an overall preference for SNPs. A statistical reason for this preference is that the profile of the process $Z_t$ seemed to give a more precise picture of the location of the gene with the more closely spaced markers. A scientific reason is that the slight discrepancy between the locations of the peaks for the 10 cM and the 0.3 cM scans seemed to suggest that the latter provided a better estimate of the location of the appropriate gene, which could be regarded as known from earlier studies.

# 9.5 Parametric Methods

The approach described above is frequently called *nonparametric* to distinguish it from the *parametric* approach pioneered by Morton [56] and systematically described by Ott [57]. While parameters appear in both approaches, in the parametric approach the latent *genetic* parameters – penetrances and allele frequencies – play an explicit role in statistics to detect linkage, whereas in the nonparametric approach one concentrates on *statistical* parameters, which can in principle be estimated directly from experimental data without a specific genetic model. Examples are the frequency of a trait, or the probability that a particular relative of an affected is also affected, which can be estimated from population samples of phenotypes, or noncentrality parameters of statistics to detect linkage, which can be estimated from a combination of phenotypic and genotypic data. The parametric approach was developed at a time when the diseases under consideration involved a single gene, often showed a clear mode of inheritance, dominant or recessive, and had essentially complete penetrance, so it might be reasonable to assume $g_0 \approx 0$, with $g_1 = g_2 \approx 1$ for a dominant trait or $g_1 \approx 0, g_2 \approx 1$ for a recessive trait. Moreover, the number of markers available was extremely small, so the limiting factor in one's ability to map a disease gene was not the genetic effect on the trait, which was quite pronounced, but the recombination distance of the nearest marker to the gene.

For example, suppose we see the disease occurring in successive generations of pedigrees with approximately 50% of the offspring of affected individuals also having the disease. This suggests that the trait is dominant and fully penetrate without sporadic cases ($g_0 = 0$), and under an assumption of Hardy-Weinberg equilibrium we can estimate the frequency $p$ of the disease gene from the population prevalence, $2p(1 - p) + p^2 = 2p - p^2$, of the trait.

The simplest illustration of a parametric analysis arises from considering a three generation pedigree of an affected grandparent, say the grandfather, the intervening parent, say the father, and an affected grandchild. We assume the grandmother and mother are unaffected, so under the assumption of full penetrance the father is also affected. Assume there is a marker that has recombination $\theta$ with the trait, and assume that all marker alleles in the pedigree are unique, so we know exactly which marker allele passes from the affected grandfather, to the father, and the allele that the father then passes to the grandchild. If in addition to the disease allele, the father passes to his child the marker allele he received from the grandfather, then by definition there is no recombination between the disease allele and the marker allele. This happens with probability $1 - \theta$. If the father passes the allele he inherited from the grandmother, there is recombination with the disease allele, which happens with probability $\theta$. If we let $R$ denote the number of recombinations, 0 or 1, we have a Bernoulli variable with $\Pr(R = r) = \theta^r(1 - \theta)^{1-r}$. If we have a sample of $n$ such grandchildren, the total number of recombinations in the sample would be binomial with probability $\theta$, and we could test the hypothesis $\theta = 1/2$ by counting these recombinations. Observe that the grandfather and grandson share one allele IBD at the marker locus if and only if $R = 0$. Hence a test based on the number of recombinations is equivalently a test based on the number of alleles shared IBD, so the parametric analysis is in this case equivalent to a nonparametric analysis based on number of alleles shared IBD. Observe also that in this scenario there can be multiple affected grandchildren in a pedigree without any essential changes, and that unaffected grandchildren also provide linkage information. For them, however, recombinations have probability $1 - \theta$ and non-recombinations have probability $\theta$, since they are assumed not to have inherited the disease allele. Finally note that this simple scenario would become substantially more complex if there are sporadic cases and/or the penetrance of the disease allele is less than one, since then we could not be sure that the grandchild has the disease allele nor that it comes from the grandfather. In that case an appropriate parametric likelihood function would involve the penetrances $g_0$, $g_1$, $g_2$ and the allele frequency $p$ in addition to $\theta$.

Returning to the case of affected sib pairs, to put parametric and non-parametric analysis on similar footing we assume a two generation pedigree, but parental phenotypes are unknown. If there are few sporadic cases in the population and the disease is rare (and dominant), we would be willing to suppose that there is exactly one copy of the disease allele in the parental generation. But we do not know which of four marker alleles present in the parents and lying on the same chromosome as the disease related locus (we assume as above that markers are completely informative) is actually linked to the disease allele itself. (Genetic terminology is that the *linkage phase* is unknown.) Hence we consider the four possibilities to be equally likely. The analysis is more complicated but similar in principle to that given above. It leads to the number of alleles shared IBD by the siblings, so we eventually arrive at the probabilities in (9.4), or more generally the corresponding probabilities, say $\pi_i(\theta)$ for a marker at recombination distance $\theta$ from the disease gene. These probabilities would be assumed known except for the parameter $\theta$. To test the null hypothesis $\theta = 1/2$ (no linkage) against a specific alternative value $\theta < 1/2$, we could use the log likelihood ratio statistic

$$\sum_{j=0}^{2} N_j \log[\pi_j(\theta)/\pi_j(1/2)] \,, \tag{9.8}$$

which is often maximized with respect to $\theta$ to reflect the fact that the true $\theta$ is almost always unknown. Under the conditions described above, the non-centrality parameter of this statistic is typically large if $\theta$ is small, but would be small if the true $\theta$ is close to $1/2$.

A detailed development of this approach, especially the modifications required to deal with the fact that for complex diseases the allele frequencies and penetrances are essentially never known, is beyond the scope of this book. The most important strength of a parametric approach, to which we return in

Chap. 11, is that, subject to being able to perform the required calculations, it generalizes directly to pedigrees having varying numbers and configurations of affecteds and to arbitrary combinations of pedigrees. This property has played an important role in dealing with single gene traits of large penetrance, where large pedigrees with multiple affecteds are common. It is less important for dealing with complex diseases, involving small penetrances where pedigrees with large numbers of affecteds are rare.

The weakness of a parametric approach is that for complex diseases there may be multiple genes of incomplete penetrance that may interact with each other or with the environment, as well as non-genetic cases of the disease ($g_0 > 0$), with the result that one has no clear idea of the number or values of the relevant penetrances and allele frequencies. In addition, since modern DNA analysis has made available a large number of mapped markers, the emphasis on testing an hypothesis about $\theta$ seems misplaced. We are prepared to assume that *some* markers are closely linked to the relevant genes. However, the complexity of the genetics can lead to small noncentrality parameters, even at tightly linked markers, so the true signal at a linked marker may be small compared to the apparent signals arising from chance fluctuations at spurious markers throughout the genome. Hence in our outlook we have emphasized a null hypothesis to the effect that at the marker locus under consideration there is effectively no departure from Mendelian segregation of genotypes, so the noncentrality parameter of any test statistic is zero.

To gain somewhat more insight into the nature of a parametric analysis and prepare for a related discussion in Chap. 11, suppose (to simplify calculations) that $\check{\delta} = 0$ and $\check{\alpha}$ is small. By using the Taylor series approximation $\log(1 + x) \approx x - x^2/2$, valid for small $|x|$, one can show that the log likelihood ratio statistic at a marker locus $t$ assumed to be a recombination distance of $\theta$ from the trait locus is approximately

$$\xi(1 - 2\varphi)Z_t - \xi^2(1 - 2\varphi)^2/2 \tag{9.9}$$

where $Z_t$ is the approximately normal statistic defined above and $\xi = (n/2)^{1/2}(\check{\alpha}/Q_2)$ is its noncentrality at the trait locus $\tau$. So far we have regarded $\xi$ as known. If we admit that it is unknown, the parameters $\xi$ and $\varphi$ cannot be estimated separately if we only observe $Z_t$. Only the parameter $\eta = \xi(1 - 2\varphi)$ can be estimated.

At this point there are different possibilities for proceeding. (i) If we maximize the preceding expression with respect to $\eta = \xi(1 - 2\varphi) \geq 0$, we get $[\max(Z_t, 0)]^2/2$, where the nonnegativity restriction arises from the fact that the parameter $\eta$ cannot be negative. This would be equivalent to the statistic $Z_t$ (for one-sided alternatives), but as we shall see, the equivalence for this simple problem turns out to be the exception, not the rule. (See Prob. 9.11 for an example and the related discussion in Chap. 11.) (ii) If we take the attitude that markers are reasonably dense, so the distance from the nearest marker to the trait locus is likely to be small, we might simply set $\varphi = 0$ in (9.9) and scan the genome for maxima with respect to $t$. The result is a monotonic function of $\max Z_t$, hence is again equivalent to using $\max Z_t$ directly. (iii) If we take into consideration that we have a collection of mapped markers, the situation is similar to that in Chap. 7. For the asymptotic Gaussian model, where the log likelihood at $\tau$ is given by (9.9) with $t = \tau$ and $\varphi = 0$, we can also compute the likelihood function for a trait locus $\tau$ lying between markers $t_i$ and $t_{i+1}$. Using the notation of Chap. 7 (but with $\varphi$ in place of $\theta$), we have that $\mathrm{E}_0[Z_\tau | Z_{t_i}, Z_{t_{i+1}}] = \sigma'_\tau W Z$, $\mathrm{var}_0[Z_\tau | Z_{t_i}, Z_{t_{i+1}}] = 1 - \sigma'_\tau W \sigma_\tau$, and hence the likelihood function equals

$$\mathrm{E}_0[\exp(\xi Z_\tau - \xi^2/2) | Z_{t_i}, Z_{t_{i+1}}] = \exp(\xi \sigma'_\tau W Z - \xi^2 \sigma'_\tau W \sigma_\tau / 2) .$$

For $\xi$ regarded as known, the likelihood ratio statistic for a genome scan would be the maximum over $\tau$ of the expression appearing in the exponent. If $\xi$ is regarded as unknown, we can maximize over $\xi \geq 0$ as well. This leads to a one-sided version of the statistic in Chap. 7. Based on the results of Chap. 7, it seems unlikely that these statistics will be substantially more powerful than the simpler statistics studied earlier in this chapter.

## 9.6 Estimating the Number of Alleles Shared IBD

In general, IBD status, which is the basis for the statistics discussed above, is not observed directly. It needs to be inferred from genotype information. Assume that both parents of the siblings were recruited and the genotypes of all the given members of the family were obtained. If multi-allelic markers are used, one may be able to observe that at a particular locus the two parents have four distinct alleles. This favorable scenario enables the precise determination of the IBD status of the siblings. At the other extreme, if both parents are homozygous, then the specific marker provides no information regarding the IBD status of the siblings at that locus. The marker is then said to be *uninformative*. There may also exist intermediate cases, e.g., where one parent is homozygous while the other is heterozygous, or when both parents are heterozygous for the same two alleles. Such markers are denoted *partially informative*. However, even when a marker is partially informative or totaly uninformative, there may be other markers nearby which are either informative or at least partially informative. If these markers are sufficiently close, so there is little chance of recombination, one may attempt to infer the IBD status at the given locus based on the genotypes at those nearby loci and then conduct a genome scan with reconstructed IBD statistics. The problem of partial information regarding IBD relations among the affected siblings and the need to exploit genotype information from nearby markers in order to reconstruct the IBD becomes even more acute if the parents are not available for genotyping. This is commonly the case in late onset diseases, such as Alzheimer disease, in which the participating affected siblings are typically older and are less likely to have living parents.

Recall that markers fall into two main classes: SNPs, which are bi-allelic, and various classes of multi-allelic markers, e.g., SSRs, which often have 4–10 alleles. While SNPs are much more numerous and more easily genotyped, they are individually less informative. In most of the following we concentrate on SNPs and find that because each one by itself is relatively uninformative, there is a considerable loss of information unless they are reasonably dense. The programs are easily adapted to multi-allelic markers and show that SSRs can be more widely separated without a corresponding loss of information.

In this section we will investigate the effect of partial information regarding the IBD relations on the statistical properties of the test statistics. We will substitute for the unknown IBD statistic its conditional expectation, given the genotypic information at hand. This is similar to the case of missing genotypes that was discussed in Chap. 7. However, the computation of the conditional expectation is more complex and will require application of algorithms that were originally developed in the context of what are called *hidden Markov models*, or HMM in short.

The section is divided into three subsections. In the first subsection we will develop R code for the simulation of pedigrees. The HMM algorithms will be presented in the second subsection. In the third subsection we will explore the statistical properties of a genome scan with affected sib pairs when only their genotypes are available. The tools that were developed in the first two subsections will be used in that exploration.

### 9.6.1 Simulating Pedigrees

Our first goal is to develop R code that will enable us to simulate affected sib pairs. The programs we develop are similar to those developed in the chapters that dealt with experimental genetics. However, there is a major difference between the situation we previously considered and the current one. In the experimental designs that we considered before, subjects were not preselected based on their phenotypes. In particular, the segregation of genetic material from one generation to the next followed Mendel's segregation rules. In the case at hand, however, the subjects are selected because they express the trait (a disease). This selection rule results in a distortion of the segregation of alleles in loci linked to trait-related genes. In fact, it is exactly this distortion that allows us to detect such loci. As a result, we now need to rewrite our programs in order to allow for distortion in the segregation in the presence of a trait locus.

Start with an adaptation to the new setting of the function "`meiosis.chr`", which simulates the gamete being segregated from a parent to an offspring:

```
> meiosis.link <- function(GF,GM,markers,qtl,inhe)
+ {
+     n <- nrow(GF)
+     GS <- GF
```

```
+      loci <- sort(c(qtl, markers))
+      rec.frac <- (1-exp(-0.02*diff(loci)))/2
+      index <- 1:length(markers)
+      from.GM <- inhe
+      for (i in index[markers >= qtl])
+      {
+          rec <- rbinom(n,1,rec.frac[i])
+          from.GM <- from.GM*(1-rec) + (1-from.GM)*rec
+          GS[from.GM==1,i] <- GM[from.GM==1,i]
+      }
+      from.GM <- inhe
+      for (i in rev(index[markers < qtl]))
+      {
+          rec <- rbinom(n,1,rec.frac[i])
+          from.GM <- from.GM*(1-rec) + (1-from.GM)*rec
+          GS[from.GM==1,i] <- GM[from.GM==1,i]
+      }
+      return(GS)
+ }
```

Observe that in the default application in the definition of the function "mating" below, which corresponds to the null case of no trait locus and which obeys Mendel's laws of segregation, the inheritance vector at the first marker consists of realizations of independent 0-1 random variables. The segregation of the rest of the markers is determined in the first loop according to the process of recombination in exactly the same way it was done in the function "meiosis.chr". The second loop is not activated.

If a trait related locus does exist at a locus denoted "qtl", then the selection rule may distort the distribution of the inheritance indicator at that locus. This distorted distribution will be generated externally, and the resulting vector of the inheritance indicator at the trait locus may be imported in the argument "inhe". The distortion is reflected at the markers on both sides of the trait locus due to linkage and the process of recombination. The process to the right of the trait locus is generated in the first loop and the process to the left is generated in the second loop.

Subjects' pairs of parental gametes are stored in a list. This list contains two matrices, one for each gamete. The columns of the matrices correspond to the markers and the rows to independent copies. The function "mating" is an adaptation to the new setting of the function "cross" which was applied in the context of experimental genetics. It takes as input two subjects (a father and a mother) and returns as output a new subject (an offspring):

```
> mating <- function(fa,mo,markers,qtl=markers[1],
+      inhe.fa=rbinom(nrow(fa$pat),1,0.5),
+      inhe.mo=rbinom(nrow(mo$pat),1,0.5))
+ {
```

```
+        pat <- meiosis.link(fa$pat,fa$mat,markers,qtl,inhe.fa)
+        mat <- meiosis.link(mo$pat,mo$mat,markers,qtl,inhe.mo)
+        return(list(pat=pat, mat=mat))
+ }
```

As an illustration of the application of the code let us generate the processes of IBD on a given chromosome for 10 independent pedigrees with markers spaced 20 cM apart:

```
> n.ped <- 10
> markers <- seq(0,140,by=20)
> n.mark <- length(markers)
> fa <- list(pat=matrix(1,n.ped,n.mark),
+        mat=matrix(2,n.ped,n.mark))
> mo <- list(pat=matrix(3,n.ped,n.mark),
+        mat=matrix(4,n.ped,n.mark))
> sib1 <- mating(fa,mo,markers)
> sib2 <- mating(fa,mo,markers)
> ibd <- (sib1$pat==sib2$pat)+
+        (sib1$mat==sib2$mat)
> ibd
        [,1] [,2] [,3] [,4] [,5] [,6] [,7] [,8]
  [1,]    0    1    0    0    0    1    1    1
  [2,]    2    2    1    0    0    0    1    1
  [3,]    1    1    0    0    0    1    1    1
  [4,]    2    2    1    2    0    0    2    2
  [5,]    2    2    2    1    1    1    1    1
  [6,]    1    1    1    1    2    1    0    1
  [7,]    0    0    0    1    1    1    2    2
  [8,]    1    1    2    1    1    1    1    2
  [9,]    1    2    2    2    2    1    1    0
 [10,]    1    1    2    1    0    0    0    0
```

Here the markers are fully informative and the IBD process can be computed directly from the alleles of the offspring.

Next we turn to the simulation of pedigrees under the alternative distribution. In Sect. 9.2 we investigated the distribution of the IBD status at the trait locus as a function of the allele and penetrance frequencies, given that both siblings are affected. The distribution of the inheritance vectors is a reflection of the IBD distribution. The inheritance vector has four components, indicating the parental source of the gamete segregated from (i) the father and (ii) the mother of the first sibling and (iii) the father and (iv) the mother of the second sibling. Observe that the marginal probability of each of the grandpaternal origins in components (i)–(iv) is 1/2. The IBD relation introduces dependence between the components. When the IBD status equals zero, then the parental sources for (i) and (iii) and for (ii) and (iv) are opposite. When the IBD status equals two, the parental sources are the same. When the IBD status equals one, then for one pair the parental source is the same and for the other it is opposite:

```
> inhe.vector <- function(ibd.prob,n.ped)
+ {
+     ibd.qtl <- sample(0:2,n.ped,replace=TRUE,prob=ibd.prob)
+     sib1.pat <- rbinom(n.ped,1,0.5)
+     sib1.mat <- rbinom(n.ped,1,0.5)
+     pat.equal <- rbinom(n.ped,1,0.5)
+     sib2.pat <- sib1.pat*pat.equal+(1-sib1.pat)*(1-pat.equal)
+     sib2.mat <- sib1.mat*(1-pat.equal)+(1-sib1.mat)*pat.equal
+     inhe <- cbind(sib1.pat,sib1.mat,sib2.pat,sib2.mat)
+     inhe[ibd.qtl==0,3:4] <- 1-inhe[ibd.qtl==0,1:2]
+     inhe[ibd.qtl==2,3:4] <- inhe[ibd.qtl==2,1:2]
+     return(inhe)
+ }
```

The program works by simulating first the independent components (i) and (ii) from the marginal Bernoulli distribution. Using random assignment, some of the rows are set to have (iii) match (i) and (iv) opposite to (ii) while the other rows are set in the opposite way. This corresponds to having IBD equal to one. Subsequently, the relations between the components are changed to the appropriate relations in the rows where IBD is equal to zero and in the rows where it is equal to two. The IBD itself is generated from the appropriate alternative distribution, which is provided in the argument "ibd.prob". To illustrate consider the additive model in which we take $p = 0.2$, $g_0 = 0.1$, $\alpha = 0.4$:

```
> n.ped <- 10^5
> ibd.prob <- DistIBD(0.2,0.1,0.5,0.9)
> qtl <- 80
> inhe.qtl <- inhe.vector(ibd.prob,n.ped)
> fa <- list(pat=matrix(1,n.ped,n.mark),
+               mat=matrix(2,n.ped,n.mark))
> mo <- list(pat=matrix(3,n.ped,n.mark),
+               mat=matrix(4,n.ped,n.mark))
> sib1 <- mating(fa, mo, markers,
+               inhe.fa=inhe.qtl[,"sib1.pat"],
+               inhe.mo=inhe.qtl[,"sib1.mat"],qtl=qtl)
> sib2 <- mating(fa, mo, markers,
+               inhe.fa=inhe.qtl[,"sib2.pat"],
+               inhe.mo=inhe.qtl[,"sib2.mat"],qtl=qtl)
> ibd <- (sib1$pat==sib2$pat)+(sib1$mat==sib2$mat)
> sum(ibd.prob*0:2)
[1] 1.137339
> 2*mean(inhe.qtl[,1:2]==inhe.qtl[,3:4])
```

```
[1] 1.1369
> round(apply(ibd,2,mean),4)
[1] 0.9996 1.0098 1.0261 1.0622 1.1369 1.0605 1.0288 1.0108
```

A susceptibility locus is present 80 cM from the telomere, next to the 5th marker. Note that the average IBD at the marker is about equal to the expectation computed from the IBD probabilities. The expected IBD is elevated in the vicinity of the trait locus and it gradually decreases to the null expectation as markers become more distant from that locus.

Now consider the replacement of the fully informative markers by partially informative ones. The information provided by markers is in the form of the classification of pedigrees based on the genotypes of those members for which the genotypes are obtained. For example, we will assume in the sequel that genotypes are obtained for both siblings but not for their parents. We will also assume that the markers have n.al distinct alleles, with the default value of two. Genotype measurement for an individual returns the combined reading of its pair of homologous chromosomes, without distinguishing the parental source. Hence, for bi-allelic markers one may obtain three distinct genotypes. More generally, for markers with n.al alleles the total number of distinct genotypes is n.al(n.al+1)/2. The total number of genotypes for pair of siblings is the square of the number of individual genotypes.

We will find it easier to simulate and compute the distribution of the four parental alleles of the two siblings. However, it should be realized that these alleles are not observable. Instead, what one gets to observe are the genotypes, which are a many-to-one mapping of the four alleles. As a first step we introduce a function that maps alleles to genotypes:

```
> genotype <- function(a1,a2,a3,a4,n.al=2)
+ {
+     a.m <- pmin(a1,a2)
+     a.M <- pmax(a1,a2)
+     g1 <- a.M + (a.m-1)*(n.al-a.m/2)
+     a.m <- pmin(a3,a4)
+     a.M <- pmax(a3,a4)
+     g2 <- a.M + (a.m-1)*(n.al-a.m/2)
+     g <- g1 + (g2-1)*n.al*(n.al+1)/2
+     return(g)
+ }
```

In the specific case of a bi-allelic markers the function returns the combined genotypes coded as an integer between one and $3^2 = 9$.

The function "ped.geno" takes as input a pair of siblings and a vector of population allele distribution. It produces as an output a matrix of coded genotypes. Each column of the matrix corresponds to a marker and each row corresponds to a pedigree. Markers are assumed to be identically distributed and in linkage equilibrium, and pedigrees are assumed to be unrelated (which means statistical independence):

```
> ped.geno <- function(sib1,sib2,f=rep(1/2,2))
+ {
+       n.ped <- nrow(sib1$pat)
+       n.mark <- ncol(sib1$pat)
+       n.al <- length(f)
+       par.al <- list()
+       for(par in 1:4) par.al[[par]] <-
+           matrix(sample(1:n.al,n.ped*n.mark,
+               replace=TRUE,prob=f),n.ped,n.mark)
+       a <- inhe <- c(sib1,sib2)
+       for (v in 1:4) for (par in 1:4)
+           a[[v]][inhe[[v]]==par] <-
+               par.al[[par]][inhe[[v]]==par]
+       geno <- genotype(a[[1]],a[[2]],a[[3]],a[[4]],n.al)
+       return(geno)
+ }
```

The function works by simulating alleles (integers in the range between 1 and
n.al) for each of the four parental gametes. The function "sample" is used in
order simulate the alleles from the population distribution of marker alleles.
The lists "sib1" and "sib2" store two matrices each with the index of the
parental source of the gamete. The resulting inherited alleles are computed
and sorted in the list "a". Finally, the function "genotype" is applied in order
to compute the resulting genotype codes.

### 9.6.2 Computing the Conditional Distribution of IBD

The goal in this subsection is to reconstruct the unobserved process of IBD
in a pedigree using the marker genotypes. This will be conducted by the
calculation of the conditional expectation of the full-information statistic –
the total number of alleles inherited IBD for the two siblings – given the
genotypic information at hand. This conditional expectation is straightforward
to compute once the conditional distribution of IBD, given the genotypes, is
known. The calculation of the latter is the subject of this subsection.

The probabilistic structure of the observations may be modeled using a
"hidden" process. The hidden process is the process of IBD at the markers.
This process may not be observed directly, but it does have an effect on the
distribution of the observed genotypes. This effect may be exploited in order to make inference on the underlying hidden process. In particular, when
the underlying hidden process is Markovian and the distribution of an observation is determined by the state of the hidden process at the location of
the observation, independently of its values at other locations, the model is
called a hidden Markov model (HMM). Our case fits into this setting since
the process of IBD is Markovian and since markers were assumed to be in
linkage equilibrium.

**Table 9.1.** Conditional distribution of genotypes of ASP, given the IBD status.

| | IBD=0 | IBD=1 | IBD=2 |
|---|---|---|---|
| 1=(0,0) | $(1-f)^4$ | $(1-f)^3$ | $(1-f)^2$ |
| 5=(1,1) | $4f^2(1-f)^2$ | $f(1-f)$ | $2f(1-f)$ |
| 9=(2,2) | $f^4$ | $f^3$ | $f^2$ |
| 2=(0,1) | $2f(1-f)^3$ | $f(1-f)^2$ | 0 |
| 4=(1,0) | $2f(1-f)^3$ | $f(1-f)^2$ | 0 |
| 6=(1,2) | $2f^3(1-f)$ | $f^2(1-f)$ | 0 |
| 8=(2,1) | $2f^3(1-f)$ | $f^2(1-f)$ | 0 |
| 3=(0,2) | $f^2(1-f)^2$ | 0 | 0 |
| 7=(2,0) | $f^2(1-f)^2$ | 0 | 0 |

The distribution of a HMM is fully determined by the initial distribution and the transition matrices of the underlying Markov process and by the conditional distribution of the observations, given the states of the underlying process. The latter refers in our case to the conditional distribution of the pair genotypes of the siblings, given the IBD status at the marker.

The conditional distribution of the genotypes of the siblings, given the IBD process, is a function of the frequency of the alleles in the population. If there is no identity-by-descent among the alleles of the siblings (IBD=0), the two genotypes are independent and follow the Hardy-Weinberg distribution. In the case where exactly one pair of alleles has a common source (IBD=1), then the other two alleles (one in each sibling) are independent of each other and of the IBD allele. Finally, when each of the two alleles in one sibling has a matching IBD allele in the other sibling (IBD−2), then the genotypes of the two siblings fully match. The distributions of the siblings' genotypes for a bi-allelic marker and for each of the IBD situations is given in Table 9.1. To avoid confusion we have used the letter $f$ to represent the population frequency of the allele of the marker. In contrast, we have used the letter $p$ to represent the frequency of the allele $D$ of the trait locus.

The function "geno.given.ibd" computes this table for allele frequencies denoted by the vector f (with the default of uniformly distributed bi-allelic marker):

```
> geno.given.ibd <- function(f=c(0.5,0.5))
+ {
+     n.al <- length(f)
+     P.0 <- outer(outer(f,f),outer(f,f))
+     P.1 <- P.2 <- array(0,dim=rep(n.al,4))
+     for(a2 in 1:n.al) for(a1 in 1:n.al)
+     for(a3 in 1:n.al) for(a4 in 1:n.al)
+     {
+         if (a1==a3 & a2==a4)
+         {
+             P.2[a1,a2,a3,a4] <- f[a1]*f[a2]
```

```
+              P.1[a1,a2,a3,a4] <- f[a1]*f[a2]*(f[a1]+f[a2])/2
+          }
+          if (a1==a3 & a2!=a4)
+          {
+              P.1[a1,a2,a3,a4] <- f[a1]*f[a2]*f[a4]/2
+          }
+          if (a1!=a3 & a2==a4)
+          {
+              P.1[a1,a2,a3,a4] <- f[a1]*f[a3]*f[a2]/2
+          }
+      }
+      a1 <- rep(1:n.al,n.al^3)
+      a2 <- rep(rep(1:n.al,rep(n.al,n.al)),n.al^2)
+      a3 <- rep(rep(1:n.al,rep(n.al^2,n.al)),n.al)
+      a4 <- rep(1:n.al,rep(n.al^3,n.al))
+      geno <- genotype(a1,a2,a3,a4,n.al)
+      P.0 <- tapply(as.vector(P.0),geno,sum)
+      P.1 <- tapply(as.vector(P.1),geno,sum)
+      P.2 <- tapply(as.vector(P.2),geno,sum)
+      P <- cbind(P.0,P.1,P.2)
+      colnames(P) <- paste("State=",0:2,sep="")
+      return(P)
+ }
```

The function works by computing the distribution of the vector of four alleles
for the two siblings in each of the IBD situations. The results are stored
in three vectors, each of length n.al to the fourth power. The probabilities
for the different genotypes are computed by summation of the four allelic
probabilities according to the level of the genotype. This is carried out with
the aid of the function "tapply". The function "tapply" takes as input a
vector, a factor, and a function. The function is applied to the collection of
values of the vector that correspond to each given level of the factor.

The other component is the distribution of the unobserved IBD process.
This process is generated in the case under consideration as a function of
the four inheritance indicators. Each of these indicators can be viewed as
an independent process with two states (0 or 1). The states of the processes
may change from one marker to the next, depending on the recombination
fraction. Under a model of no crossover interference (the Haldane model)
these independent inheritance processes are Markovian. As it turns out, for
the specific pedigree structure of two siblings the process of IBD is Markovian
as well. The transition matrix of going from one marker to the next, which
was discussed on our analysis of $J$ in Sect. 9.3, is computed in the function
"trans.mat":

```
> trans.mat <- function(theta)
+ {
+     phi <- 1-theta^2-(1-theta)^2
+     Tr <- matrix(c((1-phi)^2,2*phi*(1-phi),phi^2,
+              phi*(1-phi),phi^2+(1-phi)^2,phi*(1-phi),
+              phi^2,2*phi*(1-phi),(1-phi)^2),3,3,byrow=TRUE)
+     colnames(Tr) <- paste("to.IBD=",0:2,sep="")
+     rownames(Tr) <- paste("from.IBD=",0:2,sep="")
+     return(Tr)
+ }
```

For example, when the distance between two markers is 20 cM, then the fraction of recombination and the transition matrix are given by:

```
> theta <- 0.5 - 0.5*exp(-0.02*20)
> round(trans.mat(theta),3)
           to.IBD=0 to.IBD=1 to.IBD=2
from.IBD=0   0.525    0.399    0.076
from.IBD=1   0.200    0.601    0.200
from.IBD=2   0.076    0.399    0.525
```

It can be shown that the null IBD distribution: $(1/4, 1/2, 1/4)$ is the stationary distribution of such a matrix, i.e., multiplication of the transition matrix on the left by this row vector produces exactly the same vector:

```
> Pr <- c(0.25,0.5,0.25)
> Pr%*%trans.mat(theta)
       to.IBD=0 to.IBD=1 to.IBD=2
[1,]     0.25      0.5      0.25
```

The initial distribution and transition matrices of the Markov process and the conditional distribution of the observations determine the joint distribution of the observations and of the hidden process in a straightforward way. In principle, the determination of the conditional distribution of the hidden process, given the observation is a straightforward application of Bayes' formula. However, a naïve attempt to apply the formulas will face severe computational problems when the number of markers is even moderately large. Indeed, the sample space of the possible paths of the IBD processes over the set of markers grows exponentially fast in powers of three as a function of the number of markers. An attempt to compute directly the posterior distribution of the paths may require the manipulation of an extremely large number of terms and turns out to be impractical if more than a score of markers is considered.

As a remedy, clever algorithms have been developed for computation in HMM scenarios where the unobserved process possesses a Markovian structure and the observations are conditionally independent, given the states of the unobserved process. These algorithms exploit the sequential independence of the components of the process in order to subdivide the task of summing an exponential number of products into a sequence of multiplications of sums with a fixed number of summands. Below we apply two basic algorithms that were developed for computation in such a setting: The *forward* and the *backward* algorithms.

Denote by $J(t_m)$ the IBD process at the $m$th marker and by $\Pr(G_m \mid j)$ the probability of the observed genotypes at that marker, given that the IBD status is $j$. Denote the transition probability for the IBD process by $T_{ij} = \Pr(J(t_m) = j \mid J(t_{m-1}) = i)$, which also equals $\Pr(J(t_{m-1}) = j \mid J(t_m) = i)$, since the process of recombination does not depend on the way we order the markers, but only on the distance between them. In principle $T_{ij}$ will also depend on $m$ when distances between markers vary. The forward algorithm computes recursively the quantity $F_m(j) = \Pr(G_1, G_2, \ldots, G_m, J(t_m) = j)$, which is the joint distribution of the genotypes up to locus $t_m$ and IBD status at that locus. It does so by conditioning on the states of hidden process at the locus $t_{m-1}$ and exploiting the Markovian structure and independence in order to obtain the relation:

$$F_m(j) = \sum_i \Pr(G_1, G_2, \ldots, G_m, J(t_{m-1}) = i, J(t_m) = j)$$

$$= \sum_i \left\{ F_{m-1}(i) \times T_{ij} \times \Pr(G_m \mid j) \right\}.$$

Summation in the relation extends over the three possible values of the IBD process at locus $t_{m-1}$. Applying this relation recursively, starting with the initial relation $F_1(j) = \Pr(G_1 \mid j)\Pr(J(t_1) = j)$, allows for the computation of these quantities for all $m$ and $j$. The number of elements that needs to be manipulated is proportional to the product of the number of markers with the number of possible states of the underlying process (which equals three in our setting).

Denote by $\tilde{m}$ the index of the last marker on the given chromosome. The backward algorithm is used in order to compute the quantity $B_m(j) = \Pr(G_{m+1}, G_{m+2}, \ldots, G_{\tilde{m}} \mid J(t_m) = j)$, namely the conditional distribution of the genotypes beyond a locus, given the IBD status at the locus. A recursive relation between the quantities can be identified. This time the relation involves a sum over the states of the process at $t_{m+1}$ and takes the form

$$B_m(j) = \sum_i \Pr(G_{m+1}, \ldots, G_{\tilde{m}}, J(t_{m+1}) = i \mid J(t_m) = j)$$

$$= \sum_i \left\{ B_{m+1}(i) \times T_{ji} \times \Pr(G_{m+1} \mid i) \right\}.$$

The starting values for the recursion are $B_{\tilde{m}}(j) = 1$.

Let $G = (G_1, \ldots, G_{\tilde{m}})$ be the genetic information over the chromosome for the given pedigree. Since $\Pr(G, J(t_m) = j) = \Pr(G_1, \ldots, G_{\tilde{m}}, J(t_m) = j) = F_m(j)B_m(j)$, it follows from the definition of conditional probabilities that

$$\Pr\big(J(t_m) = j \,|\, G\big) = \frac{F_m(j)B_m(j)}{\sum_i F_m(i)B_m(i)} \; . \tag{9.10}$$

Consequently, the conditional distribution of IBD, given the genotype information, can be computed at each locus as a function of the $F$ and $B$ quantities.

The functions "forward" and "backward" apply the forward and backward algorithms in order to compute the forward an backward joint distributions of the genotypes and the IBD states. The first argument to these functions is an array "G.I" with the conditional probabilities of the observed genotypes for each of the pedigrees, each of the markers and each of the IBD states. The second and third arguments are the transition matrix of the IBD process and its initial distribution, respectively. The output are arrays that contain the forward and backward probabilities, respectively:

```
> forward <- function(G.I,Tr,Pr)
+ {
+       n.samp <- dim(G.I)[1]
+       n.mark <- dim(G.I)[2]
+       F <- G.I
+       F[,1,] <- sweep(G.I[,1,],2,Pr,"*")
+       for (i in 2:n.mark)
+       {
+           F[,i,] <- G.I[,i,]*(F[,i-1,]%*%Tr)
+           S <- F[,i,1] + F[,i,2] + F[,i,3]
+           F[,i,] <- sweep(F[,i,],1,S,"/")
+       }
+       return(F)
+ }
> backward <- function(G.I,Tr,Pr)
+ {
+       n.samp <- dim(G.I)[1]
+       n.mark <- dim(G.I)[2]
+       B <- G.I
+       B[,n.mark,] <- 1
+       for (i in seq(n.mark-1,1))
+       {
+           B[,i,] <- (G.I[,i+1,]*B[,i+1,])%*%t(Tr)
+           S <- B[,i,1] + B[,i,2] + B[,i,3]
+           B[,i,] <- sweep(B[,i,],1,S,"/")
+       }
+       return(B)
+ }
```

Note that in each iteration of the evaluation, the currently computed quantities in F and B are re-scaled to sum to one. This re-scaling increases the numerical stability of the algorithm, which would otherwise involve the manipulation of terms that become vanishingly small as the algorithm progresses.

Round off errors would have been a serious concern if that were the case. Owing to the re-scaling, the terms are no longer the probabilities per se, but are only proportional to such probabilities. Nonetheless, the constants of proportionality do not depend on the IBD status at a locus and are therefore canceled out of both the numerator and the denominator when (9.10) is applied in order to obtain the target distribution. The actual computation of the conditional distribution of the states, given the genotypes, is carried out in the function "marginal.post". This function takes as input the output arrays of the functions "forward" and "backward" and produces an array of the same type with the posterior probabilities of the states:

```
> marginal.post <- function(F,B)
+ {
+     P <- F*B
+     S <- P[,,1]+P[,,2]+P[,,3]
+     P <- sweep(P,1:2,S,"/")
+     return(P)
+ }
```

### 9.6.3 Statistical Properties of Genome Scans

With the tools developed for the simulation of random pedigrees and a function for the computation of the estimated identity-by-descent probabilities from the genotypic information, we can start investigating the statistical properties of mapping in the more realistic setting of partial information. Let us initiate our investigation by the determination of the expectation and covariance properties of the scanning statistic under both the null and the alternative hypotheses. Later, we will consider Gaussian processes with the same covariance and mean structure and obtain significance thresholds and power curves. Our investigation will include several inter-marker spacings.

The first simulation is conducted under the null distribution. We simulate $10^5$ independent pedigrees and use them in order to compute the covariance and mean structure of the reconstructed IBD process. We also assess the accuracy of the reconstruction at a central locus. The simulation is split into 100 batches in order to avoid running out of memory:

```
> n.rep <- 10^2
> n.ped <- 10^3
> Delta <- c(35,20,10,5,1)
> ibd.est.null <- matrix(nrow=3,ncol=length(Delta))
> colnames(ibd.est.null) <- paste("Delta=",Delta,sep="")
> rownames(ibd.est.null) <- c("mean","var","mse")
> cor.ibd <- vector(mode="list",length=length(Delta))
> names(cor.ibd) <- paste("Delta=",Delta,sep="")
> cor.est <- cor.ibd
```

```
> P <- geno.given.ibd()
> for(i in 1:length(Delta))
+ {
+       markers <- seq(0,140,by=Delta[i])
+       n.mark <- length(markers)
+       locus <- ceiling(n.mark/2)
+       theta <- 0.5 - 0.5*exp(-0.02*Delta[i])
+       Tr <- trans.mat(theta)
+       fa <- list(pat=matrix(1,n.ped,n.mark),
+                     mat=matrix(2,n.ped,n.mark))
+       mo <- list(pat=matrix(3,n.ped,n.mark),
+                     mat=matrix(4,n.ped,n.mark))
+       ibd.est <- ibd <- NULL
+       G.I <- array(dim=c(n.ped,n.mark,3))
+       for (rep in 1:n.rep)
+       {
+           sib1 <- mating(fa,mo,markers)
+           sib2 <- mating(fa,mo,markers)
+           geno <- ped.geno(sib1,sib2)
+           for(k in 1:3) G.I[,,k] <-
+                 matrix(P[geno,k],n.ped,n.mark)
+           F.P <- forward(G.I,Tr,Pr)
+           B.P <- backward(G.I,Tr,Pr)
+           I.G <- marginal.post(F.P,B.P)
+           ibd.est <- rbind(ibd.est,2*I.G[,,3]+I.G[,,2])
+           ibd <- rbind(ibd,(sib1$pat == sib2$pat)+
+                 (sib1$mat == sib2$mat))
+       }
+       ibd.est.null["mean",i] <- mean(ibd.est[,locus])
+       ibd.est.null["var",i] <- var(ibd.est[,locus])
+       ibd.est.null["mse",i] <-
+                 mean((ibd[,locus]-ibd.est[,locus])^2)
+       cor.ibd[[i]] <- cor(ibd)
+       cor.est[[i]] <- cor(ibd.est)
+ }
```

Observe that mean, variance, and mean-square distance between the reconstructed and actual IBD processes are stored in a matrix called "ibd.est.null". The columns of this matrix correspond to the different inter-marker spacings. The correlation matrices are stored in a list named "cor.est". For later comparison, we also store the correlation structure of the actual IBD process in the list "cor.ibd".

Let us examine the mean, the variance, and the quality of reconstruction under the null distribution:

```
> round(ibd.est.null,3)
       Delta=35 Delta=20 Delta=10 Delta=5 Delta=1
mean     1.001    1.002    1.001    1.002   0.998
var      0.126    0.155    0.217    0.292   0.425
mse      0.374    0.343    0.284    0.208   0.075
```

Several insights emerge. First, it can be seen that the expected value of the estimated IBD is about equal to the expectation of the true IBD, which is one. As a matter of fact, it can be shown mathematically that the expectations of the estimated and the true IBD coincide. This follows from the fact that the estimated expression is a conditional expectation and the mathematical fact that the expectation of a random variable is equal to the expectation of its conditional expectation. (Symbolically, $E(J) = E[E(J|G)]$.)

Second, one can conclude that the variance of the estimated IBD is less than the variance of the actual IBD, which is equal to $1/2$. Moreover, the denser the markers are, the closer the variance is to $1/2$. Mathematically the relation between variances of the actual IBD $J$ and the estimated IBD, which we denote by $\hat{J}$, is given by the relation

$$\mathrm{var}(J) = \mathrm{var}(E(J\,|\,G)) + E(\mathrm{var}(J\,|\,G)) = \mathrm{var}(\hat{J}) + E(\mathrm{var}(J\,|\,G)) > \mathrm{var}(\hat{J}).$$

The more informative $G$ is, the more similar $\hat{J}$ is to $J$ and the closer their variances are to each other.

Third, a closer examination of the numbers in the second and third rows reveals that their sum equals one-half – the variance of the actual IBD statistic. In mathematical terms one can express this relation in the form:

$$E[(\hat{J} - J)^2] = E[\mathrm{var}(J\,|\,G)].$$

Substituted into the previous equation, this relation shows that the closer the variance of the reconstructed IBD is to the variance of the actual IBD, the more accurate the reconstruction is, as measured by its mean squared error.

In the computation of a scanning statistic, standardization is carried out with the standard deviation of the estimated IBD. The resulting statistic has a zero mean and a unit variance under the null distribution. The Gaussian limiting distribution of the process of statistics is determined by the correlation structure. In Fig. 9.3 the correlation function between a statistic computed for a central marker and the statistics computed at the flanking markers is plotted. The spacing between markers is 10 cM. The code that produced the figure is:

```
> markers <- seq(0,140,by=10)
> plot(markers,cor.ibd$"Delta=10"[8,],pch=19,ylab="cor.")
> points(markers,cor.est$"Delta=10"[8,])
> legend(1,0.99,legend=c("Actual","Estimated"),pch=c(19,1))
```

The correlation values for the actual IBD process are plotted in *solid black* and the values for the estimated IBD process are plotted in *black and white*.

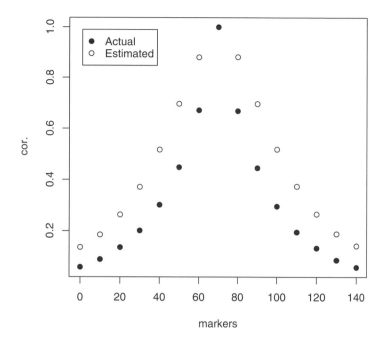

**Fig. 9.3.** Correlation functions between markers for IBD processes (for an inter-markers spacing of $\Delta = 10$ cM).

Observe that in general the correlations for the estimated process are larger than for the actual IBD process. Indeed, the same set of genotypes is used in order to infer the IBD status at the different markers. This increases the correlation beyond the correlation that results from the recombination process. A byproduct of this increase in the correlation will be a decrease in the size of the threshold that should be used to ensure a given significance level in a genome scan. Indeed, the first step in the investigation of the statistical properties of a genome scan involves a slight readjustment of the threshold to the new setting. We will get to that below.

Before considering the process of test statistics let us examine the properties of the reconstructed IBD when a susceptibility gene is present. We consider the same inter-markers spacings and assume that the locus is perfectly linked to the central marker. We assume an additive model for the trait with $p = 0.1$, $g_0 = 0.05$, and $\alpha = 0.225$. The means, variances, and mean square errors are computed for the marker linked to the trait locus. The simulation applies, with some obvious modifications, the same code as before:

```
> ibd.prob <- DistIBD(0.1,0.05,0.275,0.5)
```

```
> ibd.est.alt <- ibd.est.null
> for(i in 1:length(Delta))
+ {
+       markers <- seq(0,140,by=Delta[i])
+       n.mark <- length(markers)
+       qtl <- ceiling(n.mark/2)
+       theta <- 0.5 - 0.5*exp(-0.02*Delta[i])
+       Tr <- trans.mat(theta)
+       fa <- list(pat=matrix(1,n.ped,n.mark),
+                   mat=matrix(2,n.ped,n.mark))
+       mo <- list(pat=matrix(3,n.ped,n.mark),
+                   mat=matrix(4,n.ped,n.mark))
+       G.I <- array(dim=c(n.ped,n.mark,3))
+       ibd.est <- ibd <- NULL
+       for (rep in 1:n.rep)
+       {
+           inhe.qtl <- inhe.vector(ibd.prob,n.ped)
+           sib1 <- mating(fa, mo, markers,qtl=markers[qtl],
+                   inhe.fa=inhe.qtl[,"sib1.pat"],
+                   inhe.mo=inhe.qtl[,"sib1.mat"])
+           sib2 <- mating(fa, mo, markers,qtl=markers[qtl],
+                   inhe.fa=inhe.qtl[,"sib2.pat"],
+                   inhe.mo=inhe.qtl[,"sib2.mat"])
+           geno <- ped.geno(sib1,sib2)
+           for(k in 1:3) G.I[,,k] <-
+                   matrix(P[geno,k],n.ped,n.mark)
+           F.P <- forward(G.I,Tr,Pr)
+           B.P <- backward(G.I,Tr,Pr)
+           I.G <- marginal.post(F.P,B.P)
+           ibd.est <- c(ibd.est,2*I.G[,qtl,3]+I.G[,qtl,2])
+           ibd <- c(ibd,((sib1$pat == sib2$pat)+
+               (sib1$mat == sib2$mat))[,qtl])
+       }
+       ibd.est.alt["mean",i] <- mean(ibd.est)
+       ibd.est.alt["var",i] <- var(ibd.est)
+       ibd.est.alt["mse",i] <- mean((ibd-ibd.est)^2)
+ }
```

The matrix "ibd.est.alt" stores the mean, the variance, and the mean square error for the marker fully linked with the trait:

```
> round(ibd.est.alt,3)
      Delta=35 Delta=20 Delta=10 Delta=5 Delta=1
mean     1.042    1.052    1.073   1.096   1.144
var      0.116    0.147    0.218   0.297   0.420
mse      0.353    0.325    0.267   0.193   0.068
```

As expected, the mean of the estimated IBD is elevated. The elevation is more apparent when markers are denser. The variance and mean square error are, however, hardly affected by the change in distribution from unlinked to linked. This last observation is consistent with the mathematics of local alternatives, where one varies the mean but not the covariance structure.

It is more convenient to interpret the effect that the missing information may have on the statistical power by considering the noncentrality parameter. This parameter is equal to the difference between the alternative and the null expectations, multiplied by the square root of the sample size and divided by the null standard deviations. If we take, for example, a trial with 400 pedigrees, then we obtain

```
> n <- 400
> ncp.ibd <- sqrt(2*n)*(sum(ibd.prob*0:2)-1)
> ncp.app <- ncp.ibd*sqrt(ibd.est.null["var",]/0.5)
> ncp.sim <- sqrt(n)*(ibd.est.alt["mean",]-1)/
+       sqrt(ibd.est.null["var",])
> ncp <- rbind(ncp.sim,ncp.app)
> round(ncp.ibd,3)
[1] 4.744
> round(ncp,3)
          Delta=35 Delta=20 Delta=10 Delta=5 Delta=1
ncp.sim      2.351    2.640    3.139   3.552   4.412
ncp.app      2.383    2.638    3.128   3.626   4.372
```

The noncentrality parameter for a fully informative marker with complete linkage is 4.744. The same parameter for the same marker when only partial information is available was computed in two different ways. In the variable "ncp.sim" it was computed directly based on the results of the simulations. In the variable "ncp.app" it was approximated using (9.11) given below.

Looking at the numbers we see that the noncentrality parameter is severely deflated if the inter-marker spacing is more than 5 cM. If the spacing is 1 cM or less, then one recovers most of the noncentrality parameter. Another observation is the similarity between the actual noncentrality values as computed by simulations and the following approximation for these values. This approximation takes the general form:

$$E(\hat{Z}_t) \approx \mathrm{cor}\big(\hat{J}(t), \hat{J}(\tau)\big) \times \left[\frac{\mathrm{var}(\hat{J}(\tau))}{\mathrm{var}(J(\tau))}\right]^{1/2} \times E(Z_\tau) \,, \qquad (9.11)$$

where $E(Z_\tau) = \xi$ is the noncentrality parameter which was considered in the previous section and computed at the trait locus under the assumption of completely informative markers, while the statistic $\hat{Z}_t$ is the test statistic computed at a marker and based on the reconstructed IBD. The term $\mathrm{var}(J(\tau))$ is the variance of the actual IBD process, 1/2 in this case, and $\mathrm{var}(\hat{J}(t))$ is the variance of the reconstructed IBD, computed at the marker.

The term $\text{cor}\big(\hat{J}(t), \hat{J}(\tau)\big)$ corresponds to the correlations between elements of the reconstructed process.

The noncentrality parameter conveniently decomposes into three factors. The rightmost factor measures the contribution of the genetic effect, combined with the sample size. The central factor measures the reduction in the noncentrality parameter due to missing information. The leftmost factor measures the effect of using a marker that is only partially correlated with the trait locus.

Although we learn something from examining the noncentrality parameter, we now carry out a more comprehensive investigation of the statistical properties of genome scans when only the siblings are available for genotyping. The first stage involves finding the appropriate thresholds for the different inter-marker spacings. For the actual IBD process we use the analytical approximations that were developed in Chap. 5. For the reconstructed IBD we use simulations. Observe that the correlation structure of a scanning process formed by the summation of independent copies of reconstructed IBD processes is the same at the correlation structure of a single such process. Hence, we can use the correlation matrices that we found for the reconstructed processes as inputs for the function that simulates the Gaussian processes.

```
> library(MASS)
> Delta <- c(35,20,10,5,1)
> z <- matrix(nrow=2,ncol=length(Delta))
> colnames(z) <- paste("Delta=",Delta,sep="")
> rownames(z) <- c("ibd","ibd.est")
> n.rep <- 10^2
> n.iter <- 10^4
> for(d in 1:length(Delta))
+ {
+       z["ibd",d] <- uniroot(OU.approx,c(3,4),beta=0.04,
+           Delta=Delta[d],length=140*23,chr=23,center=0.05,
+           test="one-sided")$root
+       n.mark <- 140/Delta[d] + 1
+       Z.max <- NULL
+       for(i in 1:n.rep)
+       {
+           Z <- mvrnorm(n.iter,rep(0,n.mark),cor.est[[d]])
+           Z.max <- c(Z.max,apply(Z,1,max))
+       }
+       z["ibd.est",d] <- sort(Z.max)[n.rep*n.iter*(1-0.05/23)]
+ }
> round(z,3)
          Delta=35 Delta=20 Delta=10 Delta=5 Delta=1
ibd          3.338    3.461    3.603   3.722   3.907
ibd.est      3.322    3.425    3.515   3.587   3.742
```

As expected from our comparison of correlations, the threshold levels for the reconstructed IBD are slightly smaller than the threshold levels for the actual IBD process. We will use these lower thresholds in order to determine the power.

For the actual IBD process we use the analytical approximation that was developed in Chapter 6. For the estimated IBD we again use simulations. Motivated by the theory of local alternatives, we simulate the process under the alternative distribution using the same covariance structure that was used under the null. The mean function is computed based on the approximation (9.11) and is added to each row of the random part using the function "sweep":

```
> xi <- seq(3,6,by=0.1)
> n.rep <- 10
> power.ibd <- matrix(nrow=length(xi),ncol=length(Delta))
> colnames(power.ibd) <- names(z)
> power.est <- power.ibd
> for (d in 1:length(Delta))
+ {
+     power.ibd [,d] <-
+         power.marker(z["ibd",d],0.04,Delta[d],xi)
+     n.mark <- 140/Delta[d] + 1
+     qtl <- ceiling(n.mark/2)
+     rho <- sqrt(ibd.est.null["var",d]/0.5)*
+         ((cor.est[[d]])[qtl,])
+     Z <- mvrnorm(n.iter,rep(0,n.mark),cor.est[[d]])
+     for (i in 1:length(xi))
+     {
+         ncp <- xi[i]*rho
+         Z1 <- sweep(Z,2,ncp,"+")
+         power.est[i,d] <-
+             mean(apply(Z1,1,max) >= z["ibd.est",d])
+     }
+ }
```

Let us plot the power functions:

```
> plot(range(xi),c(0,1),type="n",xlab="xi",ylab="power")
> gr <- gray(0.75*(1:length(Delta))/length(Delta))
> for(d in 1:length(Delta))
+     {
+         lines(xi,power.ibd[,d],col=gr[d])
+         lines(xi,power.est[,d],col=gr[d],lty=2)
+     }
> legend(xi[1],1,bty="n",legend=colnames(z),
+     lty=rep(1,ncol(z)),col=gr)
```

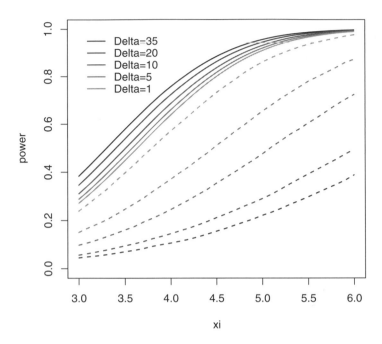

**Fig. 9.4.** Power functions for an IBD process (*solid line*) and an inferred process (*broken line*). The QTL in complete linkage with the central marker. Bi-allelic markers are used.

The output is presented in Fig. 9.4. Observe the dramatic effect on the power of failing to reconstruct the IBD process accurately enough. When markers are placed one cM apart, then most of the power is retained. However, a substantial fraction of the power is lost when markers are set 5 cM apart and becomes worse as the markers become more spread apart.

The exact same programs that were used to generate Fig. 9.4, which refers to SNPs, can be used in order to generate power curves for SSR markers. Consider, for example, markers with five uniformly distributed alleles. The only changes that are needed are the replacement of line "P <- geno.given.ibd()" by the line "P <- geno.given.ibd(rep(1/5,5))" in one location and the line "geno <- ped.geno(sib1,sib2)" by the line "geno <- ped.geno(sib1,sib2,rep(1/5,5))" in two locations.

After making these changes and rerunning the programs one gets new significance thresholds and new power curves that are given in Fig. 9.5:

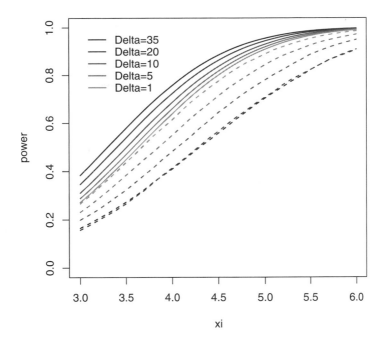

**Fig. 9.5.** Power functions for an IBD process (*solid line*) and an inferred process (*broken line*). The QTL in full linkage with the central marker. Uniformly distributed five-allelic markers are used.

|         | Delta=35 | Delta=20 | Delta=10 | Delta=5 | Delta=1 |
|---------|----------|----------|----------|---------|---------|
| ibd     | 3.338    | 3.461    | 3.603    | 3.722   | 3.907   |
| ibd.est | 3.321    | 3.442    | 3.563    | 3.663   | 3.841   |

Clearly, for markers separated by more than a few cM, the SSRs provide more information than the SNPs. Indeed, SSRs at 5 cM appear to be roughly comparable to SNPs at 1 cM. If markers are multi-allelic or if parents are also genotyped, then one can recover more information and mitigate to some extent the effect of separated markers. Nevertheless, when one takes account of the effect of a value of $\beta$ twice as large as one finds in backcross or intercross designs and the additional loss of information due to incompletely informative markers, one should plan to place markers substantially closer in human than in experimental genetics. With microsatellite markers it is common to use an inter-marker spacing of about 10 cM, and 5 cM would be substantially better. For SNPs, which are generally less informative but more common than microsatellites, the marker spacing should be about 1 cM, or less.

## 9.7 Bibliographical Comments

The idea of using affected sib pairs to map disease genes goes back to Penrose [58]. James [42] observed that the regression models used for quantitative traits could also be used for qualitative traits if one treated the (unobserved) penetrance of a qualitative trait as a quantitative phenotype. Risch [65] provided an elegant theoretical framework for single marker applications. For discussion of genome scans, see Feingold, Brown, and Siegmund [28] and Lander and Kruglyak [48].

The standard paradigm of parametric linkage analysis is due to Morton [56], and is described in detail by Ott [57].

Considerable effort has gone into the problem of reconstructing IBD relations from marker data. One of the first and still widely used programs for this purpose is based on the Elston-Stewart [24] algorithm, implemented in the software SAGE. As befits its vintage, this algorithm is particularly adept at dealing with large pedigrees and a relatively small number of markers. Two widely used and freely available programs that are built around a hidden Markov model of the kind discussed in the text are GENEHUNTER (Kruglyak *et al.* [46]) and MERLIN (Abecassis *et al.* [2]). Both of these programs can handle essentially any number of markers, but have problems in dealing with large pedigrees. ALLEGRO [33] began its existence as something of an offspring of GENEHUNTER, but lately [34] seems to have developed an independent life of its own.

These programs also provide suites of statistical methods that utilize the IBD reconstruction for gene mapping. The reconstruction is computationally intensive, so the programs should be optimized; and since the goals are reasonably clear, one would usually want to use programs prepared by professionals. The statistical applications, on the other hand, are comparatively easy to program using available statistical software, and the appropriate methods are not so generally agreed on. Hence one might want to use the IBD reconstruction of, for example, GENEHUNTER or MERLIN as a preliminary step to one's own statistical analysis.

## Problems

**9.1.** (a) Plot the mean function of the $Z$ statistic for an additive model with $g_0 = 0.05$, for various values of $p$, $\alpha$, and $n$.

(b) For inter-marker spacing of $\Delta = 5$ cM, plot the power function for various values of $p$, $\alpha$, and $n$.

(c) For $p = 0.1$ and $\alpha = 0.4$, how large a value of $n$ is required to have 90% power to detect a gene perfectly linked to a marker?

**9.2.** Use (9.4) to find the probabilities of IBD for an additive model with $g_0 = 0$. Note, in particular, that these probabilities depend only on $p$ and not on the penetrances $g_1 = \alpha$, $g_2 = 2\alpha$. Do the same for the dominant model $g_0 = 0$, $g_1 = g_2 = \alpha$. Make numerical comparisons of the noncentrality parameters of $(N_2 - N_0)/(n/2)^{1/2}$ for the two models for different values of $p$. Do you find your numerical results surprising?

**9.3.** The probabilities of IBD for the recessive model $0 = g_0 = g_1 < g_2$ are given by

$$\pi_0 = \frac{p^2}{(1+p)^2}, \quad \pi_1 = \frac{2p}{(1+p)^2}, \quad \pi_2 = \frac{1}{(1+p)^2},$$

where $p$ is the frequency of the allele $D$ in the population.

(a) Write an R function that computes the power to detect a susceptibility gene as a function of the relevant parameters.
(b) Plot the power function for various values these parameters.

**9.4.** The statistic $Z_1 = (N_2 - N_0)/(n/2)^{1/2}$ is designed for an additive model. For a recessive model, where $\sigma_D^2 \gg \sigma_A^2$, an alternative is $Z_2 = (N_2 - n/4)/(3n/16)^{1/2}$.

(a) Explain why this is a reasonable statistic (for example, by calculating its noncentrality parameter).
(b) For the additive and recessive models of the preceding two problems, for which $g_0 = 0$, compare the noncentrality parameters of $Z_1$ and $Z_2$ as functions of $p$. For which values of $p$ does $Z_2$ have a substantially larger noncentrality parameter than $Z_1$? (See Chap. 11 for additional discussion of alternative statistics when the dominance variance may be important.)

**9.5.** Consider the model (2.3) for the penetrance $y$, where $e$ is assumed to contain environmental effects and possibly the effect of other genes that are unlinked to the trait locus at $\tau$. Since two siblings are expected to share genetic material and are also likely to share a common environment, we allow for a covariance $r = \text{cov}(e_1, e_2)$. How should the assumption (9.1) be changed? What is the effect of this generalization on (9.2) and (9.4)?

**9.6.** Simulate significance thresholds at various inter-marker distances when markers are only partially informative and compare these thresholds to those obtained (either by simulation or by theoretical approximations) for fully informative markers. Now make similar comparisons of the power of genome scans based on the appropriate thresholds.

**9.7.** An alternative to the additive model, which is also particularly tractable, is the multiplicative relative risk model. The penetrances are given by $g_0 > 0$, $g_1 = g_0 R$, $g_2 = g_0 R^2$, where $R > 1$ is called the relative risk. Assuming Hardy-Weinberg equilibrium, for a sib pair find expressions for $\text{Pr}(\text{Both affected})$ and the conditional probability $\text{Pr}(\text{Both affected}|J(\tau) = j)$ for $j = 0, 1, 2$. Generalize this model to consider two (unlinked) genes acting multiplicatively between loci.

*Remark 9.2.* It is possible to reduce this to the canonical form given in Sect. 9.1 and apply the results given in the text; but the properties of this model are sufficiently simple that one can also carry out the desired calculations from first principles.

**9.8.** Let $\sigma_{\mathrm{D}}^2 = 0$ in (9.4). The log likelihood function for the data $(N_0, N_1, N_2)$ parametrized by $\check{\alpha}/Q_2$ is $\ell(\check{\alpha}/Q_2) = N_2 \log(1 + \check{\alpha}/Q_2) + N_0 \log(1 - \check{\alpha}/Q_2)$. Show that the statistic $Z$ introduced in the text is the score statistic for testing the hypothesis $H_0 : \check{\alpha} = 0$, which is of the form $\dot{\ell}(0)/\{\mathrm{E}_0[\dot{\ell}(0)]^2\}^{1/2}$. (We use the dot notation for derivatives. See Chap. 1 for a discussion of the score statistic.)

**9.9.** Let $\pi_i(\theta) = \Pr(J(t) = i|A)$ for a marker $t$ at a recombination fraction $\theta$ from a trait locus $\tau$. Show that $\pi_i(\theta)$ is of the same form as (9.4), but with $\check{\alpha}$ replaced by $\check{\alpha}(1 - 2\varphi)$ and $\check{\delta}$ replaced by $\check{\delta}(1 - 2\varphi)^2$.

**9.10.** The affected sib pair method is based on the observation that affected sibs are likely to share an excess of alleles IBD at a locus that increases susceptibility to the trait. A sib pair of whom one affected and one unaffected is likely to share a deficit of alleles IBD at a trait locus. Assuming there is no dominance variance, develop a model for using affected/unaffected sib pairs for gene mapping. Find the score statistic or an otherwise reasonable statistic and evaluate its noncentrality parameter. Can you give conditions where these sib pairs would be as useful as affected sib pairs? Does it seem plausible that these conditions might sometimes be satisfied?

**9.11.** Suppose that our sample consists of $n_1$ affected/unaffected sib pairs, as in the preceding problem, and $n_2$ affected sib pairs. Assuming a parametric model, where we regard the penetrances and allele frequencies as known, and the validity of (9.8), propose an approximate Gaussian log likelihood ratio statistic to combine these two kinds of sib pairs. Observe that this involves a specific linear combination of the two $Z$ statistics, with weights that depend on the hypothesized values of the noncentrality parameters. (This issue is discussed again in Chap. 11.)

**9.12.** Give an analysis along the lines of Sects. 9.1–9.3 for samples of (i) an affected grandparent and grandchild (ii) affected half-sibling pairs, (iii) affected first cousins pairs.

*Remark 9.3.* Determining the recombination parameter for first cousins involves a more difficult argument than the other two cases.

**9.13.** Compute the probability distribution of the total number of alleles shared IBD by a trio of siblings.

**9.14.** Show that (9.3) can be re-written in the form

$$\mathrm{E}(y_1 y_2 | J) = \mathrm{E}(y_1 y_2) + (J - 1)\check{\alpha} - (I_{\{J=1\}} - 1/2)\check{\delta} .$$

**9.15.** Recall that a trait is called recessive if $\delta = -\alpha$. Consider a child whose parents are related to each other, e.g., siblings or first cousins. Let $F$ denote the coefficient of relatedness of the parents. Using the model (2.3) together with the assumption that $E(e) = 0$, show that the probability the child is affected is $Q_1 = m + F\sigma_D$. In particular, the probability that an inbred child is affected is larger than the probability $m$ that a child of unrelated parents is affected. An individual is said to be homozygous by descent (HBD) at the locus $t$ if the two alleles at that locus are IBD. Hence the the probability of HBD at a random locus is the coefficient of relatedness of the parents. Explain how you could use a sample of inbred affected individuals, e.g., a sample of affected children of first cousins, to map a recessively acting gene. What would be the noncentrality parameter at a trait locus in a large sample in terms of the sample size $n$, $Q_1$, $F$, and $\sigma_D^2$? How would your analysis change if you want to consider the possibility of a second gene, which is unlinked to and does not interact with the first gene?

# 10

# Admixture Mapping

Admixture occurs when two or more populations merge to form a new population. A classical example in humans is the African-American population. The African population and the European Caucasian population have diverged during thousands of years of largely separate evolution. During the late eighteenth and early nineteenth centuries, substantial numbers of Africans were brought to the region that became the United States of America, where most lived as slaves of the Caucasian population. Although interracial marriages were rare, substantial genetic blending between those two populations did occur (in addition to some blending with the Native American population). It is estimated that about 20% of the genetic material in today's African-American population originated from a non-African, predominantly a Caucasian, source.

We discuss admixture mapping in this chapter in the context of human genetics, which presents the greatest challenges. Similar methods can be applied to gene mapping in outbred plants and animals, e.g., dogs, which exist in different breeds that can be phenotypically quite different and can be crossed to produce data suitable for gene mapping. In some cases of interest inbred strains may not exist, and in other cases one may simply prefer to use outbreds or a mixture of outbreds and inbreds to have a richer source of genetic and phenotypic variability than would be likely to occur in two inbred strains. In experimental genetics one can often control or at least document mating patterns, which makes this case simpler than human genetics, where mating patterns cannot be controlled and often are incompletely documented.

From a genetic point of view, there is some similarity between populations that are formed by admixture and crosses of inbred strains in experimental genetics. In both cases, the starting point is two (or more) genetically separated populations, which are merged through several generations of mating to form a new population. The genetic structure of the new population is a mixture of the genetic material of the founding populations. In the experimental genetics of inbred strains, phenotypically distinct strains are used. Correlations between the source of the genetic material and the expression of the phenotype are used in order to identify genomic regions which may be linked to the trait. One hopes that a similar approach will work in an admixed population for traits that are differentially expressed in the two founding populations, so by correlating phenotypes with the population origin of genotypes one can obtain useful insights.

In spite of this resemblance, major differences exist between mapping in an admixed population and mapping with crosses of inbred strains. The founding populations are outbred and are genetically heterogeneous. In natural populations the blending of the genetic material is typically not a result of a controlled and documented process. Consequently, admixture mapping often involves complications that can be avoided in carefully planned experiments. The main aim of this chapter is to discuss some of these complexities and the statistical means that can be used to address them.

We start the next section by proposing a working model for the investigation of the statistical properties of admixture mapping. This model will have simplifying assumptions regarding the distribution of the population source of genetic material in a random subject from the admixed population. In Sect. 10.1.2 we will describe a statistic that measures the correlation between the founder population source and expression of the phenotype in various settings. In particular, we will determine the noncentrality parameter of the statistic and examine factors that may affect it.

Unlike the situation in experimental genetics, molecular markers in admixture mapping do not uniquely identify the population source of the genetic material. In that regard, the situation in admixture mapping is more like the situation for linkage analysis in humans, where identity-by-descent relationships between subjects have to be inferred statistically from genotypic information. Likewise, in admixture mapping the founding population source must usually be indirectly inferred. In Sect. 10.2 we will describe available tools and discuss their effect on the noncentrality parameter. Broadly speaking, the same tools of hidden Markov models (HMM) that were applied in Chap. 9 can be adapted to the new setting. However, the current situation diverges from the situation in Chap. 9 in one major point. Unlike the case of affected sib pairs, where the distribution of the underlying Markov process comes directly from Mendel's laws and a model for recombination, here the distribution of the underlying process may be unknown. In the last section we will describe a statistical tool that can be used to estimate the unknown distribution of this process. That tool is derived from the *expectation maximization* (EM) algorithm, which is commonly used for the computation of the maximum likelihood estimates in settings that involve missing information.

Like some other chapters in this book, our main goal in this one is more to introduce problems and tools to study them than to produce solutions ready for application. Indeed, the applicability of gene mapping based on admixed populations depends more than anything else on finding suitable populations, which have been isolated for a long enough time that their gene frequencies and phenotypes have diverged, and have become admixed sufficiently rapidly to provide a suitable sample of individuals to study. As mentioned already, the most suitable area of application may turn out not to be humans, but animals that exist in different breeds, chosen to have differing phenotypes of interest. If one controls experimentally the mating of these animals, the question of the differing degrees of individual admixture vanish, and one can ask questions about the breeding designs that would be most powerful for one purpose or another.

## 10.1 Testing for the Presence of a QTL

In this section we describe the basic features associated with scanning the genome for the location of a QTL in an admixed population. These features are discussed in the context of a quantitative trait but the same principles apply when a qualitative trait is considered. We start by proposing a simple model for the formation of the admixed population. The scanning test statistic, its noncentrality parameter, and its correlation structure are then discussed.

### 10.1.1 A Model for an Admixed Population

In order to simplify the analysis of admixture mapping we will assume that the target population was formed by the blending together of two founding populations several generations before the present. Denote these founding populations by the numbers 1 and 2, and consider a random subject in the current admixed population. The genetic composition of that subject is a mixture of the genetic material that originated from the two founding populations. A gamete is a mosaic of consecutive segments with an alternating origin: a segment of random length inherited from one founding population is followed by a segment inherited from the other population. The homologous gamete is of the same general structure, although the points of recombination will differ. Later, especially in Sect. 10.1.3 and in Sects. 10.2 and 10.3, we will return to the mosaic structure of a random chromosome. However, for the sake of the discussion in this and in the next subsections we concentrate on a specific locus, which we denote by $\tau$.

It may occur that both homologous copies of locus $\tau$ originated from population 1. Alternatively, one copy may have originated from population 1 and the other from population 2. A third possibility is that both copies were inherited from population 2. In order to examine the distribution of the number of copies we consider the two parents of the subject. In all likelihood, in populations similar to our motivating example of the African-American population, the parents themselves originated from the same admixed population, one generation before the present. Hence, each parent may contribute to the offspring a copy of the locus inherited from population 1 or a copy inherited from population 2. Let $\chi_M$ be the number copies (0 or 1) of the allele at locus $\tau$ that originated from population 2 and was contributed by the mother and let $\chi_F$ be the number of copies contributed by the father. If we assume random mating of the parents, the probability of transmitting a copy that originated from population 2 is the same, say $\pi$, for each parent and the transmissions of the two parents are independent. Under these assumptions one may conclude that the total number of copies from population 2 in the offspring, $\chi = \chi_M + \chi_F$, has a binomial $B(2, \pi)$ distribution.

If admixture occurred very recently, one may have much more information. For example, one may know that each set of grandparents, say, consisted of a first generation admixture, with one grandparent coming from population 1 and the other from population 2. In this case a member of the present generation is just like an intercross, with $\pi = 1/2$. In the case that one set of grandparents is first generation admixed and both grandparents of the other set are from population 1, say, then a member of the current generation is similar to a backcross, since one allele definitely comes from population 1, while its homolog can be from population 1 or 2 with probability $1/2$ each. Often the parameter $\pi$ will vary from one individual to another and may be unknown because it depends on the mating patterns of several generations before the present. We will return to this point in Sect. 10.3, but here we assume that $\pi$ is a known constant.

Admixture mapping uses information regarding the population source of an allele as the tool for mapping traits that differ phenotypically between the founder populations. Consider, in particular, a quantitative trait and let us assume the case where $\tau$ is a QTL, which is taken to be bi-allelic. Denote by $A$ and $a$ the two alleles. The target in admixture mapping are QTLs that diverge in their allele frequencies between the two population, and hence presumably contribute to the observed phenotypic difference. Assume $\tau$ is such a locus and denote by $p_1$ the frequency of allele $A$ in population 1 and by $p_2$ the frequency of the same allele in population 2.

Recall the term $x_M$, which was used in Chap. 2 to denote the number of $A$ alleles (0 or 1) inherited from the mother and the term $x_F$, which is the number inherited from the father. Observe that under our assumptions the two terms are independent of each other, each obtaining the value one with probability $\bar{p} = \pi p_1 + (1 - \pi)p_2$, where $p_i$ is the frequency of allele A in population $i$. In order to describe the relations between the genotype and the phenotype let us reuse the model that was introduced in Chap. 2. For convenience, we write it down again:

$$y = m + \tilde{\alpha}[(x_M - \bar{p}) + (x_F - \bar{p})] + \tilde{\delta}[(x_M - \bar{p})(x_F - \bar{p})] + e , \qquad (10.1)$$

where $\tilde{\alpha} = \alpha + (1 - 2\bar{p})\delta$ and $\tilde{\delta} = -2\delta$. We will be interested in the conditional expectation of the phenotype given $\chi_M$ and $\chi_F$. For that it is useful to note that

$$\mathrm{E}(x_F - \bar{p} \,|\, \chi_F) = p_1\chi_F + p_2(1 - \chi_F) - p_1\pi - p_2(1 - \pi) = (p_1 - p_2)(\chi_F - \pi) ,$$

with a similar representation holding when the subscript $F$ is replaced by the subscript $M$. Under our assumptions regarding the formation of the target population it is also true that $\chi_F$ is independent of $x_M$ and $\chi_M$ is independent of $x_F$. It can be concluded that

$$E(y \mid \chi_F, \chi_M) = m + \tilde{\alpha}(p_1 - p_2)[(\chi_F - \pi) + (\chi_M - \pi)] + \tilde{\delta}(p_1 - p_2)^2[(\chi_M - \pi)(\chi_F - \pi)] . \tag{10.2}$$

The statistics, developed in the next section, are obtained by regressing the phenotype on the quantities given in the square brackets in (10.2), of the indicators of the population source $\chi_F$ and $\chi_M$. The noncentrality parameters of those statistics involve in the numerator the correlation coefficients associated with the appropriate regressors, while the variance of the phenotype appears in the denominator. Since the different components of (10.1) are uncorrelated, it follows that the phenotypic variance can be represented as a sum of the three elements $\sigma_A^2$, $\sigma_D^2$, and $\sigma_e^2$, where

$$\sigma_A^2 = 2\bar{p}(1 - \bar{p})\tilde{\alpha}^2, \quad \sigma_D^2 = \bar{p}^2(1 - \bar{p})^2\tilde{\delta}^2, \quad \text{and} \quad \sigma_e^2 = \text{var}(e) . \tag{10.3}$$

## 10.1.2 Scanning Statistics and Noncentrality Parameters

In earlier chapters we have used regression statistics for gene mapping, and admixture mapping will be similar. The guiding principle in the admixture approach for mapping is the observation that relevant genetic information can be obtained from the population source of the genetic material, which we have denoted by $\chi_F$ and $\chi_M$. We have seen in (10.2) that the conditional expectation of the phenotype is a combination of the two uncorrelated quantities $\chi = \chi_F + \chi_M$ and $\bar{\chi} = (\chi_F - \pi)(\chi_M - \pi)$. Moreover, for either parent $P = F$ or $P = M$, $E[y(\chi_P - \pi)] = E[E(y|\chi_F, \chi_M)(\chi_P - \pi)]$, so the actual phenotype and the conditional expectation of the phenotype given the population origins of the inherited alleles have exactly the same covariance with the population origins. It follows that for a sample of $n$ unrelated individuals with phenotypes $y_1, \ldots, y_n$, test statistics for determining the presence of a QTL at locus $\tau$ can be obtained from the standardized regression (correlation) coefficients:

$$Z_\alpha = \frac{\sum_{i=1}^n \left\{(y_i - \bar{y})(\chi_i - 2\pi)\right\}}{[n\hat{\sigma}_y^2 2\pi(1 - \pi)]^{1/2}} \tag{10.4}$$

$$Z_\delta = \frac{\sum_{i=1}^n \left\{(y_i - \bar{y})\bar{\chi}_i\right\}}{[n\hat{\sigma}_y^2 \pi^2(1 - \pi)^2]^{1/2}} . \tag{10.5}$$

Under the null hypothesis of no linkage (and under local alternatives) these statistics are asymptotically independent and normally distributed. A scanning statistic for the detection of an additive effect may be obtained by considering the absolute value of the statistic in (10.4) (or, equivalently, its squared value) and a scanning statistic for combined additive and dominant effects may be obtained by summing together the square of this statistic and the square of the statistic in (10.5) in order to produce a statistic with a marginal chi-square distribution on two degrees of freedom.

The noncentrality parameter at the QTL associated with the statistics (10.4) and (10.5) can be obtained by the consideration of the covariance between $y$ and $\chi$ and between $y$ and $\bar{\chi}$, which by (10.2) leads to

$$E[Z_\alpha] \approx n^{1/2} \tilde{\alpha} (p_1 - p_2) \{2\pi(1 - \pi)\}^{1/2} / \sigma_y \tag{10.6}$$

$$E[Z_\delta] \approx n^{1/2} \tilde{\delta} (p_1 - p_2)^2 \pi(1 - \pi) / \sigma_y . \tag{10.7}$$

Observe, in particular, that the parameters are functions of the difference between the frequencies of the QTL in the founding populations. The power to detect a trait locus using admixture mapping is greatly reduced when this difference is small, which would lead to small phenotypic differences between the populations. In the extreme case that $|p_1 - p_2| = 1$ and $\pi = 1/2$, the noncentrality parameters given in (10.6) and (10.7) (as well as the variance components in (10.3)) reduce to the corresponding results for an intercross.

Another interesting point is the fact that even if the QTL has a purely dominant effect, in the sense that $\alpha = 0$, the noncentrality parameter for the additive effect is non-zero unless $\bar{p} = 1/2$. This follows from the fact that $\tilde{\alpha}$ is a function of both $\alpha$ and $\delta$, and not only of $\alpha$. As a result, the statistic (10.4) may have power to detect both additive and dominance effects although one could not estimate both effects from the value of that statistic alone. This phenomenon is characteristic of mapping in natural population as opposed to mapping in crosses. We saw this already in Chap. 9 and will return to it in Chap. 11 when we discuss the issue of quantitative trait mapping using human pedigrees.

We consider next some numerical examples. The function "`ncp.admixed`" computes additive and dominance noncentrality parameters. The inputs are the parameters of the model and the output is a data frame containing the noncentrality parameters:

```
> ncp.admixture <- function(alpha,delta,p1,p2,pi1=0.8,
+                     sig=1,n=1)
+ {
+     p.bar <- pi1*p1 + (1-pi1)*p2
+     a <- alpha + (1-2*p.bar)*delta
+     d <- -2*delta
+     v.a <- 2*p.bar*(1-p.bar)*a^2
+     v.d <- p.bar^2*(1-p.bar)^2*d^2
+     v.y <- v.a + v.d + sig^2
+     ncp.a <- a*(p1-p2)*sqrt(2*pi1*(1-pi1)*n/v.y)
+     ncp.d <- d*(p1-p2)^2*pi1*(1-pi1)*sqrt(n/v.y)
+     return(data.frame(ncp.a=ncp.a,ncp.d=ncp.d))
+ }
```

Next we use the function to explore the behavior of the noncentrality parameters.

```
> round(cbind(p1,p2,delta,ncp.admixture(alpha,delta,p1,p2)),3)
    p1   p2 delta ncp.a   ncp.d
1 0.70 0.5    -1 0.106   0.009
2 0.90 0.5    -1 0.271   0.037
3 0.99 0.5    -1 0.385   0.060
4 0.70 0.5     0 0.094   0.000
5 0.90 0.5     0 0.199   0.000
6 0.99 0.5     0 0.254   0.000
7 0.70 0.5     1 0.065  -0.011
8 0.90 0.5     1 0.077  -0.048
9 0.99 0.5     1 0.059  -0.075
```

Note that the (absolute value of the) noncentrality parameter usually increases as a function of $|p_1 - p_2|$. The exception is the noncentrality of $Z_\alpha$ in the last line. In this case the role of the dominance deviation is quite pronounced. For the two degree of freedom statistic, for which the noncentrality parameter is the sum of the squares of the individual noncentrality parameters for $Z_\alpha$ and $Z_\delta$, there is an increase in all cases. The reader is encouraged to re-run the program for different values of $\pi$, $p_1$, and $p_2$, and reflect on the results.

The noncentrality parameters given in (10.6) and (10.7) for a sample of size $n$ is the product of $n^{1/2}$ and the noncentrality parameters given by the program for a sample of size 1. Recalling from earlier analyses that a total noncentrality of 4 to 5 is required to have reasonable power to detect linkage in a genome scan, we see that in some of these examples a sample size of, say, $n = 400$ would appear to be adequate, while in others a substantially larger sample size would be required.

### 10.1.3 A Model for the Covariance Structure

The noncentrality parameters in (10.6) and (10.7) measure the expectation at a trait locus. In order to measure the expectation of a statistic at other loci it is relevant to obtain the correlation coefficient between test statistics computed for two different loci. Indeed, we have seen in other examples that the noncentrality parameter at a marker is a product of the noncentrality parameter computed at the trait locus and the correlation coefficient between the test statistic at the marker and the test statistic at the trait locus, where the correlation coefficient is computed under the null hypothesis of no genetic effect. The same correlation coefficient is also relevant in the determination of the threshold for significance, when we account for multiple testing.

In the case of admixture mapping the correlation coefficient is a function of the transition rates of the population origin of DNA as one moves from one locus to the next along a random gamete. Let $s$ and $t$ be two loci on the genome and let $\chi_P(s)$ and $\chi_P(t)$ be the indicators that the population origin for a gamete inherited from parent $P$ at loci $s$ and $t$, respectively, is, say, population 2. It can be shown that, when $Z = Z_\alpha$, then

$$\text{cor}(Z_s, Z_t) = \text{cor}(\chi_P(s), \chi_P(t)) \, ,$$

and when $Z = Z_\delta$,

$$\text{cor}(Z_s, Z_t) = \left[\text{cor}(\chi_P(s), \chi_P(t))\right]^2 .$$

Hence, the the correlation at two loci of statistics for the additive effect is the same as the correlation of the indicator of population origin for a random gamete, while the correlation of the statistic for the dominance effect is the square of that quantity.

The term $\text{cor}(\chi_P(s), \chi_P(t))$ reflects the process by which the admixed population was formed and may be quite complex. However, a rough idea of its magnitude and its relation to the genetic distance between the two loci may be obtained if one is willing to tolerate additional simplifying assumptions.

Suppose that the population was formed $g+1$ generations ago by merging two infinite founding populations at proportions $1 - \pi$ and $\pi$, respectively. Assume that random mating occurs between members of the two populations, so in the first generation the proportion of matings between two individuals from population 1 is $(1 - \pi)^2$, the proportion of matings between an individual from population 1 and an individual from population 2 is $2\pi(1 - \pi)$, etc. From the same reasoning introduced in Sect. 3.3 to discuss Hardy-Weinberg equilibrium, it follows that in all generations, at a given locus, the proportion of gametes from population 2 is $\pi$. One can also apply the more detailed argument from Sect. 3.3, which was used there to analyze the decay in linkage disequilibrium in a randomly mating population. Replacing the notion of haplotype, which was used there, by the notion of "population source" and the notion of linkage disequilibrium by the notion of covariance, one finds that for two loci separated by a recombination fraction $\theta$, in the present generation

$$\Pr[\chi_P(s) = 1 = \chi_P(t)] - \pi^2 = \pi(1 - \pi)(1 - \theta)^g \, ,$$

which after some algebraic rearrangement yields

$$\text{cor}(\chi_P(s), \chi_P(t)) = (1 - \theta)^g \, . \tag{10.8}$$

Hence the log of the correlation between loci is proportional to the number of generations of random mating since admixture.

Note that unlike the noncentrality parameters given above, which in the extreme case that both $|p_1 - p_2| = 1$ and $\pi = 1/2$ are the same as for an intercross, the correlation for $g = 1$ is not the same. In fact, since the maximum value of $\theta$ is $1/2$, in this case there is correlation between unlinked markers. The difference is that with random mating, some individuals in the population are descended from only one of the founding populations, even after several generations. Although this fraction is $\pi^{2(g+1)} + (1 - \pi)^{2(g+1)}$, which becomes rapidly smaller as $g$ increases, it never vanishes completely.

In the next two sections we revisit the issue of inference based on partial information. For most loci in admixture mapping, as in linkage analysis in human pedigrees, one does not observe the population source directly. The genotypic information generally provides only indirect evidence. In order to reconstruct the unobserved process it is useful to make assumptions regarding the statistical properties of transition between states of that process. It is natural to use in the current context the same tools of hidden Markov models that were used for linkage analysis. For that purpose we will take the liberty of treating the process of population source as if it were a Markov process. The reader should observe that for the simplified model of random mating the process is actually *not* Markovian. However, the hope is that for the sake of the reconstruction of the statistic a Markov process with the same marginal distributions and one-step transition probabilities provides an adequate approximation of the actual process.

For later reference we record the transition probabilities of the two-state stationary Markov process describing the population source of the allele contributed by parent $P$. This result can be derived either by referring again to the argument of Sect. 3.3 or by observing that in a two-state Markov chain the transition probabilities are determined by the stationary probabilities and the correlation. Hence the transition probabilities are

$$\Pr(\chi_P(s) = 0 \,|\, \chi_P(t) = 1) = \pi\big[1 - (1 - \theta)^g\big]\,,$$
$$\Pr(\chi_P(s) = 1 \,|\, \chi_P(t) = 0) = (1 - \pi)\big[1 - (1 - \theta)^g\big]\,, \qquad (10.9)$$

where $\pi$ is the stationary probability of state 2, and $\theta$ is the recombination fraction between the two loci.

## 10.2 Inferring Population Origin from Molecular Markers

The analysis suggested above requires that we know the population source of each marker allele, but like IBD relations between siblings it is rare that we know the population source with certainty. It is possible at markers that have only one allele in one of the founding populations and a different allele in the other population. However, it is much more common to have markers that are represented in both populations, but with different allele frequencies. For such a marker, observing an allele that is more frequent in one founding population increases the likelihood that the associated gamete originated from that founding population. The accumulation of evidence from a collection of markers in the vicinity of a given locus may allow one to infer the population source of that locus with probability close to one. The reconstructed values of the process may then be used for the computation of the scanning statistic.

A computationally convenient way of accumulating evidence from an observed process in order to infer the state of another unobserved but correlated process is to use the backward and forward algorithms, which were introduced in Chap. 9. In order to use those algorithms, one needs to assume a hidden Markov model. According to this structure the unobserved process is Markovian and the components of the observed process are conditionally independent, given the states of the hidden process. In the current setting, if we consider for example the reconstruction of the statistic $Z = Z_\alpha$, the modeling assumption is that for each subject $i$ the process of total counts from population 2, denoted here by $\chi_i = \chi_{Mi} + \chi_{Fi}$, is a Markov process. Another assumption is that, conditional on the states of $\chi_i$, the genotypes at different markers are independent, and their distribution at each marker is determined by the state of the process $\chi_i$ at that marker.

Consider first the Markovian assumption. The distribution of the process $\chi_i$ is determined by the transition matrix. Under our general assumption that the parents are not related and originate from the same admixed population, this process is created by summing two independent and identically distributed two-state Markov processes. The transition matrix is a function of the one-step transition probabilities from one state of the two-state process to the other. Let $\beta = (\beta_1, \beta_2)$ denote the vector of this pair of probabilities, so for $j = 1$ or $2$, $\beta_j$ denotes the probability that if a given marker is derived from population $j$, then the next marker on the chromosome is derived from the other population. (Under the very special random mating assumption of the preceding section, $\beta_1$ and $\beta_2$ are given in (10.9). Although this is the specific example we use throughout this chapter, the same methods apply more generally.) The function "transition.matrix" computes the transition matrix of the process $\chi$ as a function of $\beta$:

```
> transition.matrix <- function(beta)
+ {
+     Tr <- matrix(0,3,3)
+     Tr[1,] <- c((1-beta[1])^2,2*beta[1]*(1-beta[1]),
+         beta[1]^2)
+     Tr[2,] <- c(beta[2]*(1-beta[1]),
+         prod(beta)+prod(1-beta),beta[1]*(1-beta[2]))
+     Tr[3,] <- c(beta[2]^2,2*beta[2]*(1-beta[2]),
+         (1-beta[2])^2)
+     return(Tr)
+ }
```

The stationary distribution of a Markov process is determined by the transition matrix and is therefore also a function of the transition probabilities $\beta$. Consider the computation of the the stationary distribution:

```
> stationary.probability <- function(beta)
+ {
+     p <- rev(beta)/sum(beta)
+     Pr <- c(p[1]^2,2*p[1]*p[2],p[2]^2)
+     return(Pr)
```

```
+ }
```

Specifically, consider an admixed population, with admixture frequencies for populations 1 and 2 of 20% and 80%, respectively. Assume the population was formed 9 generations ago, so $g = 8$, and adopt the random mating assumptions that led to the transition probabilities given in (10.9). If we assume that markers are placed 2 cM apart, then we obtain:

```
> g <- 200/25
> pi2 <- 0.8
> delta <- 2
> theta <- 0.5 - 0.5*exp(-0.02*delta)
> true.beta <- c(pi2,1-pi2)*(1-(1-theta)^g)
> Tr <- transition.matrix(true.beta)
> Pr <- stationary.probability(true.beta)
> Tr
              [,1]        [,2]       [,3]
[1,]  0.7793474497  0.20691825  0.0137343
[2,]  0.0258647814  0.86037537  0.1137598
[3,]  0.0008583937  0.05687992  0.9422617
> Pr
[1] 0.04 0.32 0.64
```

Note that unlike the situation in linkage analysis using affected sib pairs, the stationary distribution is not symmetric.

This transition matrix and stationary distribution will be used throughout this section. In particular, they will be used in order to simulate the hidden process $\chi$ over the entire chromosome. For that simulation we will be using the function "sim.pop":

```
> sim.pop <- function(Pr,Tr,n.mark,n.subj)
+ {
+       pop <- matrix(ncol=n.mark,nrow=n.subj)
+       states <- 0:(length(Pr)-1)
+       pop[,1] <- sample(states,n.subj,rep=TRUE,prob=Pr)
+       for (i in 2:ncol(pop))
+       {
+            for(x in states)
+            {
+                 initial.state <- (pop[,i-1] == x)
+                 prob <- Tr[x+1,]
+                 pop[initial.state,i] <- sample(states,
+                      sum(initial.state),rep=TRUE,prob=prob)
+            }
+       }
+       return(pop)
+ }
```

This function takes as input an initial distribution "Pr" and a transition matrix "Tr" and returns a matrix. Each row in the matrix corresponds to an independent path of a Markov process over "n.mark" points. The total number of independent paths is "n.subj".

The distribution of the hidden process is one component. The other component is the conditional distribution of the observed process, given the state of the hidden process. These conditional distributions are determined by the allele frequencies of markers in the two founding populations. Indeed, let $f_1$ be the frequency of one of the two alleles of a given marker in population 1 and let $f_2$ be the frequency of the same allele in population 2. Hence, given that both gametes originated from population 1 ($\chi = 0$), one obtains that the distribution of the genotype of the marker is $B(2, f_1)$. On the other hand, if both gametes originated from population 2 ($\chi = 2$), then the distribution is $B(2, f_2)$. If one gamete originated from population 1 and the other from population 2 ($\chi = 1$), then the distribution of the genotype is slightly more complex. The probability of an homozygote of the identified allele is $f_1 f_2$, the probability of a homozygote of the other allele is $(1 - f_1)(1 - f_2)$, and the probability of an heterozygote is the complementary probability: $f_1(1 - f_2) + f_2(1 - f_1)$. Finally, these genotypes are conditionally independent given the process $\chi(t)$ if the marker alleles are assumed to be in linkage equilibrium in each founding population before admixture occurs.

The function "geno.given.pop" simulates the observed genotypes. It takes as input the hidden process "pop" and the frequency of the allele in both populations and returns a matrix of genotypes. The dimensions of the returned matrix are the same as the dimensions of "pop". For convenience we will take the frequencies of the allele to be the same for each marker. Straightforward modifications will allow the introduction of different frequencies for different loci. (See Prob. 10.9.)

```
> geno.given.pop <- function(pop,f1,f2)
+ {
+   geno <- pop
+   geno[pop==0] <- rbinom(sum(pop==0),2,f1)
+   geno[pop==2] <- rbinom(sum(pop==2),2,f2)
+   P.1 <- c((1-f2)*(1-f1),(1-f2)*f1+f2*(1-f1),f1*f2)
+   geno[pop==1] <- sample(0:2,sum(pop==1),rep=TRUE,prob=P.1)
+   return(geno)
+ }
```

In principle, we now have the tools to carry out for admixture mapping the same type of investigation that was conducted in Chap. 9 for linkage analysis, where we asked how much information one loses by the process of reconstruction as a function of the density and informativeness of the markers.

Instead, we would like here to study a different aspect of the problem and investigate the issue of the effect of the estimation of unknown parameters. Even in our very simplistic formulation of the problem, the distribution is still a function of the actual values of the Markov chain transition probabilities and the allele frequencies $f_1$ and $f_2$ in the two founding populations. In principle, the parameters of the Markov chain may vary from one individual to the next, depending on the specific admixture history associated with that particular individual; and the allele frequencies may vary between markers. In general, these parameters are unknown and need to be estimated. To illustrate some aspects of the problem, in this section we will deal specifically with the issue of the estimation of allele frequencies while we assume the transition probabilities are known. In the next section we will discuss the estimation of the transition probabilities and assume that the allele frequencies are known.

Estimating the allele frequencies in the founding population has by no means a straightforward solution. People that reside today in the regions where the founding populations originated may not adequately represent the population at the time the admixture took place. Nonetheless, we will assume that representing samples are available for both founding populations and that they produce estimates $\hat{f}_1$ and $\hat{f}_2$ for each marker used in the investigation.

For the sake of the reconstruction of the underlying process, estimates of the frequencies will be used. The function "prob.geno.given.pop" takes as input the matrix of genotypes and two matrices of estimated allele frequencies "f1.hat" and "f2.hat" and return an array of conditional distributions of genotypes, given the states of the hidden process. This array is used as input for the application of the forward and backward algorithms:

```
> prob.geno.given.pop <- function(geno,f1.hat,f2.hat)
+ {
+      n.subj <- nrow(geno)
+      n.mark <- ncol(geno)
+      G.nu <- array(dim=c(n.subj,n.mark,3))
+      G.0 <- G.1 <- G.2 <- geno
+      for (g in 0:2)
+      {
+          G.0[geno==g] <- dbinom(g,2,f1.hat[geno==g])
+          G.2[geno==g] <- dbinom(g,2,f2.hat[geno==g])
+      }
+      G.1[geno==0] <- (1-f1.hat[geno==0])*(1-f2.hat[geno==0])
+      G.1[geno==1] <- (1-f1.hat[geno==1])*f2.hat[geno==1] +
+          (1-f2.hat[geno==1])*f1.hat[geno==1]
+      G.1[geno==2] <- f1.hat[geno==2]*f2.hat[geno==2]
+      G.nu[,,1] <- G.0
+      G.nu[,,2] <- G.1
+      G.nu[,,3] <- G.2
+      return(G.nu)
+ }
```

The application of the forward and backward algorithms will produce a reconstructed process $\hat{\chi}_i$ for each subject. The test statistic is obtained from the covariance $\sum_{i=1}^{n} \{(y_i - \bar{y})\hat{\chi}_i\}$, which must be standardized by dividing by the square root of its variance, or by a suitable estimator. In the computation of the variance, however, there is a possibly important difference between the situation when $\chi$ is known (and also when $\chi$ is unknown but $f_1$ and $f_2$ are known) and the situation when $\chi$ is unknown and estimates of allele frequency are used for its reconstruction. The difference stems from the fact that the summands are no longer independent since the same estimates of allele frequencies are used for all subjects. As a result, the variance includes an extra term that results from the correlation between the summands. Since $[\sum_{i=1}^{n} (y_i - \bar{y})]^2 = 0$, we obtain that $\sum_{i \neq j} (y_i - \bar{y})(y_j - \bar{y}) = -\sum_{i=1}^{n} (y_i - \bar{y})^2$, which together with some calculation leads to the equation:

$$\frac{1}{n} \text{var}\left( \sum_{i=1}^{n} [y_i - \bar{y}]\hat{\chi}_i | y_1, \cdots y_n \right) = \hat{\sigma}_y^2 \left( \text{var}[\hat{\chi}_1] - \text{cov}[\hat{\chi}_1, \hat{\chi}_2] \right) . \qquad (10.10)$$

The term $\text{cov}[\hat{\chi}_1, \hat{\chi}_2]$ is the covariance between the reconstructed processes for two different individuals that share the same estimated allele frequencies in the reconstruction. Although this extra term in principle might not be negligible, we will see below some numerical evidence that it is small and positive. Consequently we will then ignore it.

We turn now to the actual simulation of reconstructed counts of the population source when the allele frequencies at the founding populations are estimated using samples from these populations. Specifically, we will take equal sample sizes from both populations and consider three possibilities. A sample size 20 corresponds to taking 10 individuals from each population (each individual contributing a pair of independent homologous chromosomes) and represents a small sample. A sample size of 100 corresponds to 50 individuals and represents a medium-sized sample. And a sample size of 1000 represents a large sample. The distribution of frequency of the allele for the sample is the appropriate binomial. The relative frequency is used for an estimate. Linkage equilibrium is assumed among markers. Hence, the binomial frequencies are independent between markers. For the simulation the actual allele frequencies are 0.75 and 0.25 for all markers. However, these frequencies are assumed to be unknown, and the estimated frequencies are used in the reconstruction.

In the following code we generate population-source processes and their genotypes. Allele frequencies are obtained by simulating binomial random variables. The estimated frequencies are used in order to compute the conditional distribution of the genotypes given the states of the underlying process. The backward and the forward algorithms are used in order to reconstruct the hidden process. For the computation of the covariance term from (10.10) the reconstruction is carried out again for independent processes but with the same estimated allele frequencies. The mean, variance, and covariance terms at the central locus are stored in the matrix "pop.est.null". The covariance structure of the reconstructed process and the covariance of that process with the hidden process are stored in a list for later use. The simulation involves a total of $10^5$ iterations:

```
> samp <- c(20,100,1000)
> f1 <- 0.75
> f2 <- 0.25
> n.mark <- round(140/delta)+1
> locus <- ceiling(n.mark/2)
> n.rep <- 10^2
> n.subj <- 10^3
> pop.est.null <- matrix(nrow=3,ncol=length(samp))
> colnames(pop.est.null) <- paste("Sample=",samp,sep="")
> rownames(pop.est.null) <- c("mean","var","cov")
> cor.est <- vector(mode="list",length=length(samp))
> names(cor.est) <- paste("Sample=",samp,sep="")
> cor.est.pop <- cor.est
> for(i in 1:length(samp))
+ {
+     n1 <- n2 <- samp[i]
+     pop.est <- pop.est.1 <- pop <- NULL
+     for (rep in 1:n.rep)
+     {
+         f1.hat <- rbinom(n.subj*n.mark,n1,f1)/n1
+         f1.hat <- matrix(f1.hat,nrow=n.subj,ncol=n.mark)
+         f2.hat <- rbinom(n.subj*n.mark,n2,f2)/n2
+         f2.hat <- matrix(f2.hat,nrow=n.subj,ncol=n.mark)
+         pop.temp <- sim.pop(Pr,Tr,n.mark,n.subj)
+         pop <- c(pop,pop.temp[,locus])
+         geno <- geno.given.pop (pop.temp,f1,f2)
+         G.pop <- prob.geno.given.pop(geno,f1.hat,f2.hat)
+         P.F <- forward(G.pop,Tr,Pr)
+         P.B <- backward(G.pop,Tr,Pr)
+         P <- marginal.post(P.F,P.B)
+         pop.est <- rbind(pop.est,P[,,2]+2*P[,,3])
+         pop.temp <- sim.pop(Pr,Tr,n.mark,n.subj)
+         geno <- geno.given.pop (pop.temp,f1,f2)
+         G.pop <- prob.geno.given.pop(geno,f1.hat,f2.hat)
+         P.F <- forward(G.pop,Tr,Pr)
+         P.B <- backward(G.pop,Tr,Pr)
+         P <- marginal.post(P.F,P.B)
+         pop.est.1 <- c(pop.est.1,P[,locus,2]+2*P[,locus,3])
+     }
+     pop.est.null["mean",i] <- mean(pop.est[,locus])
+     pop.est.null["var",i] <- var(pop.est[,locus])
+     pop.est.null["cov",i] <- cov(pop.est.1,pop.est[,locus])
+     cor.est[[i]] <- cor(pop.est)
+     cor.est.pop[[i]] <- cor(pop.est,pop)
+ }
```

After running the simulations we obtain the following results:

```
> round(pop.est.null,3)
      Sample=20 Sample=100 Sample=1000
mean      1.560      1.594       1.599
var       0.188      0.173       0.171
cov       0.005      0.001       0.001
```

The expected value of $\chi$ is $2 \times 0.8 = 1.6$. The reconstruction of the hidden process based on estimated allele frequencies introduces some bias when the sample sizes are small or moderate. The variance of $\chi$ is $2 \times 0.8 \times 0.2 = 0.32$. Compare this to the variance of $\hat{\chi}$ when the sample size is large (0.171). Assuming that using a large sample size corresponds, more or less, to actual knowledge of the allele frequencies, we see that the ratio of the noncentrality parameter of the reconstructed process (when we use estimated allele frequencies in the reconstruction, but assume they are known) to that of the ideal process is approximately $(0.171/0.32)^{1/2} \approx 0.73$. Since the noncentrality parameter is proportional to $n^{1/2}$, one can conclude that under the given idealized conditions, compared to using the ideal unobservable processes $\chi(t)$, using the reconstructed process requires a sample size about $1/0.73^2 \approx 2$ times as large. We would like to understand how much more power is lost due to variability in the estimates of the allele frequencies. Finally, from the the covariance term, one sees that the effect of the dependence between reconstructed processes on the variance of the statistic is small. As a result, we will ignore this dependence from now on.

In the process of obtaining power curves we should get the appropriate significance thresholds that take into account multiple testing. For that end we assume normality and simulate the test statistic under the null hypothesis of no linkage to a trait locus. Observe that we are using a two-sided test:

```
> library(MASS)
> samp <- c(20,100,1000)
> z <- samp
> names(z) <- paste("n1=n2=",samp,sep="")
> n.rep <- 10^2
> n.iter <- 10^4
> for(i in 1:length(samp))
+ {
+     Z.max <- NULL
+     for(rep in 1:n.rep)
+     {
+         Z <- mvrnorm(n.iter,rep(0,n.mark),cor.est[[i]])
+         Z.max <- c(Z.max,apply(abs(Z),1,max))
+     }
+     z[i] <- sort(Z.max)[n.rep*n.iter*(1-0.05/23)]
+ }
```

```
> round(z,3)
  n1=n2=20   n1=n2=100 n1=n2=1000
     3.972      3.927       3.926
```

Although the approximating Gaussian model is very similar to that used in linkage analysis for affected sib pairs, the resulting thresholds are higher for two reasons. First, the test is two-sided. Second, since there are eight generations of meiotic recombination involved, recombinations are more frequent, so the process $\chi(t)$ is less correlated than the IBD process for sib pairs. The effective value of the correlation parameter (denoted by $\beta$ in Chaps. 5 and 6) is $0.01g = 0.08$. With perfect information about the population origins of each allele, this would lead to a threshold $z = 4.1$, about 4–5% larger than the simulated thresholds. The most important reason for this difference between the mathematical approximation and the simulated thresholds is the reconstruction process. Since we use information from nearby markers in our reconstruction of $\chi(t)$, the process $\hat{\chi}(t)$ is more correlated, which in turn makes it less likely to take on extreme values. This factor also operates in the case of affected sib pairs, where it had a similar effect on the significance thresholds. Finally, we also use estimated allele frequencies for the reconstruction. Since this increases variability, it seems plausible that it makes extreme values more likely, hence works to increase the threshold. This is consistent with larger simulated thresholds for smaller sample sizes (hence greater variability).

The distribution of the test statistic under the alternative involves a nonzero mean. This noncentrality parameter can be obtained using the relation:

$$\mathrm{E}(\hat{Z}_l) \approx \mathrm{cor}\big(\hat{\chi}(t), \chi(\tau)\big) \times \mathrm{E}(Z_\tau) = \mathrm{cor}\big(\hat{\chi}(t), \chi(\tau)\big) \times \xi. \tag{10.11}$$

(Observe that this relation reduces to (9.11) when the correct distribution is used for the reconstruction and $\hat{\chi}(t) = \mathrm{E}[\chi(t)|G]$, but not in general.) We use this approximation in order to generate the Gaussian process of the statistics under the alternative distribution and plot the power curves:

```
> n.rep <- 10
> power.est <- matrix(nrow=length(xi),ncol=length(samp))
> colnames(power.est) <- names(z)
> for (i in 1:length(samp))
+ {
+     qtl <- ceiling(n.mark/2)
+     rho <- (cor.est.pop[[i]])
+     Z <- mvrnorm(n.iter,rep(0,n.mark),cor.est[[i]])
+     for (j in 1:length(xi))
+     {
+         ncp <- xi[j]*rho
+         Z1 <- sweep(Z,2,ncp,"+")
+         power.est[j,i] <- mean(apply(abs(Z1),1,max) >= z[i])
+     }
+ }
```

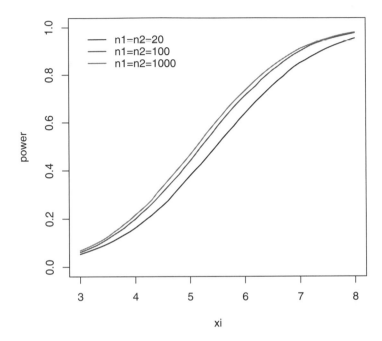

**Fig. 10.1.** The power function for different sample sizes for estimating frequency of alleles.

```
> plot(range(xi),c(0,1),type="n",xlab="xi",ylab="power")
> gr = gray(0.75*(1:length(samp))/length(samp))
> for(i in 1:length(samp))
+       lines(xi,power.est[,i],col=gr[i])
> legend(xi[1],1,bty="n",legend=names(z),
+     lty=rep(1,length(z)),col=gr)
```

The resulting power curves are given in Fig. 10.1. For our simulations the power curve for a moderate sample size coincides with the curve for a large sample size. Only when the samples are extremely small can we identify a noticeable decrease in power. We can conclude that, at least in the example we considered, the fact that the allele frequencies are estimated has only a minor effect on the statistical properties. However, one should be careful not to extrapolate this finding too generally. We have used independent data in order to estimate the allele frequencies of the markers. In linkage analysis in humans, allele frequencies of markers are also relevant for the reconstruction of hidden processes and are typically unknown. However, the common practice is to use the same data to estimate the allele frequencies and for subsequent linkage analysis. Issues of dependencies, as well as issues associated with biased sampling, may or may not have a larger effect in that case.

The simple formula of Chap. 6 for approximating the power can be used here. If we take as known the factor 0.73 by which the noncentrality parameter must be reduced, it is easy to see that the power approximation for $\xi = 6$ (hence noncentrality 4.38) is about 0.7, and the power does not get into the range of 0.8–0.9 until $\xi \approx 6.5$–7. These numbers are roughly consistent with the simulations given in Fig. 10.1.

One can also explore the effect of changing parameters. For example, if $f_1 = 0.2$, $f_2 = 0.8$, which provides better information for reconstructing the process $\chi(t)$, and other parameters remain the same as before, the significance threshold increases only slightly to 3.96, but the power when $\xi = 6$ increases to slightly more than 0.8. Going in the other direction, if $f_1 = 0.33$, $f_2 = 0.67$, the significance threshold decreases slightly to 3.89, while the power falls to about 0.4.

Other important parameters are the length of time since admixture, over which one has only limited control, and the distance between markers, which is easily controlled. The power will increase if either the time since admixture or the distance between markers decreases. For example, suppose $f_1 = 0.33$, $f_2 = 0.67$, but the time since admixture is only 5 generations, so $g = 4$. Then the significance threshold is 3.74, and for $\xi = 6$ the power jumps to slightly more than 0.7. If $g = 8$, but the interval between markers is reduced to 1 cM, the significance threshold is 3.93, and for $\xi = 6$ the power is about 0.66.

## 10.3 Estimating Parameters of the HMM

In the discussion of the reconstruction of the test statistic, we have treated the parameters of the underlying Markov process as known quantities. This may be reasonable in the case of interest in Chap. 9 involving linkage analysis in human pedigrees, where the parameters are determined by the familial relationships, or in admixture mapping of outbred strains in experimental genetics, where one controls the degree of admixture through experimental design. In admixed human populations, however, the distribution of the underlying process is a function of a complex history of the subject's ancestry. For admixed populations that have evolved over several generations it is unlikely that one can obtain a reliable assessment of the process in question from recorded ancestral history. An alternative approach, which is the subject of this section, is to estimate the parameters of the process using the genotypic information collected throughout the genome. We will start with the case where the process $\chi$ is directly observable and present the maximum likelihood estimator (MLE) of the parameters that determine the distribution of the process. We will then turn to the more realistic setting where the process is observed indirectly and consider the calculation of the MLE. The algorithm that we will present for carrying out the calculation is a close relative of the *expectation maximization* (EM) algorithm. A small simulation study will be conducted in order to explore the accuracy of the estimators.

The model for admixture we have used so far has the simplifying feature that the probability $\pi$ is the same for everyone. Another simple model with quite different properties is described in Prob. 10.7. In that model there are different values of $\pi$ depending on the number of generations since initial admixture. We can generalize our basic model to allow $\pi$ to be specific to the individual. Then equation (10.4) should be modified by letting $\pi$ in the numerator depend on $i$ and by replacing $n\pi(1 - \pi)$ in the denominator by $\sum_i \pi_i(1 - \pi_i)$. Equations (10.5), (10.6), (10.7), and (10.11) are similarly modified. In the discussion that follows there is no specific model for admixture – only the (convenient but mathematically incorrect) assumption that the processes giving the allelic population sources are Markov chains. Note, however, that the numerical examples are restricted to the simpler model having a single value of $\pi$ and a fixed number of generations since admixture.

Hence we assume that the process $\chi$ is a sum of two independent and identically distributed stationary two-state Markov processes over equally spaced markers. Recall that the distribution of such a process is determined by the two transition probabilities of the two-state process from state 1 to state 0 and from 0 to 1. We denoted the pair of probabilities by $\beta = (\beta_1, \beta_2)$ and the stationary distribution of the process $\chi$ by $\pi = \pi(\beta)$. We write the transition probabilities of the process as a $3 \times 3$ matrix $T = T(\beta)$ with elements $T_{ij}(\beta)$, which give the probability of moving from state $i$ to state $j$ in one step. It follows from the Markovian assumption that the probability of a path of the process on an autosome in a random subject is given by:

$$\Pr_\beta\big(\chi(t_1), \ldots, \chi(t_{\tilde{m}})\big) = \prod_{j=0}^{2} \pi_j^{I_{\{\chi(t_1)=j\}}} \prod_{m=2}^{\tilde{m}} \prod_{i=0}^{2} \prod_{j=0}^{2} T_{ij}^{I_{\{\chi(t_{m-1})=i,\chi(t_m=j)\}}} \, ,$$

where $I_{\{\chi(t_1)=j\}}$ indicates that the initial value of the process is $j$ and $I_{\{\chi(t_{m-1})=i,\chi(t_m)=j\}}$ is the indicator of the event that the process moved from state $i$ to state $j$ between the markers $t_{m-1}$ and $t_m$. Observe that the dependence of the probability on the parameter $\beta$ is through the dependence of $\pi$ and $T$ on $\beta$.

Information regarding the value of $\beta$ can be accumulated from several chromosomes. Certainly, the other chromosomes within a subject can be used. If there is good reason to believe that the properties of the process are the same for a group of subjects, then the information from all such subjects can be pooled together. However, in this investigation we will use only the information collected from a single subject in an attempt to assess the properties of the subject-specific process $\chi$. We will ignore the special properties of the sex chromosomes and, for the sake of simplifying the notation and programming, we consider 23 pairs of autosomes of equal length. Taking the log of the probability and summing over the different chromosomes one obtains the log-likelihood function of the parameter:

$$\ell(\beta; \chi) = \sum_{j=0}^{2} \left[ \sum_{c=1}^{23} I_{\{\chi(t_1)=j\}} \right] \log \pi_j$$

$$+ \sum_{i=0}^{2} \sum_{j=0}^{2} \left[ \sum_{c=1}^{23} \sum_{m=2}^{\tilde{m}} I_{\{\chi(t_{m-1})=i, \chi(t_m)=j\}} \right] \log T_{ij} . \quad (10.12)$$

When the process $\chi$ is observed, one may evaluate the indicators and obtain a data-dependent function of the parameter. The MLE is obtained as a maximizer of this function.

In order to illustrate the computation of the maximum likelihood estimator consider the function "log.lik" that computes the *negative* log-likelihood. The first argument to the function is a vector of length two containing the parameters. The second argument is a vector of length 3 with the sums of indicators that appear inside the first brackets of (10.12). The third argument is a matrix of dimension $3 \times 3$ with the elements inside the other square brackets. The output of the function is minus the log-likelihood. Observe that parameters are corrected to fall within the permitted range:

```
> log.lik <- function(beta,MP.sum,JP.sum)
+ {
+     beta <- pmax(beta,0); beta <- pmin(beta,1)
+     lt <- log(transition.matrix(beta))
+     lp <- log(stationary.probability(beta))
+     return(-sum(c(lp*MP.sum,lt*JP.sum)))
+ }
```

The maximum likelihood estimator is calculated by minimization of the (minus) log-likelihood function. For that we apply R's general optimization function "optim". The calculation of the MLE is implemented in the function "MLE.beta". The input is a starting value of $\beta$, which is passed on to the optimization function and a matrix that contains the observed paths of the process. Cross-tables of counts are calculated with the aid of the function table. The levels of the factors are specified since some levels may have a frequency of zero in the sample:

```
> MLE.beta <- function(initial,pop)
+ {
+     M.s <- table(factor(pop[,1],0:2))
+     J.s <- table(factor(pop[,1:(n.mark-1)],0:2),
+         factor(pop[,2:n.mark],0:2))
+     beta.hat <- optim(initial,log.lik,MP.s=M.s,JP.s=J.s)$par
+     return(beta.hat)
+ }
```

For an illustration we consider the same process parameters that were used in the previous section:

```
> pi2 <- 0.8
> delta <- 2
> theta <- 0.5 - 0.5*exp(-0.02*delta)
> true.beta <- c(pi2,1-pi2)*(1-(1-theta)^8)
```

We simulate 23 independent copies of the process and compute the estimate. The initial values of the parameters correspond to the same marginal distribution as the true values but a different number of generations of random mating:

```
> Tr <- transition.matrix(true.beta)
> Pr <- stationary.probability(true.beta)
> n.mark <- round(140/delta)+1
> n.chr <- 23
> pop <- sim.pop(Pr,Tr,n.mark,n.chr)
> initial.beta <- c(pi2,1-pi2)*(1-(1-theta)^6)
> MLE.beta(initial.beta,pop)
[1] 0.12437302 0.02980149
> true.beta
[1] 0.11719342 0.02929836
```

Observe that the estimated values of $\beta$ are not identical to the true values but they are reasonably close, with a relative error of about 4–6%.

The situation is more complex when the process $\chi$ is not directly observable. In that case the likelihood (10.12) cannot be calculated, so alternative approaches are required.

The expectation-maximization (EM) algorithm was developed in order to compute the maximum likelihood estimator of a parameter in situations where the computation of the log-likelihood is relatively simple when full information is available but is difficult when only part of the information is observed. This algorithm may be used in order to estimate the parameters of the unobserved process on the basis of the observed genotypes.

Denote the observed genotypes by $G$. The conditional distribution of the genotypes given the process is independent of the parameter $\beta$. It has been ignored so far. However, if one considers the log of that likelihood and adds it to the right-hand side of (10.12), then one gets the joint log-likelihood of $G$ and $\chi$ as a function of $\beta$. Denote this log-likelihood by $\ell(\beta; \chi, G)$. When only the genotypes are observable, then the function which needs to be maximized in order to obtain the MLE is the marginal log-likelihood of the genotypes which is equal to:

$$\ell(\beta; G) = \log \sum_\chi e^{\ell(\beta; \chi, G)} \ ,$$

with the sum extending over all paths of the $\chi$ process. Since the number of paths is extremely large, it is not feasible to maximize this function directly.

In the EM algorithm the maximization of $\ell(\beta; G)$ is replaced by an iterative process of maximization, each of a function which is much easier to compute compared to the actual log-likelihood. The basic function defined in the algorithm is a function of two variables:

$$Q(\beta \,|\, \beta^*) = \mathrm{E}_{\beta^*}(\ell(\beta; \chi, G) \,|\, G) \,. \tag{10.13}$$

Here $\beta^*$ is our current best estimate of the value of $\beta$. The principle behind (10.13) is to use instead of the unobservable quantity $\ell(\beta; \chi, G)$ its conditional expectation given the observed $G$. The function $Q$ is maximized with respect to $\beta$. The resulting maximizer becomes our new best estimate and replaces the value of $\beta^*$ for an iteration of the entire procedure. In each step the maximizer replaces the previous value of $\beta^*$. The iterations proceed until the values of $\beta^*$ converge.

Consider the evaluation of the function $Q$ in our setting. The term involving the conditional distribution of the genotypes given the states of $\chi$ can be ignored since it does not involve $\beta$, and hence will not change the maximizing value of $\beta$. Taking a conditional expectation, given the values of the genotypes and $\beta^*$, corresponds to replacing the indicators in the square brackets of (10.12), which we do not observe, by their conditional expectations:

$$Q(\beta \,|\, \beta^*) = \sum_{j=0}^{2} \Big[ \sum_{c=1}^{23} \mathrm{Pr}_{\beta^*}(\chi(t_1) = j \,|\, G) \Big] \log \pi_j +$$

$$\sum_{i=0}^{2} \sum_{j=0}^{2} \Big[ \sum_{c=1}^{23} \sum_{m=2}^{\tilde{m}} \mathrm{Pr}_{\beta^*}(\chi(t_{m-1}) = i, \chi(t_m) = j \,|\, G) \Big] \log T_{ij} \,.$$

The required conditional probabilities can be computed by a slight modification of the forward and backward algorithms. Indeed, the conditional probability of the initial state given the genotypes is a direct application of the algorithm. The joint distribution of consecutive states can be computed from the forward and backward probabilities using the formula:

$$\mathrm{Pr}_{\beta^*}(\chi(t_{m-1}) = i, \chi(t_m) = j \,|\, G) = \frac{F_{m-1}(i) \, T_{ij} \, \mathrm{Pr}(G_m \,|\, j) \, B_m(j)}{\sum_{k,l} \{ F_{m-1}(k) \, T_{kl} \, \mathrm{Pr}(G_m \,|\, l) \, B_m(l) \}} \,. \tag{10.14}$$

This formula is implemented in the function "joint.post":

```
> joint.post <- function(F,B,G.pop,Tr)
+ {
+       n.samp <- dim(G.pop)[1]
+       n.mark <- dim(G.pop)[2]
+       n.state <- dim(G.pop)[3]
+       GB <- G.pop*B; GB <- GB[,-1,]
+       FF <- F[,-n.mark,]
+       JP <- array(dim=c(n.samp,n.mark-1,n.state,n.state))
+       for (i in 1:n.state) for (j in 1:n.state)
+           JP[,,i,j] <- FF[,,i]*GB[,,j]*Tr[i,j]
```

```
+        S <- apply(JP,1:2,sum)
+        JP <- sweep(JP,1:2,S,"/")
+        return(JP)
+ }
```

Observe, therefore, that the EM algorithm for the computation of the maximum likelihood estimators of the transition rates can be implemented iteratively by the computation of the conditional probabilities based on the current estimates of the rates followed by application of a maximization procedure. The target function for the maximization is the same log-likelihood function that was considered in the case of full information, but with counts replaced by sums of conditional probabilities.

The function "EM.beta" implements the EM algorithm for genotypes. The input is an initial value of $\beta$ and an array with the conditional distribution of the genotypes given the states of the hidden process. The maximal number of iterations and a tolerance level for monitoring convergence can be set. In each iteration the conditional probabilities are computed using the backward and the forward algorithms. The function "optim" is used in order to obtain the new value of the parameter. The process is stopped when the ratio between the old and the new values of the parameters are close enough to 1 or when the maximal number of iterations is reached.

```
> EM.beta <- function(initial,G.pop,max.iter=1000,tol=0.001)
+ {
+      beta.new <- initial
+      beta.old <- rep(-Inf,2)
+      iter <- 1
+      while((iter<max.iter)&max(abs(beta.old/beta.new-1)>tol))
+      {
+         beta.old <- beta.new
+         Tr <- transition.matrix(beta.new)
+         Pr <- stationary.probability(beta.new)
+         P.F <- forward(G.pop,Tr,Pr)
+         P.B <- backward(G.pop,Tr,Pr)
+         MP <- marginal.post(P.F,P.B)
+         MP.s <- apply(MP[,1,],2,sum)
+         JP <- joint.post(P.F,P.B,G.pop,Tr)
+         JP.s <- apply(JP,3:4,sum)
+         beta.new <- optim(beta.new,log.lik,
+               MP.s=MP.s,JP.s=JP.s)$par
+         iter <- iter+1
+      }
+      return(list(beta=beta.new,iter=iter))
+ }
```

Let us run a small simulation in order to explore the statistical properties of the MLE of the transition rates. We consider the same level of informativeness of the genotypes that was assumed in the previous section, but this time we take the allele frequencies to be known:

```
> f1 <- 0.75
> f2 <- 0.25
> f1.true <- matrix(f1,nrow=n.chr,ncol=n.mark)
> f2.true <- matrix(f2,nrow=n.chr,ncol=n.mark)
```

In each repeat we will generate a new subject from the given distribution and compute for the subject the estimate of the transition rates, once by assuming that the 23 processes $\chi$ are observable and then by assuming that they are not. We will use the simulated values in order to assess the accuracy of the estimators in both cases. However, owing to the long time it takes the EM algorithm to run, we simulate only 100 independent copies:

```
> locus <- ceiling(n.mark/2)
> n.rep <- 10^2
> beta.full <- beta.hat <- matrix(nrow=n.rep,ncol=2)
> iter <- vector(length=n.rep)
> pop.true <- pop.est <- NULL
> for(i in 1:n.rep)
+ {
+     pop <- sim.pop(Pr,Tr,n.mark,n.chr)
+     beta.full[i,] <- MLE.beta(initial.beta,pop)
+     geno <- geno.given.pop(pop,f1,f2)
+     G.pop <- prob.geno.given.pop(geno,f1.true,f2.true)
+     out <- EM.beta(initial.beta,G.pop)
+     beta.hat[i,] <- out$beta
+     iter[i] <- out$iter
+     Tr.est <- transition.matrix(out$beta)
+     Pr.est <- stationary.probability(out$beta)
+     P.F <- forward(G.pop,Tr.est,Pr.est)
+     P.B <- backward(G.pop,Tr.est,Pr.est)
+     P <- marginal.post(P.F,P.B)
+     pop.true <- rbind(pop.true,pop[,locus])
+     pop.est <- rbind(pop.est,P[,locus,2]+2*P[,locus,3])
+ }
```

With this simulation we hope to gain some insight regarding the statistical properties of the MLE as an estimator of the rate and the efficiency of the EM algorithm in its computation as well as the effect of not knowing the transition rates on the noncentrality of the scanning statistic. Observe that we simulate the unobserved population source and the observed genotypes. The estimator of the transition rates that would have been obtained had we known the population source is computed and stored. The EM algorithm is applied to the observed genotypes in order to produce the actual estimator of the rates. The population source at the center of each chromosome is reconstructed based on the genotypic information and the estimated values of the rates. The true and the reconstructed quantities are stored for later use.

Compare the true rates to the average of the estimated values:

```
> signif(true.beta,3)
[1] 0.1170 0.0293
> signif(apply(beta.full,2,mean),3)
[1] 0.1190 0.0302
> signif(apply(beta.hat,2,mean),3)
[1] 0.121 0.031
```

It can be seen that the maximum likelihood estimator, both when the process $\chi$ is observable and when it is not, is basically unbiased. If we considered the standard deviation of these estimates, we obtain:

```
> signif(apply(beta.full,2,sd),3)
[1] 0.01220 0.00306
> signif(apply(beta.hat,2,sd),3)
[1] 0.02880 0.00864
```

Hence, the variability of the estimator when the hidden process is inferred is more than doubled. Yet, it seems that even when partial information is used for estimating the rate and when the rates are subject-specific, the results are reasonable. They will improve further if more informative markers can be used or if the information from several subjects can be pooled together.

Although estimation of the transition rates involves some error, our primary interest is in the noncentrality parameter. From (10.11) we see that the reduction in this parameter can be summarized in terms of the correlation between the reconstructed process $\hat{\chi}$ and the true process $\chi$. We compute the correlation for each chromosomes and then average the results across chromosomes:

```
> r.est <-  NULL
> for (i in 1:n.chr)
+     r.est[i] <- cor(pop.true[,i],pop.est[,i])
> mean(r.est)
[1] 0.7110186
```

The new correlation is only slightly lower than the value of 0.73 that was obtained when the transition rates were known (although the allele frequencies were estimated and hence slightly misspecified). Consequently it appears that, under the conditions we have examined, not knowing the transition rates has a minor effect compared to not knowing the population source.

In order to understand why it took so long to run so few iterations let us consider the distribution of the number of iterations in the application of the EM algorithm:

```
> signif(c(mean(iter),sd(iter)),3)
[1] 30.4 11.0
```

Convergence is slow. This is a known drawback of the algorithm, which requires many iterations. Each iteration involves the computation of the conditional probabilities, which is a computationally intensive task in its own right. Hence, it should not come as a surprise that a long time is needed to obtain estimates of the rates.

## 10.4 Discussion

By comparing the noncentrality parameter derived in this chapter with, for example, that given in Chap. 11 for mapping QTL using sib pairs, one sees that use of admixed populations in human gene mapping is potentially very useful, but only if two conditions are satisfied: the base populations differ substantially in the frequencies $p_i$ for the QTL, and the admixture process is such that one can recover population sources of marker alleles with reasonable efficiency. This dependence on population history, over which one has no control, makes it difficult to know how useful admixture mapping is likely to be in any specific case. Population history will be seen in Chap. 12 to be an issue there as well.

Although this chapter used several basic ideas that come directly from experimental genetics, we also saw another application of hidden Markov models, which appeared for the first time in Chap. 9, and an introduction to the EM algorithm, which will reappear in a different context in Chap. 13.

## 10.5 Bibliographical Comments

McKeigue [53, 54] has recommended use of admixed populations to map disease genes in humans. The simple models for admixture discussed in this chapter were first suggested by Long [50]. Zhu *et al.* [93] study these models of population history, but their recommended test statistics are oriented toward testing the null hypothesis $\theta = 1/2$, as in classical parametric linkage analysis.

For a general introduction to the EM algorithm and applications to a variety of problems in statistical genetics, see Lange [49].

## Problems

**10.1.** Suppose that the approach in this chapter is to be used for a designed experiment involving an intercross of two (outbred) breeds of dogs. What if any of the basic theory developed in Sect. 10.1 and summarized in (10.1) to (10.9) must be changed? Modify the programs in Sect. 10.2 accordingly and evaluate significance thresholds and power for different sets of parameters.

**10.2.** Develop statistics comparable to (10.4)–(10.5) and noncentrality parameters as in (10.6)–(10.7) for the case of a 0-1 trait, based on a sample of $n$ unrelated individuals who have the trait. Assume that markers are fully informative with regard to population origin.

**10.3.** Consider the model (2.2) with, for simplicity, $\delta = 0$, for the *penetrance* of a qualitative trait. Consider the offspring of a mother from population 1 and a father from population 2, who is "backcrossed" to someone from, say, population 1 to produce an offspring who has one set of chromosomes that definitely originates from population 1 and homologous chromosomes that have some DNA derived from population 1 and some from population 2. Assume that markers are fully informative, so one knows with certainty the population origin of each marker allele.

(a) Suppose that this individual is affected. Given a sample of size $n$ of such (unrelated) individuals, what are an appropriate statistic, significance threshold, and noncentrality parameter to detect linkage? Does it make any difference for the statistical power whether $p_1 > p_2$ or $p_2 > p_1$?

(b) Suppose now that regardless of the phenotype of the individuals from the first generation of backcrosses, the backcross to population 1 is repeated to produce a second generation. For a sample of $n$ affecteds from the second generation, what would be an appropriate statistic, a genome-wide significance threshold and noncentrality parameter?

**10.4.** (a) Spell out the details of the arguments that produce (10.8) and (10.9).
(b) Use (10.8) and the approximation $\exp(-x) \approx 1 - x$, which is valid for small $|x|$, to verify the claim in the text that the correlation parameter to use in the approximations from Chaps. 5 and 6 is $\beta = 0.01g$ per cM.

**10.5.** Assume the probability that an arbitrary allele in the $i$th individual comes from population 2 is $\pi_i$, which may depend on $i$. What should be the form of (10.5), (10.6), (10.7) and (10.11)?

**10.6.** Admixture creates linkage disequilibrium, even if there is none in the individual populations before admixture. Consider the population model of admixture discussed in Sect. 10.1.3 and two bi-allelic loci. Let the allele frequency of a particular allele at locus $i$ in population $j$ be denoted by $p_{ij}$. Let $\delta_i = p_{i1} - p_{i2}$ be the difference in allele frequencies at locus $i$, and let $D_g$ denote the measure of disequilibrium defined in Chap. 3 (the difference between the joint frequency of two alleles and the product of their marginal frequencies). Show that $D_g = \pi(1 - \pi)\delta_1\delta_2(1 - \theta)^g$. (See Chap. 12 for an example of the difficulty linkage disequilibrium can cause when admixed populations are used for case-control studies.)

**10.7.** * A different model of admixture from the one discussed in Sect. 10.1.3 is to assume that in the first generation a relatively small fraction $\lambda$ of chromosomes from population 2 are mixed in with the fraction $1 - \lambda$ of population

1 to form a base population, and in each successive generation a fraction $\lambda$ of new population 2 chromosomes are mixed with the fraction $1 - \lambda$ of chromosomes from the current base population (where mating otherwise occurs at random). Let $\bar{\pi}_g = \Pr[\chi_g(s) = 0]$ and $p_g = \Pr[\chi_g(s) = 0, \chi_g(t) = 0]$, where $s$ and $t$ are two loci separated by a recombination fraction $\theta$. The probability that an arbitrary allele in the admixed population after $g$ generations originated from population 1 is $\bar{\pi}_g = (1 - \lambda)^g$.

(a) Show that $p_g = (1 - \lambda)[(1 - \theta)p_{g-1} + \theta\bar{\pi}_{g-1}^2]$. Solve this recursion together with the appropriate initial condition for $p_1$ to show that

$$\text{cor}[\chi_g(s), \chi_g(t)] = \frac{\lambda[(1 - \theta)^g - (1 - \lambda)^g]}{(\lambda - \theta)[1 - (1 - \lambda)^g]} .$$

For large $g$ and small $\theta$, this is approximately $(1-\theta)^g$, in agreement with (10.8). (b) Write a program to simulate the process $\chi_g(t)$ and check numerically the formula for the correlation found in (a).

**10.8.** In this exercise you are asked to check some of the effects of replacing the actual process of the population source by a Markov process that approximates it.

(a) Write a function that simulates gametes in a population that is admixed followed by several generations of random mating.

(b) Try running the HMM algorithm twice, once with the original function "sim.pop" that generates a Markov process and then with the function you wrote in (a) replacing that function. Compare the results of the two runs.

**10.9.** In the estimation of recombination rates, we have assumed that the transition probabilities $T_{ij}$ are stationary, i.e., they are the same for all (adjacent) pairs of markers. This would be reasonable if markers are equally spaced, which is unlikely to be the case in practice. Assuming that the genetic distance between adjacent markers is known, explain how one can use the theory of continuous time Markov chains to represent $T_{ij}$ as a function of two parameters and the distances between markers, so that the approach developed in Sect. 10.3 can still be used.

# 11

# Mapping Complex and Quantitative Traits with Data from Human Pedigrees

In this chapter we first study QTL mapping in human pedigrees and then consider for qualitative traits some more complicated issues that were not covered in Chap. 9.

The basic model goes back to Chap. 2, while the methods developed for affected sib pairs in Chap. 9 also play an important role. Putting the two sets of ideas together becomes complicated with large pedigrees, so we start our detailed discussion with the study of sib pairs.

## 11.1 Model and Covariances

We use the model (2.3):

$$y = m + \tilde{\alpha}[(x_M - p) + (x_F - p)] + \tilde{\delta}[(x_M - p)(x_F - p)] + e , \qquad (11.1)$$

where $\tilde{\alpha} = \alpha + (1 - 2p)\delta$ and $\tilde{\delta} = -2\delta$. Recall also from Chap. 2 that

$$\sigma_y^2 = \sigma_A^2 + \sigma_D^2 + \sigma_e^2 , \qquad (11.2)$$

where $\sigma_A^2 = 2pq[\alpha^2 + (1 - 2p)\delta]^2$, $\sigma_D^2 = 4p^2q^2\delta^2$, $\sigma_e^2 = \mathrm{Var}(e)$. Whereas in experimental genetics involving, e.g., an intercross, we knew that $p = 1/2$ from the experimental design, in human genetics we do not know the allele frequency $p$ and in fact have no reason to suppose that the QTL has only two alleles. Hence we tend to regard the variance components, which can also be defined for multi-allelic loci (cf. Prob. 11.5), as the important parameters of the model, not the specific formulas given for them in terms of $p$, $\alpha$, and $\delta$ in the preceding set of equations.

Suppose now we have a pair of siblings with phenotypes $y_1$ and $y_2$. Exactly the same analysis that led to (9.2) yields

$$\mathrm{cov}(y_1, y_2) = \sigma_A^2/2 + \sigma_D^2/4 + r\sigma_e^2. \qquad (11.3)$$

We have included the last term, which does not appear in (9.2), to allow for the possibility that $e_1$ and $e_2$ are correlated, either because the siblings are likely to live in similar environments, and hence have correlated environmental contributions to their phenotypes, or because there are other genes that contribute to the trait, as in Chap. 8, but which we have not included explicitly in the model. It is convenient to introduce a parameter $\rho = \mathrm{cor}(y_1, y_2) = \mathrm{cov}(y_1, y_2)/\sigma_y^2$ to denote the phenotypic correlation between sibs.

In affected sib pairs, the evidence for linkage is based on the observation that the conditional probability that both sibs are affected given the IBD count at a trait locus is different from the unconditional probability. Similarly, the evidence for linkage of a quantitative trait is that the overall phenotypic covariance and the covariance conditional on the IBD status at a QTL are different. To compute the conditional covariance, we follow the the argument that gave us equation (9.3). After some tedious algebra we obtain

$$\mathrm{cov}[y_1, y_2 | J(\tau)] = \mathrm{cov}(y_1, y_2) + \check{\alpha}[J(\tau) - 1] - \check{\delta}(I_{\{J(\tau)=1\}} - 1/2) , \qquad (11.4)$$

where $\check{\alpha} = (\sigma_A^2 + \sigma_D^2)/2$ and $\check{\delta} = \sigma_D^2/2$. We have used the labels $\check{\alpha}$ and $\check{\delta}$ for the parameters appearing in (11.4) because they play a role in the analysis to follow that is very similar to the role played by $\alpha$ and $\delta$ in (11.1) when that equation (with $p = 1/2$) is used as the model for an intercross. Note, however, that $0 \leq \check{\delta} \leq \check{\alpha}$, while $\alpha$ and $\delta$ have no such restrictions.

The important difference between (11.1) and (11.4) is that (11.1) describes how the *phenotype* changes as a function of the number of $A$ alleles at the QTL, while (11.4) describes how the *covariance* of the phenotypes of two siblings changes as a function of the number of alleles they share IBD. The covariance is a quadratic function of the two phenotypes, $y_1$ and $y_2$. Other quadratic functions, which turn out to be convenient to work with, are $D^2$ and $S^2$, where $D = (y_1 - y_2)/2^{1/2}$ and $S = (y_1 + y_2 - 2\mu)/2^{1/2}$. Note that $E(D^2) = \sigma_y^2(1 - \rho)$ and $E(S^2) = \sigma_y^2(1 + \rho)$. Calculations based on (10.4) show that

$$E[D^2 | J(\tau)] = E(D^2) - \check{\alpha}[J(\tau) - 1] + \check{\delta}(I_{\{J(\tau)=1\}} - 1/2) , \qquad (11.5)$$

$$E[S^2 | J(\tau)] = E(S^2) + \check{\alpha}[J(\tau) - 1] - \check{\delta}(I_{\{J(\tau)=1\}} - 1/2) \qquad (11.6)$$

and $\mathrm{cov}[D, S | J(\tau)] = 0$. To simplify matters, assume initially that $\check{\delta} = 0$. Then the equations (11.5) and (11.6) show that the conditional expectation of $D^2$ and $S^2$ are linear functions of $J(\tau) - 1$. Hence to test for linkage we can form linear regression statistics, much like those used in Chap. 4, but now with statistics based on the variables $D^2$ and $S^2$.

## 11.2 Statistics to Detect Linkage

### 11.2.1 Regression Statistic

It turns out that $D^2$ often contains more linkage information than $S^2$, so initially we consider only $D^2$ and (11.5) with $\breve{\delta}$ assumed to equal 0. Consider a sample of $n$ sib pairs, and let $\overline{D^2} = \sum_{i=1}^{n} D_i^2/n$ denote the average value of $D^2$ in the sample. Assume that markers are fully informative, so that the number of alleles shared IBD, $J(t)$, can be observed. Then a suitable regression statistic to test for linkage at a marker $t$ is (cf. the regression statistics of Sects. 4.1.3 and 4.3):

$$Z_t = \left[ \sum_{i=1}^{n} \left( D_i^2 - \overline{D^2} \right) [J_i(t) - 1] \right] \bigg/ \left[ \sum_{i=1}^{n} \left( D_i^2 - \overline{D^2} \right)^2 / 2 \right]^{1/2} . \qquad (11.7)$$

The factor of $1/2$ in the denominator of (11.7) comes from the fact that the variance of $J(t)$ is $1/2$. If markers are not fully informative, then $J(t)$ must be replaced by the estimated number of alleles shared IBD, say $\hat{J}(t)$, as computed in Chap. 9; and the theoretical variance of $1/2$ in the denominator must be replaced by, for example, the sample variance of $\hat{J}(t)$.

By the central limit theorem, $Z_t$ is approximately normally distributed. Since the part of $Z_t$ that depends on $t$ is only the factor $J(t) - 1$, one can show that the covariance function of (11.7) is simply the covariance function of $J(t)$ given in Chap. 9. Hence we can use the formulas from earlier chapters to determine significance thresholds.

At the trait locus $\tau$, it follows from the definition of conditional expectation and (11.5) that the numerator of (11.7) has expected value $-n\breve{\alpha}E\{[J(\tau) - 1]^2\} = -n\breve{\alpha}/2$. After using the law of large numbers to evaluate the denominator approximately for large $n$, we find that the asymptotic noncentrality of (11.7) is $-(n/2)^{1/2}\breve{\alpha}/\{E(D^4) - [E(D^2)]^2\}^{1/2}$. (Since $\breve{\alpha} \geq 0$, the minus sign indicates that our test should be one-sided, with very *small* values of $Z_t$ interpreted as evidence of linkage.) The noncentrality parameter can be simplified if the phenotypes have bivariate normal distributions, so $D$ is normally distributed. Since the fourth moment of a normal distribution is 3 times the square of the variance, from the expression given above for $E(D^2)$, we find after some algebraic simplification that $E(D^4) - [E(D^2)]^2 = 2\sigma_y^4(1 - \rho)^2$. Hence in this case the asymptotic noncentrality parameter is

$$-n^{1/2}\breve{\alpha}/[2\sigma_y^2(1 - \rho)] . \qquad (11.8)$$

It turns out that the noncentrality parameter given in (11.8) is much smaller than the corresponding noncentrality parameter for a backcross or intercross with the same underlying genetic model (cf. Prob. 11.1), so much larger sample sizes are required to have reasonable power to detect linkage.

## 11.2.2 Score Statistic

More linkage information can be obtained from the pair $(D, S)$ than from $D$ alone, and substantially more can be extracted from larger pedigrees. However, to extend the ideas of the preceding section, we must go beyond the simple univariate regression model that has been the basis for most of our analysis up to now. In this section we consider an alternative analysis of sib pairs, which requires an additional argument. Dealing with pedigrees larger than sib pairs requires similar ideas with still more technical computations.

For a sib pair we make in addition to the assumptions of Sect 11.1 the new assumption that at each marker $t$, conditional on $J(t)$, the joint distribution of $y_1$ and $y_2$ is the bivariate normal distribution with mean values $m$, variances $\sigma_y^2$, and conditional covariance given by (11.4).

A consequence of the assumption that $y_1$ and $y_2$ are conditionally bivariate normal is that each one is univariate normal (conditionally or unconditionally, since the marginal distributions do not depend on the number of alleles shared identical by descent). However, some reflection will show that because the genetic contributions to the model (11.1) come in discrete amounts, depending on which of three possible genotypes an individual has, the assumption that $y$ is normally distributed cannot be exactly correct and may not even be approximately correct. One sometimes uses the phrase "working model" to connote that an assumed model, which must be viewed with some skepticism under the most favorable conditions, is admitted from the start not to be correct. This places an extra burden on the user to make sure that the consequences of the assumption are not unreasonable. (See Prob. 11.7 for a different formulation of the basic model, which has the virtue of mathematical consistency, although it may not be any more appropriate scientifically than (11.1).)

We consider a sample of $n$ sib pairs. From the normality assumption it follows that in the $m$th sibship $D_i$ and $S_i$ are bivariate normal with variances and covariance given by (11.5)–(11.6). In particular, $D_i$ and $S_i$ are conditionally independent since they are uncorrelated. Let

$$\rho(J) = \mathrm{cor}[y_1, y_2 | J(\tau)], \quad \rho = \mathrm{cor}[y_1, y_2],$$

so dividing (11.4) by $\sigma_y^2$ leads to

$$\rho(J) = \rho + \{\check{\alpha}(J(\tau) - 1) - \check{\delta}[1_{\{J(\tau)=1\}} - 1/2]\}/\sigma_y^2. \tag{11.9}$$

To simplify the notation, it will be convenient to assume initially that $\mu = 0$ and $\sigma_y^2 = 1$. The log-likelihood at $\tau$ is $\ell = \ell(\tau, \check{\alpha}, \check{\delta}, \rho)$ given by

$$\ell = -2^{-1} \sum_{i=1}^{n} \left[ \log\left(1 - \rho^2(J_i)\right) + D_i^2/(1 - \rho(J_i)) + S_i^2/(1 + \rho(J_i)) \right]. \tag{11.10}$$

For mathematical simplicity we also assume initially that $\check{\delta} = 0$. Let $C_i = C_i(\check{\alpha}, \rho)$ be defined by

$$C_i = \rho(J_i)/(1 - \rho^2(J_i)) + S_i^2/2(1 + \rho(J_i))^2 - D_i^2/2(1 - \rho(J_i))^2.$$

Momentarily regarding $\tau$ to be known, we find the components of the efficient score, $\ell_\alpha = \partial\ell/\partial\check{\alpha}$ and $\ell_\rho = \partial\ell/\partial\rho$:

$$\ell_\alpha(\tau) = \sum_{i=1}^{n}[J_i(\tau) - 1]C_i , \tag{11.11}$$

and $\ell_\rho = \sum_{i=1}^{n} C_i$.

At a given marker $t$ the score test of the hypothesis of no linkage, i.e., $\check{\alpha} = 0$, is based on $Z_t = \ell_\alpha(t)/\{\mathrm{var}[\ell_\alpha(t)]\}^{1/2}$. In evaluating the numerator we set $\check{\alpha} = 0$ and replace the parameter $\rho$ by its maximum likelihood estimator under the null hypothesis, namely $\hat{\rho} = \max\left(\frac{1}{n}\sum_{i=1}^{n} y_{1i}y_{2i}, 0\right)$. (We make $\hat{\rho} \geq 0$ because, according to (11.3), the correlation can reasonably be assumed to be nonnegative.)

For the denominator of $Z_t$, we consider $\mathrm{var}[\ell_\alpha(t)] = \sum_{i=1}^{n} \mathrm{E}[C_i^2]\mathrm{E}[(J_i(t) - 1)]^2 = n\mathrm{E}[C_i^2]/2$. If we use the hypothesis of normality, we find that

$$\mathrm{E}[C_i^2] = (1 + \rho)^2/[(1 - \rho^2)^2].$$

In carrying out this calculation, we encounter $S_m^4$ and $D_m^4$, and we use the fact that for a normally distributed random variable with mean 0 the fourth moment is three times the square of the variance. If the normality assumption is not true, this may lead to an incorrect estimate of variability. An estimator that does not depend on specific distributional assumptions is obtained by simply using the sample moment, $\frac{1}{n}\sum_{i=1}^{n} C_i^2$, to estimate $\mathrm{E}[C^2]$. Hence we redefine $Z_t$ by

$$Z_t = \left[\sum_{i=1}^{n} \{\hat{C}_i[J_i(t) - 1]\}\right]\bigg/\left[\sum_{i=1}^{n} \hat{C}_i^2/2\right]^{1/2}, \tag{11.12}$$

where we have put hats on $C_i$ to indicate that it is evaluated with $\check{\alpha} = 0$ and $\rho$ replaced by $\hat{\rho}$. Its approximate expectation when $n$ is large and $t = \tau$ is $n^{1/2}$ times

$$\xi = \frac{\check{\alpha}}{\sigma_y^2}\frac{(1 + \rho^2)/[(1 - \rho^2)^2]}{[2\mathrm{E}_0 C_i^2]^{1/2}} .$$

When the normality assumption is approximately true, we can use the expression for $\mathrm{E}_0[C_i^2]$ given above to obtain

$$\xi = \frac{\check{\alpha}}{\sigma_y^2}\{(1 + \rho^2)/2[(1 - \rho^2)^2]\}^{1/2}. \tag{11.13}$$

Since we do not know the location of the QTL $\tau$, we scan the genome using $\max_t Z_t$, where the max is taken over all marker loci $t$. Approximations for the significance level and power can be obtained from the formulas of Chaps. 5 and 6, where, as in Chap. 9, $\beta = 0.04$ and the test is one-sided. For an idealized human genome of 23 chromosomes of an average length of 140 cM and markers spaced at 1 cM, a threshold of $z = 3.91$ is required for a genome wide significance level of 0.05. A noncentrality parameter of about 5 gives power of 0.9.

For a numerical example, suppose that the allele frequency of a QTL is $p = 1/2$, while $\alpha = 1$ and $\delta = -1$, so $\sigma_A^2 = 0.5$, $\sigma_D^2 = 0.25$. Assume also that $\sigma_e^2 = 1$. Then $\sigma_y^2 = 1.75$ and the heritability of the trait is 0.43. Finally, we assume there is no residual correlation between sibs ($r = 0$). For this example the noncentrality parameter given by (11.13) is about 0.16 so a sample size of about 990 sib pairs would be required to achieve a noncentrality of $\xi = 5$.

If we were dealing with an intercross of inbred strains, the QTL would be quite easy to map. The noncentrality parameter per unit sample size would be 0.53 for the one degree of freedom statistic to detect the additive effect and 0.65 for the two degree of freedom statistic to detect both additive and dominance effects. As usual, these should be multiplied by $n^{1/2}$ to give the noncentrality parameters for a sample of size $n$. To achieve a noncentrality of about 5 for the one degree of freedom statistic, a sample size of only about 90 would be required, and the two degree of freedom statistic would be somewhat more powerful, even after accounting for the higher significance threshold.

*Remark 11.1.* The statistic (11.12) is a regression like statistic, as was (11.7). Both involve the empirical correlation between a function of the phenotypes and the number of alleles inherited identical by descent. The specific form of the function of the phenotypes, $\hat{C}_i$, arises from the assumption of normality, while the particular function of the number of alleles inherited IBD arises from the additive nature of the model. Since the question of linkage involves correlation of phenotype with genotype, the statistics (11.7) and (11.12) both pass a minimal test of reasonableness. An important question is whether a different function of the phenotypes would in some sense result in a more powerful or more robust test statistic in problems where the assumption of normality is not a reasonable approximation.

*Remark 11.2.* For the more general case, where $\sigma_y^2$, $\mu$, and $\rho$ are all unknown, the efficient score depends on all these parameters, which can be estimated by $\hat{\mu} = \frac{1}{2n} \sum_{i=1}^{n} (y_{1i} + y_{2i})$, $\hat{\sigma}_y^2 = \frac{1}{2n} \sum_{i=1}^{n} [(y_{1i} - \hat{\mu})^2 + (y_{2i} - \hat{\mu})^2]$, and $\hat{\rho} = \max\left(\frac{1}{n} \sum_{i=1}^{n} (y_{1i} - \hat{\mu})(y_{2i} - \hat{\mu}), 0\right)/\hat{\sigma}_y^2$. The asymptotic approximations for the p-value and power are unchanged.

*Remark 11.3.* We have assumed completely informative markers in order to simplify the analysis. With partially informative markers the situation is much like the corresponding case in Chap. 9. We replace $J_i(t)$ in the numerator of (11.12) by $\hat{J}_i(t) = \mathrm{E}[J_i(t)|G_i]$, where $G_i$ denotes the marker data in the $i$th pedigree, and in the denominator we replace the factor of $1/2$, which arises as $\mathrm{var}[J(t)]$ by a suitable approximation or estimate, e.g., $\frac{1}{n} \sum_{i=1}^{n} (\hat{J}_i(t) - 1)^2$.

*Remark 11.4.* The score statistic we have developed here is closely related to the regression statistic (11.7). In fact, writing down the likelihood function for $D_1, \ldots, D_n$, and follow the reasoning given above produces that regression statistic (Prob. 11.4). The likelihood-based approach seems statistically more natural for more complex problems, e.g., generalizations to larger sibships and pedigrees, although one can obtain much the same results by considering multivariate regression models.

## 11.3 A Two Degree of Freedom Statistic

As in the case of an intercross, there is with sib pairs the possibility of using a two degrees of freedom statistic, which should be particularly useful when the dominance variance, $\sigma_D^2$, is positive. The discussion below applies with only minor changes for the 0-1 traits studied in Chap. 9 as well.

Let $Z_1$ denote the statistic defined above and let $\xi_1 = EZ_{1\tau}$. If we take the score statistic for $\check{\delta}$, $\ell_\delta = \partial\ell/\partial\check{\delta}$, and follow the pattern of analysis given above, we find a second statistic, say $Z_2$, that is of the same form as $Z_1$, but has $1/2 - I\{J = 1\}$ in the place of $J - 1$ in the numerator and the appropriate variance, $1/4$, in place of $1/2$ in the denominator. Since at unlinked loci, $1/2 - I\{J = 1\}$ and $J - 1$ are uncorrelated, the statistics $Z_1$ and $Z_2$ are also. The correlation function of $Z_2$ is $\exp(-2\beta|t - s|)$, for the same reasons as the analogous result for in intercross. Finally, one can also show that the noncentrality parameter $\xi_2 = EZ_{2\tau}$ is of the same form as $\xi_1$, except that $\check{\alpha}$ is replaced by $\check{\delta}$, and 2 in the denominator is replaced by 4.

These properties are formally very similar to those of the two degrees of freedom statistic for an intercross. There is one, very important, difference that arises from the inequality $0 \leq \check{\delta} \leq \check{\alpha}$, which implies that $0 \leq \xi_2 \leq \xi_1/2^{1/2}$. For an intercross the first degree of freedom was related to the additive variance, the second to the dominance variance, and there was no necessary relation between the two variance components. Although experience suggests that the dominance variance is usually smaller than the additive variance, this is an empirical observation that can be violated. Here the relation between $\xi_1$ and $\xi_2$ is a mathematical consequence of our basic modeling assumptions.

This relation has consequences for the choice of statistic, say $U$, which can be defined as follows. Consider the point $(Z_1, Z_2)$ in the $z_1 \times z_2$ plane and the set of points $S$ in the positive quadrant formed by the positive $z_1$-axis and the ray $z_2 = z_1/2^{1/2}$ with $z_1 \geq 0$. Project the point $(Z_1, Z_2)$ onto $S$. (This projection could be the origin, for example if both coordinates are negative.) Then $U$ is the squared length of the resulting projection. This has the geometric interpretation that we impose on $(Z_1, Z_2)$ the constraints relating $\xi_1$ and $\xi_2$. Then we take the squared distance of the constrained point from the origin. In the case of an intercross, there were no constraints, so $U$ was simply the squared distance of $(Z_1, Z_2)$ from the origin.

This definition can also be expressed algebraically as follows. If the point $(Z_1, Z_2)$ satisfies the constraint $0 \leq Z_2 \leq Z_1/2^{1/2}$, then $U = Z_1^2 + Z_2^2$. If $Z_2 < 0$, but $Z_1 > 0$, then $U = Z_1^2$. If $Z_2 > 0$, and $Z_2 \geq Z_1/2^{1/2}$, then $U$ is the square of the perpendicular projection of the point $(Z_1, Z_2)$ onto the line $z_2 = z_1/2^{1/2}$, provided that the projected point is in the first quadrant. In all other cases $U = 0$.

One effect of these constraints is that the significance threshold for the two degrees of freedom statistic is only slightly higher than for the one degree of freedom statistic. See Dupuis and Siegmund [21], who give an approximation that resembles (5.3) and show that under the conditions of the preceding section the 0.05 significance threshold, which was 3.91 for a one degree of freedom statistic, would now be 4.11.

A second effect of the constraints relating $\xi_1$ and $\xi_2$ is that in most circumstances we can gain very little by considering the two-dimensional statistic. This is quite different from the situation in experimental genetics, where there are no such constraints. Consider the numerical example of the preceding section. For an intercross the ratio of the noncentrality parameter of the two degrees of freedom statistic to that of the one degree of freedom statistic would be about 1.23, so the two degrees of freedom statistic would turn out to have more power, even though it would require a substantially larger significance threshold. For sib pairs, the ratio is only about 1.03, so after correcting for the increase in threshold the power is essentially unchanged. In more extreme cases one can gain some power by considering the two degrees of freedom statistic, but these cases do not appear to be common.

## 11.4 *Sibships and Larger Pedigrees

A particularly attractive feature of components of variance models with the working assumption of multivariate normality is the ease with which they can be generalized to larger sibships and extended pedigrees. The essential technical feature is that the multivariate normal distribution is determined by the means, variances and covariances of the individual components.

As an illustration, suppose we have a sample of $n$ sibships, each of size $s$. We index sibs within a sibship by $i$ and $j$ and sibships by $k = 1, \ldots, n$. The subscript $k$ is often suppressed in our notation. To simplify the notation we assume that $m = 0$ and $\sigma_y^2 = 1$, and consider a model for a single QTL. Denote by $Y$ the vector of length $s$ of phenotypes of a given sibship. Let $J_{ij}$ denote the number of alleles shared identical by descent at a given locus by the $i$th and $j$th sibs in the $m$th sibship. Let $A_J$ denote the $s \times s$ matrix with entries $J_{ij} - 1$ for $i \neq j$ and zeroes along the diagonal. Let $\Sigma_J = \mathrm{E}(YY'|A_J)$. Under our working assumption that conditional on $A_J$ the phenotypic vector $Y$ has a multivariate normal distribution, the log-likelihood for a single QTL at $\tau$ is $\ell = \ell(\tau, \breve{\alpha}, \breve{\delta}, \rho)$ given by

$$\ell = -\frac{1}{2} \sum_{k=1}^{n} \left\{ \log |\Sigma_J| + \mathrm{tr}\Sigma_J^{-1} YY' \right\}, \tag{11.14}$$

where $J = J(\tau)$.

For a nonsingular matrix $G$ depending on $x$, it is known that $\partial \log |G|/\partial x = \operatorname{tr}(G^{-1}\partial G/\partial x)$ and $\partial G^{-1}/\partial x = -G^{-1}\partial G/\partial x G^{-1}$. Hence by differentiation of (11.14) we obtain the score equations

$$\ell_\alpha = \frac{1}{2} \sum_{m=1}^n \left\{ -\operatorname{tr}(\Sigma_J^{-1} A_J) + \operatorname{tr}(\Sigma_J^{-1} A_J \Sigma_J^{-1} YY') \right\} \tag{11.15}$$

and

$$\ell_\rho = \frac{1}{2} \sum_{m=1}^n \left\{ -\operatorname{tr}(\Sigma_J^{-1} B) + \operatorname{tr}(\Sigma_J^{-1} B \Sigma_J^{-1} YY') \right\}, \tag{11.16}$$

where $B = \partial \Sigma_\nu / \partial \rho = 11' - I$. We omit the similar expression for $\ell_\delta$.

As in the case of sib pairs, one can use (11.15), with $\check{\alpha} = 0$ and unknown segregation parameters replaced by their maximum likelihood estimates calculated under the null hypothesis of no linkage, as a test statistic. We can, as above, obtain a false positive error rate that does not depend on the true distribution of the $Y$'s if we consider the conditional distribution of $\ell_\alpha$ given the phenotypic values. This means that for large sample sizes $\ell_\alpha$ should be standardized by $\{E[\ell_\alpha^2 | Y_1, \cdots, Y_n]\}^{1/2}$. This quantity can be computed, but is algebraically complicated and is not reproduced here.

In general the noncentrality parameter of the score statistic depends on the true distribution of the phenotypes. In the case that the working hypothesis of multivariate normality is approximately correct, the square of the noncentrality parameter for a sample of $n$ sibships of size $s$ is approximately

$$\xi^2 = n \frac{\check{\alpha}^2}{2\sigma_y^4} \binom{s}{2} \frac{\{[1 + (s-2)\rho]^2 + \rho^2\}}{\{(1-\rho)[1 + (s-1)\rho]\}^2}, \tag{11.17}$$

which according to large-sample statistical theory is the largest value possible under the assumed normal model. One can see from simple numerical examples that the squared noncentrality parameter for sibships of size $s$ is roughly $\binom{s}{2}$ times that for a sibship of size 2.

A thorough development of these results is beyond the scope of this book. Nevertheless, an important conclusion that one can draw from (11.17) is that the noncentrality parameter grows very fast with the size of the sibship. Although large sibships can be difficult to find, a similar result holds for extended pedigrees. Each pairwise combination of related individuals makes a contribution, and although these can be individually small, the number of pairs can make the total noncentrality reasonably large. In view of the very large sample size required in the numerical example given above for sib pairs, one welcomes the possibility to use larger pedigrees to reduce the overall costs of a linkage study.

*Remark 11.5.* A technical point that sometimes complicates the case of larger pedigrees is that the IBD of various relative pairs within pedigrees are not independent. A consequence is that the score statistic can have a skewed distribution, and the Gaussian approximations that we have used throughout the book may no longer be adequate unless the number of pedigrees is quite large. This would mean that the significance threshold must be larger to control the false positive error rate, and hence the gain in power would be somewhat less than one would think simply by comparing noncentrality parameters. See [82] for a systematic discussion including a modified approximation to account for skewness and Prob. 11.11 for an example.

## 11.5 *Multiple Gene Models and Gene-Gene Interaction

In this section we consider models involving multiple genes, which extends to human genetics our brief discussion in Chap. 8. For this analysis, we return to the case of qualitative traits. Similar results apply to mapping multiple QTL, although a similar analysis of that case involves different details.

Suppose the penetrance involves multiple genes, for example, satisfying (8.3) (without the residual $e$) for two unlinked loci, with dominance, and an additive-additive interaction between the two genes. Direct generalization of the calculations leading to (9.2) now yields

$$Q_2 = \mathrm{E}(y_1 y_2) = \sigma_{A_1}^2/2 + \sigma_{A_2}^2/2 + \sigma_{D_1}^2/4 + \sigma_{D_2}^2/4 + \sigma_{A_1A_2}^2/4 .$$

A slightly more complex calculation shows that at, say, the first trait locus (9.4) continues to hold, but with $\check{\alpha}_1 = (\sigma_{A_1}^2 + \sigma_{D_1}^2)/2 + \sigma_{A_1A_2}^2/4$. More generally, one can obtain for the two trait loci the joint probability

$$\Pr(J(\tau_1) = i, J(\tau_2) = j|A) = 2^{-(2+|i-1|+|j-1|)}\left\{1 + \frac{1}{Q_2}[(i-1)\check{\alpha}_1 \quad (11.18)\right.$$

$$\left. +(1/2 - I_{\{i=1\}})\check{\delta}_1 + (j-1)\check{\alpha}_2 + (1/2 - I_{\{j=1\}})\check{\delta}_2 + (i-1)(j-1)\check{\gamma}]\right\} ,$$

where $\check{\gamma} = \sigma_{A_1A_2}^2/4$.

For a numerical example, which we consider again in the following section, suppose there are two unlinked genes that have only additive effects and do not interact. Let $g_0 = 0$, $\alpha_1 = 0.3$, $p_1 = 0.05$, $\alpha_2 = 0.15$, $p_2 = 0.1$. The incidence of the trait is $m = 0.06$, and each of the genes is responsible for half of the overall incidence. However, the penetrance at one of the loci is twice that at the other, with the result that the noncentrality parameters per sib pair are quite different from each other: $\xi_1 = 0.31$, $\xi_2 = 0.14$. Since the noncentrality parameters for $n$ affected sib pairs equal these quantities multiplied by $n^{1/2}$, the sample would have to be about four times as large to detect the second locus with specified power in a genome scan as to detect the first locus (with the same power).

Methods suggested in Chap. 8 for dealing with multiple genes can be adapted to the present context. Simultaneous consideration of multiple loci can be implemented exactly as in Chap. 8. Sequential detection is, however, somewhat different, although the properties that make this approach likely to succeed are much the same as in Chap. 8.

Consider the two locus model suggested above, with no interaction, and suppose we have detected one gene at a locus $\tau_1$, where a large value of $Z_t$ allows us to estimate $\tau_1$ very precisely. At that locus many sib pairs will share an excess of alleles IBD over the number expected, and a smaller number of pairs will share a deficit. Intuitively, those that share a deficit are presumably not affected by virtue of their genotypes at the first (detected) locus, and thus are affected by virtue of their genotypes at the second locus. Hence among those sharing a deficit of alleles at the detected locus, we expect to find an excess of alleles shared IBD at the second locus.

We now compute the conditional probability of IBD at $\tau_2$ given the IBD count at $\tau_1$. For simplicity we assume the loci are unlinked and there is no gene-gene interaction. Let $\alpha_i^* = \check{\alpha}_i/Q_2$ and $\delta_i^* = \check{\delta}_i/Q_2$. Then from the joint distribution given in (11.18) we obtain:

$$\Pr(J(\tau_2) = j | J(\tau_1) = i, \ A) = \frac{1}{2^{1+|j-1|}} \left[ 1 + \frac{(j-1)\alpha_2^* + (1/2 - I_{\{j=1\}})\delta_2^*}{1 + (i-1)\alpha_1^* + (1/2 - I_{\{i=1\}})\delta_1^*} \right].$$
(11.19)

Observe that this is of the same form as the unconditional probability, but with $(j-1)\alpha_2^* + (1/2 - I_{\{j=1\}})\delta_2^*$ now divided by $1 + (i-1)\alpha_1^* + (1/2 - I_{\{i=1\}})\delta_1^*$ to reflect the conditioning on the IBD count at the first trait locus.

Let $N_{ij,t}$ denote the number of affected sib pairs having $i$ alleles IBD at the trait locus $\tau_1$ and $j$ alleles IBD at a putative second trait locus $t$, which is unlinked to $\tau_1$. Let $N_{i\cdot} = \sum_j N_{ij,t}$ denote the number of pairs with $i$ alleles IBD at $\tau_1$. Conditional on $N_{i\cdot}$, consider the statistic

$$(N_{i2,t} - N_{i0,t})/(N_{i\cdot}/2)^{1/2} \ ,$$
(11.20)

which is the conditional version of (9.5) for testing that there is an excess of allele sharing at $t$ among the $N_i$ sib pairs having $i$ alleles IBD at $\tau_1$ Pursuing this analogy, we see from (9.6) and the form of (11.19) that the (conditional) noncentrality parameter at $t = \tau_2$ is:

$$(N_{i\cdot}/2)^{1/2} \alpha_2^* / [1 + (i-1)\alpha_1^* + (1/2 - I_{\{i=1\}})\delta_1^*] \ .$$

For large $n$, $N_{i\cdot}/n \approx \pi_{i\cdot} = 2^{-(1+|i-1|)}[1 + (i-1)\alpha_1^* + (1/2 - I_{\{i=1\}})\delta_1^*]$ and hence the unconditional asymptotic noncentrality parameter obtained by considering the subset of pairs sharing $i$ alleles IBD at $\tau_1$ is

$$(n/2)^{1/2} \alpha_2^* / [2^{(1+|i-1|)}\pi_{i\cdot}]^{1/2} \ .$$
(11.21)

As suggested above for $i = 0$, this can be larger than the noncentrality $(n/2)^{1/2}\alpha_2^*$ of the simple scan statistic at $\tau_2$ if the denominator is sufficiently small.

It is possible to form an optimal combination of the three statistics in (11.20), which turns out to equal

$$\frac{2^{1/2} \sum_{i=0}^{2} [N_{i2,t} - N_{i0,t}]/(2^{|i-1|} N_{i\cdot})}{\left[ \sum_{i=0}^{2} 1/(2^{2|i-1|} N_{i\cdot}) \right]^{1/2}} . \tag{11.22}$$

This statistic has an asymptotic noncentrality parameter that equals the product of the noncentrality parameter of the simple scan statistic and

$$\left\{ \sum_{i=0}^{2} 2^{-2(1+|i-1|)}/\pi_{i\cdot} \right\}^{1/2} . \tag{11.23}$$

One derivation of this result uses arguments similar to those presented below in Sect. 11.6, but the details are complicated and have been omitted.

Like the case of QTL mapping discussed in Chap 8, the impact of conditioning on the IBD status at the detected locus is most helpful when there is a very strong signal at that locus. For the example already discussed in this section, which is additive at both loci, $\alpha_1^* \approx 0.44$ and the factor given in (11.23), by which the unconditional noncentrality parameter is increased, is only about 1.06.

A very interesting example from the literature involves Type I diabetes, where there is a very large contribution from somewhere in the HLA complex of genes on chromosome 6, and there appear to be smaller contributions at several loci elsewhere in the genome. See Cordell et al. [13], whose Table 2 gives estimated joint probabilities of IBD for various two locus models, one of which is always the HLA locus (referred to as IDDM1 for Insulin Dependent Diabetes Mellitus 1). From any column of the table one can infer estimates of the marginal IBD probabilities at IDDM1 and hence estimate the factor (11.23). The result is about 1.23. This is large enough to have some impact on the attempt to detect other genes, as one can verify by examining Table 1 of [13]. (Note that the statistics given in Table 1 are in the LOD scale. To convert to the Z-scale, which we have used throughout this book, one uses the relation $|Z| = (4.6\,\text{LOD})^{1/2}$.) Since the Z-score at IDDM1 is a very large 12.6, it is disappointing that the factor (11.23) is still only slightly larger than one. A careful reading of Cordell et al. [13] suggests other related problems and shows that our discussion has just scratched the surface.

## 11.6 Using Pedigrees to Map Disease Genes

We saw earlier in this chapter that it is relatively straightforward to extend QTL mapping in humans from sib pairs to larger sibships and to pedigrees if we use the variance components approach combined with the working hypothesis of multivariate normality. Moreover, the noncentrality parameter grows rapidly with the size of the pedigree, which suggests that large pedigrees can be very valuable for QTL mapping. For mapping dichotomous traits, the corresponding jump from sib pairs to larger sibships and pedigrees presents some new difficulties. Although this subject fits quite properly into Chap. 9, we have delayed its discussion until now, in order to prepare better to deal with some comparatively technical calculations.

The relative simplicity of the QTL case arises from the fact that the multivariate normal distribution involves only pairwise correlations, while for dichotomous traits a theoretically correct analysis involves higher-order correlations. The problem becomes even more complicated if we consider pedigrees containing both affecteds and unaffecteds and samples consisting of different kinds of pedigrees and numbers of affecteds within those pedigrees. We begin with some generalities and then turn to some simple specific examples.

An important feature of dichotomous traits is that unless the penetrance is large, most of the linkage information comes from affected individuals. For example, it follows from Prob. 9.10 that for affected/unaffected sib pairs to have as much power as affected pairs one must have $m \leq 2Q_2$, where $m$ is the population frequency of the trait. Since (for additive penetrances) $m$ increases linearly with the additive effect (which is less than one), while $Q_2$ increases with the square of that effect, it will usually be the case that $m$ is substantially larger than $Q_2$ and hence affected/unaffected pairs will contribute relatively little power. This makes intuitive sense as well. Linkage information from an affected/unaffected pair comes from *decreased* sharing of alleles IBD near a trait locus. But unless the penetrance is large, the unaffected member of an affected/unaffected sib pair might well have inherited the same allele IBD as the affected member, yet still be unaffected. A simple numerical example is provided by the two locus model of the preceding section, where $m = 0.06$, while $Q_2 = 0.0099$. It follows that even at the locus having the larger penetrance the noncentrality parameter for affected/unaffected sib pairs is roughly $1/5$ as large as for affected sib pairs. Hence (since the noncentrality parameter is proportional to the square root of the sample size) about 25 affected/unaffected pairs would be required to have the same noncentrality as 1 affected sib pair.

If the penetrance is not large, as one expects will be the case for complex diseases, and if most of the linkage information comes from affected individuals, it was appropriate to focus on affected sib pairs in Chap. 9 because they will be the most easily identified pedigrees. This contrasts with the situation described at the beginning of our discussion in Sect. 9.5 of parametric methods, historically aimed at single gene traits of large penetrance, where one could find large pedigrees containing multiple affecteds and could use information from both affected and nonaffecteds.

Assume that we have different classes of pedigrees (e.g., affected sib pairs, affected sib trios, affected sib pairs together with an affected grandparent, affected sib pairs together with an affected first cousin, etc.). Also assume that for the $i$th class we have a statistic $Z_i$ that is approximately standard normal under the null hypothesis of no linkage and has noncentrality parameter at a trait locus of $\xi_i > 0$ under the alternative. (We discuss below how we obtain the statistics $Z_i$, but for the moment we take them as given.) For a parametric analysis based on an approximate Gaussian model as described in Sect. 9.5, we would regard the $\xi_i$ as known and use the log-likelihood ratio $\sum_{i=1}^{n}[\xi_i Z_i - \xi_i^2/2]$. Since the $\xi_i$ are unlikely to be known, we consider a general weighted average: $[\sum_{i=1}^{n} w_i Z_i]/(\sum_{i=1}^{n} w_i^2)^{1/2}$, which has been standardized to have unit variance. Its noncentrality parameter at the trait locus is $[\sum_{i=1}^{n} w_i \xi_i]/(\sum_{i=1}^{n} w_i^2)^{1/2}$, which is maximized with respect to the $w$'s by choosing $w_i$ to be proportional to $\xi_i$. This would give the maximal noncentrality of $[\sum_{i=1}^{n} \xi_i^2]^{1/2}$, but like the parametric analysis it would require knowledge that we do not have, in this case of the ratios $\xi_i/\xi_1$.

Still another possibility would be to form a chi-square like statistic, say, $[\sum_{i=1}^{n} \max(Z_i, 0)^2]^{1/2}$, where we have taken the square root to put this statistic on the same scale as those suggested above. This statistic would have the same noncentrality parameter as the likelihood ratio statistic, even though it does not require knowledge of the $\xi_i$. Like the two degrees of freedom statistics, which were introduced to study both additive and dominance effects in an intercross, it would require a larger threshold to control the false positive error rate. The adverse effect on the power of the higher threshold, especially for large degrees of freedom, makes this possibility not especially appealing.

It is possible that the required ratio can be estimated in some cases. Consider, for example, a mixture of $n_1$ affected/unaffected sib pairs, which were studied in Prob. 9.11, and $n_2$ affected sib pairs. Under the assumption of no dominance variance, the noncentrality parameter for the affected pairs is $\xi_2 = (n_2/2)^{1/2}\breve{\alpha}/Q_2$, while that for affected/unaffected pairs is $\xi_1 = (n_1/2)^{1/2}\breve{\alpha}/(m - Q_2)$. The ratio involves $m$, the population frequency of the trait, and $Q_2$, or equivalently $Q_2/m$, the conditional probability that the sibling of an affected is also affected. In principle these parameters might be known, at least approximately, from population data; but as indicated above, the affected/unaffected pairs are not likely to provide a substantial amount of information.

To obtain some insight into the issues involved in a common and still fairly simple case, we consider in detail the case of affected sibling trios. The analysis follows the pattern laid out in Chap. 9, with some added complications. The reader who is more interested in the consequences than in the detailed derivation can skip to (11.25) and the subsequent discussion.

We use the model (11.1), where as in Chap. 9 we regard $y$ as the penetrance of a 0-1 trait, and we assume for simplicity that $\delta = 0$. We leave the residual $e$ in the model as a reminder that there may be other genes involved or environmental contributions to the penetrance. Let $y_i$ denote the penetrance for the $i$th sibling (i = 1,2,3). Multiplying out the expressions given in (11.1) and taking expectations, one can show by some tedious algebra that the probability that three siblings are affected is given by

$$\Pr(A_1 A_2 A_3) = \mathrm{E}[y_1 y_2 y_3] = m^3 + 3m(\sigma_\mathrm{A}^2/2) + m_3/2 + \mathrm{E}(e_1 e_2 e_3) \,, \quad (11.24)$$

where we have introduced the notation $m_3 = \alpha^3 E[(x_M - p)^3]$, which comes from two terms (one for mother and one for father) of the form $\alpha^3 E[(x_{M1} - p)(x_{M2} - p)(x_{M3} - p)]$. The expectation is equal to zero by independence unless all three alleles are IBD, which happens with probability $1/4$. The second term on the right-hand side comes from the three terms of the form $m\alpha^2 E[(x_{Mi} - p)(x_{Mj} - p)] = m\sigma_A^2/4$, for $i < j$, and the corresponding terms for the paternally inherited allele.

For a given trio at the locus $t$, let $J(t)$ denote the total number of alleles inherited IBD by the three siblings. By considering $J(t)$ as the sum of the three terms consisting of the number of alleles inherited IBD by each of the three pairs of siblings, one sees that at unlinked loci $E[J(t)] = 3$. Some consideration of the maternally inherited alleles and the paternally inherited alleles shows that for any two pairs of siblings, say the pair consisting of the first and second sib and the pair consisting of the first and third sib, the numbers of alleles inherited IBD are independent random variables (although the number of alleles inherited IBD by all three pairs are not independent). It follows that $\text{var}[J(t)] = 3/2$.

Hence for a sample of $n$ affected sib trios one can use $Z_t = \sum_{i=1}^{n}[J_i(t) - 3]/(3n/2)^{1/2}$ to test for linkage of the marker $t$. By the central limit theorem this statistic is asymptotically standard normal at unlinked loci. To prepare for a genome scan, one can also show that the covariance function is exactly the same as for sib pairs, but of more concern here is the noncentrality parameter, $E(Z_\tau)$.

By conditioning on the value of $J(\tau)$ and reasoning along the lines that produced (9.3) and (11.24), one can show that

$$E[y_1 y_2 y_3 | J(\tau)] = m^3 + J(\tau)m\sigma_A^2/2 + [J(\tau) - 2]m_3/2 + E(e_1 e_2 e_3) .$$

Comparing this with (11.24), we see that

$$E[y_1 y_2 y_3 | J(\tau)] = E[y_1 y_2 y_3] + [J(\tau) - 3]\check{\alpha}_3 ,$$

where $\check{\alpha}_3 = (m\sigma_A^2 + m_3)/2$. Put $Q_3 = \Pr(A_1 A_2 A_3) = E[y_1 y_2 y_3]$. From Bayes theorem and some algebra we obtain

$$E[J(\tau)|A_1 A_1 A_3] = E[(J(\tau) - 3)^2](\check{\alpha}_3/Q_3),$$

from which some additional algebra shows that for a sample of $n_3$ affected sib trios the noncentrality parameter equals

$$\xi_3 = E(Z_\tau) = (3n_3/2)^{1/2}(\check{\alpha}_3/Q_3). \tag{11.25}$$

We now compare (11.25) with the noncentrality parameter for sib pairs given in (9.6) and discuss the implications for practice. Before doing so, we begin with several qualitative remarks.

The primary statistic discussed in Chap. 9 for affected sib pairs is the score statistic for an additive model (cf. Prob. 9.8), and it may be shown that the same is true of the statistic suggested above for affected sib trios when the mode of inheritance of the trait is additive. However, it is not obvious how to combine a sample consisting of some affected sib pairs and some affected sib trios in a likelihood analysis when the parameters are unknown. To make a more systematic comparison, it is helpful to re-write the noncentrality parameter for sib pairs given in (9.6) in the form

$$\xi_2 = (n_2/2)^{1/2}\breve{\alpha}_2/Q_2 \tag{11.26}$$

where $\breve{\alpha}_2 = \sigma_A^2/2$, $Q_2$ is the probability that both sibs are affected, and $n_2$ is the number of affected sib pairs.

Observe that *if* the factors $\breve{\alpha}_3/Q_3$ and $\breve{\alpha}_2/Q_2$ were equal, the noncentrality parameter per affected sib trio would be about $3^{1/2}$ times as large as that for affected sib pairs. This is a reflection of the fact that there are three pairwise comparisons within the sib trio (cf. the factor $\binom{s}{2} = 3$ for $s = 3$ appearing in (11.17)). In this hypothetical situation the ratio of the noncentrality parameters would be known, so the principles described above would tell us how to combine affected sib pairs and trios; and one affected sib trio would be roughly as valuable as three affected sib pairs.

For a numerical example, we use parameter values from an example in Chap. 9: $\alpha = 0.15$, $p = 0.2$, and $g_0 = 0$. Then $\xi_2 = 0.35$ and $\xi_3 = 0.45$, so the noncentrality parameter for sib trios is larger than for pairs, but by only the factor 1.29, definitely smaller than $3^{1/2} \approx 1.73$.

One possibility for weighting affected sib pairs and trios is to choose reasonable values for the unknown parameters and use the optimal weights for these hypothetical values. This is, in effect, where a parametric analysis would lead. It turns out that the resulting noncentrality parameter is fairly robust to the exact values of the weights, so this strategy frequently works reasonably well. Suppose, as is often the case, that it is substantially easier to identify affected sib pairs than trios with the result that $n_2 \gg n_3$ (and sibships with more affecteds are even rarer). Let $n_2 = 120$ and $n_3 = 30$ (which are approximately the values obtained in the study of Type II diabetes by Cox *et al.* [14]). The optimal weights would give a noncentrality of 4.55. Weights in the ratio $n_2^{1/2} : (3n_3)^{1/2}$ would give 4.51, while weights in the ratio $n_2^{1/2} : n_3^{1/2}$, which would be optimal if pairs are as valuable as trios, would give 4.53. The differences are insignificant.

As a second example consider the two locus model introduced in the preceding section. There for a single sib pair, the first locus had noncentrality 0.31, while the second had noncentrality 0.14. For sib trios these become 0.50 and 0.15. At the first locus the ratio of the noncentrality for trios to that for pairs is roughly $3^{1/2}$, but at the second locus affected sib trios and sib pairs provide essentially the same linkage information. (Intuitively, the reason is that pedigrees with larger numbers of affecteds are relatively more likely to be segregating a high penetrance susceptibility allele than a low penetrance allele.) This means that weights that are optimal at one locus cannot be optimal at the other. Nevertheless, at both loci the same choices of weights used above would again result in noncentrality parameters that differ only slightly. (See Prob. 11.8.)

For a rarer trait the noncentrality parameters are larger, and the differences are more pronounced. Let $\alpha_1 = 0.3$, $\alpha_2 = 0.15$ as before, but now let $g_0 = 0.006$, $p_1 = 0.01$, $p_2 = 0.02$. Then $m = 0.018$ and phenocopies contribute $1/3$ of the trait incidence. Noncentrality parameters per affected sib pair are 0.38 and 0.19 at the two loci, while they are 0.75 and 0.20 per affected sib trio. Detection of the first locus is substantially easier than before, while noncentrality parameters at the second locus are hardly changed. In this case weights in the ratio $n_2^{1/2} : n_3^{1/2}$ lead to an overall noncentrality parameter of 5.56, while weights in the ratio $n_2^{1/2} : (3n_3)^{1/2}$ lead to a noncentrality parameter of 5.83 (and optimal weights give 5.85). The effect of unequal weighting is greater than in the preceding case, but still increases efficiency only slightly.

A similar argument can be applied to sibships with larger numbers of affecteds, which would usually be rarer still, and hence unlikely to make a substantial difference in the noncentrality parameter whether they are optimally weighted or not. It can also be applied to other combinations of relatives. Thus, while there are no clear rules that apply in all situations, one can frequently through educated guess work obtain reasonable guidelines for combining pedigrees containing different numbers of affecteds.

There remains the question that was deferred at the beginning of this section: how to choose the statistic $Z_i$ to represent the $i$th class of pedigree. Like the problem of combining different pedigrees, this is another problem that has a definite solution for a parametric model if one assumes that penetrances and allele frequencies are known and no ideal solution otherwise.

The problem is that when there are more than two affecteds, the probabilities of IBD generally depend on more than a single parameter, even if we restrict consideration to additive models. (Affected sibling trios, discussed above, are an exception to this general rule.)

The lack of a satisfactory theoretical solution has led to a number of *ad hoc* suggestions. The simplest of these is just to sum the IBD scores for each pair of affecteds in the pedigree. Alternatives that try to capitalize on the idea that with several affecteds in a pedigree one should give particular consideration to alleles that most or all affecteds share IBD have been proposed by Whittemore and Halpern [89] and by McPeek [55]. These latter statistics seem to be most successful in detecting loci where the penetrance is relatively large. In these cases the occurrence of multiple affecteds in a pedigree is an indication that all have their phenotype for the same genetic reason. When penetrances are relatively small, multiple cases within a pedigree may not be indicative of a common genetic cause.

A simple illustrative example is provided by an affected sib pair with one affected parent, say the father. Whereas the sib pair statistic introduced in Chap. 9 does not distinguish between IBD with respect to the paternally inherited allele and the maternally inherited allele, it seems intuitively clear in this case that more weight should be attached to the IBD score of the paternally inherited alleles. Only a specific model will reveal the ideal weights. If we assume that the father is the source of the allele that contributes to the siblings' disease, an intuitively appealing statistic is to regard the siblings as half siblings related through their father. We score the siblings 1 or 0 at a marker locus $t$ according as they are or are not IBD on their paternally inherited chromosome, and ignore their relationship on their maternally inherited chromosome.

For a numerical example of the effect of this suggestion, we return to the two locus model considered above. For the first case, the noncentrality parameter per pedigree at the relatively high penetrance locus increases from 0.31 to 0.38, while at the low penetrance locus it decreases from 0.14 to 0.10. For the second case, where $p_1 = 0.01$, $p_2 = 0.02$, $g_0 = 0.006$, and the noncentralities per sib pair were 0.38 and 0.19, for the modified statistic they become 0.60 and 0.15. Again there is an increase at the high penetrance locus and a decrease at the low penetrance locus. The increase at the first locus is now large enough that it would lead to noticeably greater power. See also Prob. 11.10.

## 11.7 Discussion

We have clearly just scratched the surface of the subjects discussed in this chapter. There are many additional issues that must be studied. Some of them are relatively simple, while others can be quite complicated.

Although the analysis given here has many points in common with that in Chap. 9, there are important differences. A virtue of the multivariate normal distribution as a working model for quantitative traits is that the likelihood function is quite tractable for multi-generation pedigrees. For dichotomous traits similar calculations for large sibships or pedigrees involving more than two affecteds quickly become very complicated, and the noncentrality parameters cannot be given in terms of variance components alone.

For dichotomous traits, most of the information comes from affected individuals, so one often ignores other family members unless they are useful for reconstructing IBD information. For quantitative trait mapping, there is also evidence that one can select pedigrees where some individuals have particular phenotypes, say extremely high or extremely low, to achieve greater power (cf. Chap. 8). But this produces another problem, which is how to estimate the segregation parameters $\mu$, $\sigma_y^2$, and $\rho$, when the natural maximum likelihood estimates would be biased. The same *ascertainment* problem exists if the quantitative trait is related to a disease and one samples pedigrees according to the affected status of some members. In addition, it appears that the gain in efficiency achieved by selecting particular pedigrees decreases with the size of the pedigree. The value of recruiting particularly informative pedigrees through selective sampling and correcting for non-random ascertainment compared to recruiting large randomly sampled pedigrees is not a clearly resolved (or perhaps even resolvable) issue.

The models of this chapter allow one to replace the mean phenotype $\mu$ with a regression model involving covariates, e.g., environmental variables, that may affect the value of the phenotype. One can also adapt the models to study gene-covariate interactions, but this is a much more difficult undertaking, particularly if there is also non-random ascertainment. See, for example, Peng, Tang, and Siegmund [60].

Standardizing the score statistic by a nonparametric variance estimate, as suggested above, makes the statistic robust in the sense that its asymptotic distribution at unlinked loci will be approximately normal even if the working hypothesis of multivariate normality is violated. This contrasts with the likelihood ratio statistic, which can have an inflated false positive error rate. However, the power of the test, i.e., its performance at linked loci, is not robust, and there are various suggestions for transforming the phenotypes or otherwise trying to achieve greater power when the normality assumption is violated. See the references cited below.

## 11.8 Bibliographical Comments

A very influential pioneering paper, which proposed the regression method based on $D^2$ that is described in Sect. 11.2.1, is Haseman and Elston [37]. Variance components methods were suggested by Hopper and Mathews [39]. They have been systematically developed, with emphasis on the use of the likelihood ratio statistic, in a series of papers by Blangero and colleagues, who have also developed the software SOLAR. See, for example, Blangero and Almasy [7], Almasy and Blangero [4]. A disadvantage of the likelihood ratio statistic is its sensitivity to the working hypothesis of multivariate normality. The exposition given in the text, emphasizing the robust score statistic, is based on Tang and Siegmund [82].

The area of QTL analysis in human pedigrees is one of considerable recent research. Sham et al. [71] present a regression based method for pedigrees. Feingold and colleagues (e.g., Forrest and Feingold [31], Skatkiewicz et al. [78]) have developed methods for using selected sib pairs efficiently. Peng, Tang, and Siegmund [60] have studied methods for dealing with gene-environment interaction.

Thorny problems that have spawned substantial research with no clear resolution involve (i) strategies to mitigate the effect of violation of the working assumption that phenotypes are approximately multivariate normally distributed and (ii) suitable adjustments when some members of pedigrees are ascertained on the basis of their phenotypes. Chen, Broman, and Liang [10] suggest the use of generalized estimating equations to develop statistics that are robust against some failures of the working assumption of multivariate normality.

Other approaches to this problem are suggested by Wang and Huang [88] and by Diao and Lin [20]. These procedures involve transformations of the marginal distributions of phenotypes to make them closer to normal distributions, with no clear understanding of what this does to the joint distributions. Issues of non-random ascertainment have been discussed by Andrade and Amos [5] and Peng and Siegmund [59] among others.

Discussions of the choice of statistics for qualitative traits when there are more than two affecteds in a pedigree are given by Whittemore and Halpern [89], McPeek [55], and Teng and Siegmund [85].

The implications of the non-negativity of variance components on two-degrees of freedom statistics to detect both additive and dominance effects was first discussed by Holmans [38] in the context of testing a single marker. Dupuis and Siegmund [22] discuss the issue for genome scans.

Discussions of multi-gene models, with applications to Type I diabetes, are found in Cordell *et al.* [12, 13]. Tang and Siegmund [83] give a more theoretical discussion of multi-gene models, which points out the limitations for human genetics imposed by the non-negativity of variance components. A frequently cited example of successful modeling of gene-gene interaction in linkage detection is Cox *et al.* [14]. Because of the limitations observed in [83], Tang and Siegmund express skepticism that results like those of Cox *et al.* [14] will be found to hold more broadly.

## Problems

**11.1.** Consider model (11.1) with $\alpha = 1$, $\delta = 0$, $p = 1/2$, and $\sigma_e^2 = 1$. What is the value of the heritability, $h^2 = \sigma_A^2/\sigma_y^2$? What would the noncentrality parameter be for an intercross and a sample of size $n = 200$? Suppose that the residual correlation $r$ given in (11.3) is $1/4$. What is the value of $\rho$? Now large would $n$ have to be for the noncentrality parameter for sib pairs given in (11.8) to be as large as for the intercross with $n = 200$.

**11.2.** Evaluate the noncentrality parameter (11.13) under the conditions or Prob. 11.1. How large would $n$ have to be for the score statistic (11.12) to have the same power as for the intercross?

**11.3.** Show that the noncentrality parameter (11.13) always exceeds (11.8).

**11.4.** Under the normality assumption of Sect. 11.2.2, what is the conditional log-likelihood of $D$ given $J(\tau)$? Show that if we start from this log-likelihood, the argument leading from (11.10) to (11.12) will lead to the statistic (11.7).

**11.5.** Suppose the phenotype is given by $y = G_{M,F} + e$, where $G_{M,F}$ designates the genetic contribution of the alleles contributed by the mother (M) and father (F) at a particular trait locus. Suppose the frequency of allele $i$ in the population is $p_i$ and the population is in Hardy-Weinberg equilibrium, so $M$ and $F$ are independent with $\Pr(M = i) = \Pr(F = i) = p_i$. Denote averaging with respect to the allele frequencies by a dot as subscript, e.g., $G_{\mathrm{M},\cdot} = \sum_i G_{\mathrm{M},i} p_i$, $G_{\cdot,\cdot} = \sum_{i,j} G_{i,j} p_i p_j$, etc. Then $G_{\mathrm{M},\mathrm{F}} = m + a_{\mathrm{M}} + a_{\mathrm{F}} + d_{\mathrm{M},\mathrm{F}} + e$, where $m = G_{\cdot,\cdot}$, $a_{\mathrm{M}} = G_{\mathrm{M},\cdot} - G_{\cdot,\cdot}$, and $d_{\mathrm{M},\mathrm{F}} = G_{\mathrm{M},\mathrm{F}} - G_{\mathrm{M},\cdot} - G_{\cdot,\mathrm{F}} + G_{\cdot,\cdot}$. Show that this generalizes the representation (11.1) to multiple alleles, in the sense that $a_{\mathrm{M}}$, $a_{\mathrm{F}}$, and $d_{\mathrm{M},\mathrm{F}}$ all have mean 0 and are uncorrelated. Show that starting from the representation (2.2) and following this prescription leads to the representation (2.3).

**11.6.** Suppose that dominance is allowed into the model (8.3), both within genes and between genes. This means, for example, that there is a term in the regression equation called the additive-dominance interaction, which involves $(x_{\mathrm{M1}} + x_{\mathrm{F1}} - 2p_1)(x_{\mathrm{M2}} - p_2)(x_{\mathrm{F2}} - p_2)$. There is also a dominance-additive interaction and a dominance-dominance interaction. Find an expression for $\mathrm{cov}(y_1, y_2 \mid J(\tau_1), J(\tau_2))$ as a displacement from the unconditional covariance, $\mathrm{cov}(y_1, y_2)$. What is the appropriate value of $\check{a}$ in this case?

**11.7.** Suppose $y = m + a_{\mathrm{M}} + a_{\mathrm{F}} + \delta_{\mathrm{M},\mathrm{F}} + e$, where $a_{\mathrm{M}}$, $a_{\mathrm{F}}$, $\delta_{\mathrm{M},\mathrm{F}}$, and $e$ are independent and normally distributed with mean 0. Assume also for two siblings that $\Pr(a_{\mathrm{M1}} = a_{\mathrm{M2}}) = 1/2$, and with probability $1/2$, $a_{\mathrm{M1}}$ and $a_{\mathrm{M2}}$ are independent; similarly $\Pr(a_{\mathrm{F1}} = a_{\mathrm{F2}}) = 1/2$, while $\Pr(\delta_{\mathrm{M1},\mathrm{F1}} = \delta_{\mathrm{M2},\mathrm{F2}}) = 1/4$. Interpret these assumptions in terms of the concepts of identity by descent and Hardy-Weinberg equilibrium. Show (perhaps with additional similar assumptions) that this model satisfies the "working" assumptions of the model of this chapter.

**11.8.** Generalize the analysis of affected sib trios to allow for two unlinked genes with no gene-gene interaction. For a numerical example suppose that $\alpha_1 = 0.3$, $p_1 = 0.05$, $\alpha_2 = 0.15$, $p_2 = 0.10$, and $g_0 = 0$. Note that the two genes contribute equally to the incidence of the trait. Verify the numerical values for the noncentrality parameters and other statements about combining affected sib pairs and trios made in the text.

**11.9.** Extend the model of the previous problem to allow for additive-additive interactions. (The calculation for affected sib trios is substantially more complicated than for affected sib pairs.)

**11.10.** The results reported in the text for pedigrees consisting of two affected sibs and exactly one affected parent require a more complicated analysis than other cases considered in the text because four relatives are involved (the siblings and the affected and the unaffected parents). To avoid these technical problems, consider the case of affected sibs and an affected father, with the status of the mother left unspecified. Verify that for the case of $\alpha_1 = 0.3$, $\alpha_2 = 0.15$, $p_1 = 0.01$, $p_2 = 0.02$, $g_0 = 0.06$, the noncentrality parameters per pedigree for the statistic that treats the siblings as half siblings related through their father are 0.53 and 0.13. A comparison with the numerical results given in the text indicates that some information would be lost if the mother is unaffected, but this is unknown. What would you expect to find if both mother and father are affected?

**11.11.** For sib trios and a sample of size 1, find the exact distribution at an unlinked marker locus of the statistic $(J - 3)$ discussed in the text. (Note that if we consider $J$ to be made up of the sum of three sib pair statistics, each pairwise comparison has a completely symmetric distribution, but because they are not independent the distribution of their sum is not symmetric.) What is the skewness (third moment) of the standardized statistic $(J - 3)/(3/2)^{1/2}$? Simulate the distribution of the standardized statistic for samples of sizes $n = 30, 50, 100$. What are the 0.999 and 0.9999 quantiles of the distributions? How do these compare with the corresponding quantiles of the standard normal distribution? Repeat the entire exercise for sibships containing four siblings. Repeat the simulation for the situation described in the text, involving a mixed sample of 120 sib pairs and 30 sib trios.

# Association Studies

Association studies are an alternative to linkage analysis as tools for the discovery of genetic factors that may contribute to the susceptibility to a disease. For reasons that will become apparent below, they can also be used in conjunction with linkage analysis to give a much more precise estimate of the location of a gene that has already been mapped to a chromosomal region. Whereas linkage analysis is based on sampling pedigrees and investigating the pattern of inheritance within families, in many approaches to association studies, unrelated genomes are sampled and compared. For example, in the *case-control* design (CC), samples of affected cases and unaffected controls, unrelated to each other, are collected. The genetic composition of one group is compared with the genetic composition of the other group at a set of genetic markers. Loci which show substantial between-groups divergence are taken to be associated with susceptibility to the disease.

Association studies resemble linkage studies in experimental genetics, to the extent that we directly study the relation between an individual's phenotype and marker genotypes. By way of contrast, we saw in Chaps. 9 and 11 that linkage studies in humans proceed indirectly, starting from the observation that within families phenotypes are correlated, presumably because of the family relationships. Hence we try to find the key to the similarity of phenotypes by looking for regions of similarity in genotype, i.e., regions of identity by descent. In association studies, we assume that genotypes contributing to a trait have arisen by mutation at some time in the past and are co-inherited together with nearby markers; and we study this co-inheritance of phenotype and marker genotype. This is the same strategy we follow in experimental genetics, but in humans we do not have the opportunity to start from inbred strains and to perform controlled mating experiments. One important consequence is that associated markers must usually be very close to the relevant genes, so that they have not been separated by recombination during the many meioses that presumably have intervened since the appearance of the relevant genetic variant. Another is that the degree to which a genetic variant and associated markers are in fact associated is subject to population history, which usually is poorly understood and can be favorable or unfavorable to the success of the study.

In this chapter we investigate in some detail a version of the CC design, namely the *case-random* design (CR). In this design random controls from the general population are used, instead of controls which are actively tested for not having the disease. Though somewhat less efficient statistically, the CR design is advantageous since general DNA resources, like blood banks, can be used in order to obtain control samples. This may be important, since healthy individuals may have less an incentive to participate in a trial compared to affected ones. Nonetheless, unless the trait under investigation is very common, it is unlikely that much statistical power will be lost due to contamination of the control sample by undetected affecteds.

As mentioned above, samples in the CR studies are assumed not to be correlated, both between groups (cases and controls) and within each group. Occasionally, this assumption may be violated due to some relatedness between subjects that went undetected. Such a violation may affect the statistical properties of the samples and may lead to increased levels of false positive errors. One of the objectives of this chapter is to investigate the potential harmfulness of such relatedness. Another concern in association studies is the possibility of *population substructure*, which is often attributed to the presence of reasonably homogeneous social subclasses or ethnic subgroups that tend not to mix with one another. As we will see below, population substructures may bias the statistics and produce elevated levels of false positives. The *transmission disequilibrium test* (TDT) was introduced as a remedy for these concerns. In the simplest version of this design affected individuals are sampled with their parents. The allele that was not transmitted to the affected individual serves as a control for the allele that was. Since the analysis is conditioned on the alleles the parents possess, the pitfalls mentioned above appear to be less of a problem.

We start by presenting a model for the case-random design, develop a test statistic for QTL detection, and describe the basic properties of the test statistic. We conclude the first section with the analysis of a data set in a form that may occur in actual case-random trials. The second section deals with the issue of population substructure and cryptic relatedness in samples, which are hazards in population-based association studies. Example are given for the way these hazards may effect the statistical properties of the case-random test statistic. The last section introduces the TDT as a more robust alternative.

## 12.1 Statistical Models and Statistical Tests

The method presented below for the construction of statistical tests and for the analysis of their statistical properties will follow the same route we used previously in the context QTL mapping in experimental crosses and in human pedigrees. We start by formulating a model for the relation between an individual's genotype and phenotype. As a matter of fact, we will be using the same or very similar models as those presented in the context of linkage analysis. We will then identify an appropriate test procedure for determining the significance of a locus, assuming the observed genotypes and the trait locus coincide and investigate large-sample statistical properties of the resulting test statistic. In the next step, we will assess the reduction in power that results from the use of markers, which may be correlated with, but are not identical to, the trait locus. A major difference will emerge between this chapter and the discussion in the previous ones. When mapping with experimental crosses or with pedigrees, the source of correlation between adjacent loci is the recombination fraction. This is true, to some extent, also in admixture mapping. In the case of association studies in outbred populations the correlation is created by *linkage disequilibrium* (LD). This process involves not only recombination, but also the history of the population. It may have a complex structure, which is no longer a simple function of the recombination fraction alone. One of the unfortunate consequences of the lack of a simple model for the correlation structure is unavailability of a clean theory for the computation of p-values in genome scans and for the computation of power. A common practice is to apply the Bonferroni correction (which does not utilize correlations) to the p-value and use the power of the most highly correlated marker.

## 12.1.1 A Model for the Trait

Recall the model of a single autosomal trait locus with allele $D$, associated with the disease, and a wild-type allele $d$. In the model we allowed for sporadic cases and partial penetrance:

$$g_0 = \Pr(\text{Affected} \,|\, dd) \,,$$
$$g_1 = \Pr(\text{Affected} \,|\, Dd) \,,$$
$$g_2 = \Pr(\text{Affected} \,|\, DD) \,.$$

An alternative parametrization was given in the form:

$$g_1 = g_0 + \alpha + \delta \,; \quad g_2 = g_0 + 2\alpha \,.$$

Specifically, we identified the *additive model* by requiring that $g_1 = (g_0 + g_2)/2$, or equivalently that $\delta = 0$. A different model that we would like to emphasize here is the *multiplicative model*. According to the multiplicative model the penetrance probabilities obey the relation:

$$g_2/g_0 = R_2 = R_1^2 = (g_1/g_0)^2 \,.$$

This assumption results in simplifications in the statistical properties of the samples and will help identify appropriate test statistics.

The second component in the genetic model, apart from the penetrance probabilities, is a population model. This model describes the frequencies of the genotypes at the susceptibility locus in the general population. By definition, these frequencies directly determine the expected frequencies for the sample of the random control. We will typically assume Hardy–Weinberg equilibrium, which states that at any given locus the two alleles within each individual are selected randomly and independently from a population of alleles. In particular the Hardy-Weinberg assumption implies that the distribution of the number of $D$ alleles for a randomly selected individual follows the $B(2,p)$ distribution, where $p$ is the frequency of the allele $D$ in the population.

The Hardy-Weinberg assumption is natural in the context of random samples. Occasionally, however, it may be violated. A possible explanation for a violation may be divergence from the equilibrium in the population. An alternative explanation may be errors in the determination of the genotypes. It is important to test for a divergence from the Hardy-Weinberg equilibrium and to find a satisfactory explanation for divergences, especially if they occur among the controls. In Chap. 3 and in the problems at the end of that chapter statistical tests for the validity of Hardy-Weinberg equilibrium were introduced.

## 12.1.2 The Distribution of the Genotypes Among the Cases

The distribution of the genotypes among the cases can be easily derived from the combination of the model for the trait and the model for the population using Bayes formula:

$$\Pr(\text{Genotype}|\text{Affec.}) = \frac{\Pr(\text{Affec.}|\text{Genotype})\Pr(\text{Genotype})}{\sum_{\text{All genotypes}}\Pr(\text{Affec.}|\text{Genotype})\Pr(\text{Genotype})} .$$

Specifically, assuming the multiplicative model and Hardy-Weinberg equation, the sum in the denominator reduces to:

$$\Pr(\text{Affected}) = p^2 g_2 + 2p(1-p)g_1 + (1-p)^2 g_0 = g_0[R_1 p + (1-p)]^2 .$$

Consequently, the conditional distribution of the genotypes becomes:

$$\Pr(DD\,|\,\text{Affec.}) = \frac{R_1^2 p^2}{[R_1 p + (1-p)]^2} = p_a^2 ,$$

$$\Pr(Dd\,|\,\text{Affec.}) = \frac{2R_1 p(1-p)}{[R_1 p + (1-p)]^2} = 2p_a(1-p_a) ,$$

$$\Pr(dd\,|\,\text{Affec.}) = \frac{(1-p)^2}{[R_1 p + (1-p)]^2} = (1-p_a)^2 .$$

Inspecting the results we can see that the distribution of the genotypes is again binomial, albeit with a different probability for drawing allele $D$. Among the random controls the probability is $p$. Among the cases the probability is altered to

$$p_a = R_1 p/(R_1 p + 1 - p) = p + \frac{(R_1 - 1)p(1 - p)}{p(R_1 - 1) + 1} \ .$$

Hence, we may conclude that for the multiplicative model Hardy-Weinberg equilibrium is maintained among cases whenever it holds in the general population.

For the additive model (as well as for the recessive and dominant models) the distribution of the genotypes among cases is not binomial. However, to a first order of approximation both the additive and the dominant models agree with the multiplicative model provided the penetrance $g_0$ of the $dd$ genotype is not zero and the genetic effect of substituting a $D$ allele is small.

As an illustration consider the following comparison between the additive and the multiplicative models. Consider the genetic relative risk of an heterozygote in relation to the wild-type homozygote, which we have denoted by $R_1$. The genetic relative risk of a $DD$ homozygote is $R_1^2$ in the multiplicative model and it is $2R_1 - 1$ in an additive model. The difference between the two is equal to $R_1^2 - (2R_1 - 1) = (R_1 - 1)^2$, which is negligible when $R_1 - 1$ is small. The multiplicative model is also very convenient when it comes to creating simple models involving more than one trait locus. (See Prob. 9.7.)

### 12.1.3 Statistical Tests for Association when Genotypes at a Trait Locus are Observed

Assume that each individual in a case-random trial is classified according to the group it belongs to: cases or controls, and according to its genotype: $DD$, $Dd$, or $dd$. One can summarize the output in a $2 \times 3$ table of frequencies:

**Table 12.1.** Observed genotype frequencies

|          | $dd$     | $Dd$     | $DD$     |         |
|----------|----------|----------|----------|---------|
| Affected | $N_{a0}$ | $N_{a1}$ | $N_{a2}$ | $n_a$   |
| Random   | $N_{r0}$ | $N_{r1}$ | $N_{r2}$ | $n_r$   |
|          | $N_0$    | $N_1$    | $N_2$    | $n$     |

Observe that we denote the number of cases in the trial by $n_a$ and the number of controls by $n_r$. The total number of individuals in the sample is $n = n_a + n_r$. A natural approach for testing the association between the phenotype and the genotype is by testing independence between rows and columns in the above table. A standard tool is Pearson's chi-square (often written $\chi^2$) test, which compares the observed table of frequencies with the table of frequencies that would be expected under the assumption of independence.

**Table 12.2.** Expected genotype frequencies

|  | $dd$ | $Dd$ | $DD$ |  |
|---|---|---|---|---|
| Affected | $N_0 n_a/n$ | $N_1 n_a/n$ | $N_2 n_a/n$ | $n_a$ |
| Random | $N_0 n_r/n$ | $N_1 n_r/n$ | $N_2 n_r/n$ | $n_r$ |
|  | $N_0$ | $N_1$ | $N_2$ | $n$ |

The general form of the test statistic is

$$\chi^2 = \sum_{i,j} \frac{(O_{ij} - E_{ij})^2}{E_{ij}} , \tag{12.1}$$

where $O_{ij}$ is the observed frequency at a cell at the $i$th row and $j$th column and $E_{ij}$ is the expected frequency at that cell under the hypothesis of independence. The sum extends over all the cells of the table. In the case of Table 12.1, we would have $O_{11} = N_{a0}$, $O_{12} = N_{a1}$, etc., and the $E_{ij}$ given by the corresponding entries in Table 12.2. Note that the entries in Table 12.2 are obtained via the multiplication with each other of the marginal relative frequencies of Table 12.1 (and multiplying the results by $n$). Under the null assumption of independence the distribution of the test statistic has (approximately) a chi-square distribution. The number of degrees of freedom is the product $(C - 1) \times (R - 1)$, where $C$ is the number of columns and $R$ is the number of rows in the table, in this case $(3 - 1) \times (2 - 1) = 2$.

The statistic for testing may be simplified if the multiplicative model is assumed. In such a scenario, the problem reduces to one of testing the frequencies of the two alleles in the two groups and can be formulated as problem of testing for equality of two binomial distributions. Specifically, the sample frequencies of the alleles are computed from the genotypic data, and can be presented in the form of a $2 \times 2$ table:

**Table 12.3.** Observed allele frequencies

|  | $d$ | $D$ |  |
|---|---|---|---|
| Affected | $2N_{0a} + N_{1a}$ | $2N_{2a} + N_{1a}$ | $2n_a$ |
| Random | $2N_{0r} + N_{1r}$ | $2N_{2r} + N_{1r}$ | $2n_r$ |
|  | $2N_0 + N_1$ | $2N_2 + N_1$ | $2n$ |

The test statistic to compare two binomial probabilities given in Sect. 1.5 can be expressed as

$$Z^2 = \frac{(\hat{p}_a - \hat{p}_r)^2}{\hat{p}(1 - \hat{p})[1/(2n_a) + 1/(2n_r)]} ,$$

where $2\,n_a$ and $2\,n_r$ is the total number of alleles in the sample of affected and the sample of random controls, respectively. The term $\hat{p}_a = (2N_{2,a} +$

$N_{1,a})/(2n_a)$ is the relative frequency of the allele $D$ among affected, $\hat{p}_r = (2N_{2,r} + N_{1,r})/(2n_r)$ is the relative frequency of the allele $D$ among random controls, and $\hat{p} = (2N_2 + N_1)/(2n)$ is the relative frequency of the allele $D$ in the combined sample. Some algebraic simplification of (12.1) applied to the expressions in Table 12.3 shows that $Z^2$ is exactly the same as Pearson's chi-square statistic for this problem. As we saw in Chap. 1, the asymptotic null distribution of $Z$ is standard normal, so that of $Z^2$ is chi-square on one degree of freedom.

The test associated with Table 12.1 of genotypes is called the *genotype test* and the test associated with Table 12.3 of alleles is called the *allele test*. We will concentrate our efforts below in understanding the statistical properties of the allele test.

Since the allele test is appropriate when both Hardy-Weinberg equilibrium and a multiplicative model hold and can be misleading otherwise, its routine use cannot be recommended. However, its properties are relatively easy to understand and a detailed study of it teaches us about a number of difficulties that plague case-control studies. Some alternatives to the allele test are briefly mentioned, and a parallel discussion of the genotype test is left as an exercise.

### 12.1.4 The Statistical Properties of the Allele Test

The null distribution of the allele test results from the null distribution of the statistic $Z$, which corresponds to the standardized difference between the allele frequency in the two groups. The distribution of this statistic can be approximated by the standard normal distribution, provided that the allele is not too rare.

The power to detect a trait locus is determined by the noncentrality parameter, i.e., the expectation of the statistic $Z$ when the allele $D$ does effect the penetrance, given by

$$\mathrm{E}(Z) \approx \frac{(p_a - p)}{[p(1-p)(\frac{1}{2n_a} + \frac{1}{2n_r})]^{1/2}} = \left[2n\frac{n_a}{n}\frac{n_r}{n}\right]^{1/2} \frac{(R_1 - 1)[p(1-p)]^{1/2}}{1 + p(R_1 - 1)}.$$

(12.2)

It follows that the noncentrality parameter, and hence the power, is a function of the genetic relative risk and of the frequency of the allele $D$ in the population. It is also a function of the sample sizes. The larger the sample sizes, the larger is the power to detect the QTL. For a given total sample size $n$, the noncentrality parameter is maximized by assigning half the sample size to the cases and half to the controls. However, if a lower cost is associated with recruiting random controls compared to affected cases, then it may be more cost-effective to increase the proportion of the random controls.

Let us examine the effect of the genetic relative risk $R_1$ and the allele frequency $p$ on the noncentrality parameter. Observe that the contribution of these two terms is independent of the sample size. The following R code produces Fig. 12.1, which gives the noncentrality parameter as a function of

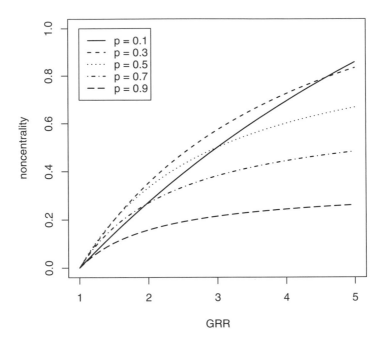

**Fig. 12.1.** The noncentrality parameter of the allele statistic (excluding the sample sizes).

the genetic relative risk, $R$, for $1 \le R \le 5$, and the allele frequency $p$, for $0.1 \le p \le 0.9$:

```
> p <- seq(0.1,0.9,by=0.2)
> R <- seq(1,5,by=0.1)
> plot(range(R),c(0,1),type="n",xlab="GRR",
+      ylab="noncentrality")
> for (i in 1:length(p))
+ {
+      ncp <- (R-1)*sqrt(p[i]*(1-p[i]))/(p[i]*(R-1)+1)
+      lines(R,ncp,lty=i)
+ }
> legend(1,1,legend=paste("p = ",p, sep=""),lty=1:length(p))
```

Observe that for values of the genetic relative risk close to one a disease gene with a common allele frequency closer to $1/2$ is easier to detect. Rarer alleles may be detected more easily when they have a stronger effect on the penetrance.

## 12.1.5 The Effect of Linkage Disequilibrium

So far we have implicitly assumed that the genetic information involves direct measurements of a functional polymorphism – a locus that effects the penetrance of the disease. In reality, however, this is unlikely. Even in the favorable scenario, where a candidate gene is correctly targeted, a functional polymorphism may be any of scores, or even hundreds, of polymorphic sites scattered within or in the vicinity of the gene. Consequently, one should assume that the genetic marker one uses for measurement is neutral with respect to the trait. The power to detect association when using such a marker results from the reflection, via linkage disequilibrium, of the effect of the functional polymorphism on the distribution of a test statistic computed at the marker.

In Chap. 3 we defined linkage disequilibrium to be the statistical dependence between different loci on a chromosome (or chromosomes) inherited from a particular parent. Here we focus our attention on the parameter that is most directly linked to the statistical power of the allele test, namely the correlation coefficient between two bi-allelic loci. (This is not the measure discussed in Chap. 3, which was simply the difference between the joint frequency of two alleles and the product of marginal frequencies of each allele separately.)

In order to define the correlation coefficient consider the two-by-two table 12.4, which describes the joint distribution of alleles in a random population:

**Table 12.4.** Joint distribution of alleles

|   | $d$ | $D$ |   |
|---|---|---|---|
| $B$ | $p_{Bd}$ | $p_{BD}$ | $f$ |
| $b$ | $p_{bd}$ | $p_{bD}$ | $1-f$ |
|   | $1-p$ | $p$ | $1$ |

Here $B$ and $b$ are the two alleles of the anonymous marker at locus $t$. The entries are the relative frequencies of gametes with alleles $B$ or $b$ at locus $t$ and allele $D$ or $d$ at the trait locus $\tau$. The correlation coefficient between the two loci is defined by:

$$r = \frac{p_{BD}p_{bd} - p_{Bd}p_{bD}}{\sqrt{p(1-p)f(1-f)}} \ . \tag{12.3}$$

This parameter turns out to be the standard correlation coefficient introduced in Chap. 1 when numerical values are assigned to the alleles (cf. Prob. 12.3). It may in principle take values between -1 and 1. An absolute value of one corresponds to perfect linkage disequilibrium. In that case, measuring the marker is statistically equivalent to measuring the functional locus. At the other extreme, a value of zero corresponds to no linkage disequilibrium (and is called *linkage equilibrium*).

The relation between the correlation coefficient $r$ and the power to detect a QTL with a marker results from the relation between the noncentrality parameter for the marker and for the QTL, respectively. This relation is given by:

$$\mathrm{E}(Z_t) = r \times \mathrm{E}(Z_\tau) = r \times \xi \,. \tag{12.4}$$

Observe that this is the same type of relation we first encountered in the context of experimental genetics and then for linkage analysis of pedigrees. The only difference is the source of the correlation. The correlations we have used previously reflected the breakdown in linkage between adjacent loci due to recombination events during meioses. The correlation coefficient we are currently using reflects the statistical dependence in random chromosomes and is a result of the accumulation of historical recombination events (cf. Sect. 3.3) as well as other events that the population may have gone through such as population bottlenecks, genetic drift, admixture, and so on.

The structure of the correlation coefficient of linkage disequilibrium is much more complex than the structure of the correlation coefficient used in linkage analysis. It is not uncommon to find a nearby locus with a lower level of linkage disequilibrium than a locus that is farther away. Hence, it cannot be expected that the level of linkage disequilibrium is a simple function of the distance between loci as was the case in linkage analysis. Moreover, as the analysis in Sect. 3.3 shows, linkage disequilibrium decays rapidly with each generation of random mating, so usually extends only for short distances. In an outbred population one does not typically find significant linkage disequilibrium between pairs of loci at distances more than a few hundred kbp (kilo-base-pairs) away, which corresponds to a fraction of a cM. The rapid decay of linkage disequilibrium, as a function of distance, provides the possibility of estimating the location of a disease gene very precisely, since markers only a short distance away will behave essentially independently from the disease locus itself. If one has a rough idea of the location of the disease locus, perhaps from a previous linkage study, one can place additional markers in the neighborhood of the gene and use association analysis to get a much more precise estimate of its location.

However, a whole-genome scan based on association requires tens or even hundreds of thousands of markers in order to ensure reasonable power to detect trait-related polymorphisms with unknown locations. With so many markers the problem of multiple testing is a serious issue. A standard approach to dealing with this issue is to apply the Bonferroni correction, where a marker is declared to be associated with the trait if its p-value multiplied by the number of markers involved in the scan is still below 0.05. Because the Bonferroni correction is usually very conservative, it cannot be regarded as an ideal solution, but a possibly more important issue is that the individual p-values themselves may be incorrectly evaluated because of unknown features of population history. We return to this point below after briefly discussing a case study.

## 12.1.6 A Demonstration

In order to demonstrate the analysis of data from case-random studies let us consider an artificial example. Assume we are investigating a target gene for association with a trait. Imagine that three SNPs from this gene were geno-typed in samples of affected cases and random controls. For sake of illustration we have created a file with the outcome of the genotypes. A link to the file can be found in `http://pluto.huji.ac.il/~msby`. Save the file in the same directory where you are running R and give it the name "`CaseRandom.csv`". Below we will test the association between each of the SNPs and phenotype.

The first step involves reading the data file into R. Observe that the data file is a comma delimited `ASCII` file. This is one of the standard ways by which one can store datasheets in a text format. Formatted text files can be read into R using the function "`read.file`":

```
> CR <- read.table("CaseRandom.csv",header=TRUE,sep=",")
> summary(CR)
   group        sex         snp1        snp2        snp3
 CASE: 634   F: 855    A/A: 556    A/A: 701    C/C: 474
 RAND:2626   M:2405    T/A:1600    G/A:1674    T/C:1585
                       T/T:1104    G/G: 885    T/T:1201
```

This function takes as input a file name and instructions regarding the structure of the file. It produces as output a dataframe. The later is an R object for storage of data in the standard matrix-like format in which columns represents variables and rows represent observations. The argument "`header=TRUE`" identifies the fact that the column names are given in the first row of the input file. The argument "`sep=","`" indicates that commas are used as field separators.

The function "`summary`" is a generic function that produces a textual summary of the content of the object in its argument. In the case of a dataframe it produces a short summary of the distribution of the variables. The two types of variables are (i) numeric, in which case the summary gives the mean and quantiles, and (ii) factors, in which case the distribution is summarized in the form of a frequency table. By factors we mean discrete, qualitative variables. The function "`read.table`" automatically transforms any variable that contains non-numeric values into factors. In the case of the data file "`CaseRandom.csv`" all variables are non-numeric and are turned into factors.

Observe that there is a total of 3260 subjects in the trial; 634 are cases and the rest are random controls. About one quarter are females. The frequencies of genotypes of the three SNPs are also given. We are interested in the joint distribution of the group status and each of the SNPs. Start with SNP1:

```
> table(CR$group,CR$snp1)

        A/A  T/A  T/T
  CASE  140  310  184
  RAND  416 1290  920
> chisq.test(table(CR$group,CR$snp1))

        Pearson's Chi-squared test

data:  table(CR$group, CR$snp1)
X-squared = 17.1176, df = 2, p-value = 0.0001919
```

The function "table" produces a frequency table that can then be passed on to the function "chisq.test", which applies Pearson's chi-square test for independence. It may be noted that variables within a dataframe may be addressed via the format "dataframe.name$variable.name". The result is what we referred to as the genotype test. In order to obtain the allelic test we should count alleles:

```
> allele <- matrix(c(2,1,0,0,1,2),3,2)
> table(CR$group,CR$snp1)%*%allele

        [,1] [,2]
  CASE  590  678
  RAND 2122 3130
```

Observe that the table created by multiplication of the frequency tables of genotypes with the matrix "alleles" contains the allele counts. Applying the chi-square test to this table produces a slightly more significant p-value:

```
> chisq.test(table(CR$group,CR$snp1)%*%allele)

   Pearson's Chi-squared test with Yates' continuity correction

data:  table(CR$group, CR$snp1) %*% allele
X-squared = 15.5287, df = 1, p-value = 8.126e-05
```

The output of the "chisq.test" function is a list. One of the objects in the list is "p.value", which contains the p-value produced by the test. Let us make a table of p-values for the three SNPs and for both the allelic and genotype tests:

```
> p <- matrix(1,2,3)
> colnames(p) <- colnames(CR)[3:5]
> rownames(p) <- c("genotype","allele")
> for(m in 1:3)
+ {
+       tab <- table(CR$group,CR[,m+2])
```

```
+      p[1,m] <- chisq.test(tab)$p.value
+      p[2,m] <- chisq.test(tab%*%allele)$p.value
+ }
> round(p,6)
              snp1      snp2      snp3
genotype 0.000192 0.034598 0.000160
allele   0.000081 0.025178 0.000542
```

Observe that snp1 is the most significant, snp2 the least significant, and that for snp3 the genotype test produces a more significant result than the allele test. Note also that elements in a list can be called with the format "list.name$object.name", which is similar to the format used for dataframes.

The p-value for the most significance SNP is about $10^{-4}$. This is clearly a significance result, given that the three SNPs were the only ones considered. However, if the trial involved examination of hundreds of markers and the markers produced the most extreme results, then the issue of statistical significance is more murky. For example, a trial that involves 500 SNPs chosen to cover a 25 cM region identified in a previous linkage study would produce a Bonferroni-corrected p-value of about 5%, of borderline significance.

## 12.2 Population Substructure and Inbreeding

Population-based association studies have many advantages and are widely applied. Typically, they are much simpler to administer than family-based studies. Family-based linkage studies call for the recruitment of pedigrees with multiple affecteds. Such pedigrees may be difficult to find; and even when a pedigree is identified, substantial effort may be required in order to assess the phenotypic status and obtain DNA samples from pedigree members. In family-based association studies, which are discussed below, parents of affected individuals need to be recruited. This may also complicate the process of collecting the sample. Consequently, it may be difficult to organize a family-based study with a large sample size. A reduced sample size means that the statistical power of the trial is compromised.

In population-based studies, on the other hand, unrelated individuals are used. Administration of the recruitment is much more straightforward. Patients that come to visit their doctors or hospital units on a regular basis are a natural resource for the trial. Recruiting can be carried out on the spot by the patient's physician, perhaps helped by a nurse or administrative staff. No extra effort is needed in order to locate and summon other family members. Control DNA samples may be obtained from public resources. Consequently, reasonably sized trials can be organized even by a group of researchers with relatively modest resources.

As an added advantage, a population-based association study with a sufficient density of markers may be more powerful than a linkage study for mapping genes with minor effects. Careful examination of the noncentrality parameter for affected sib pairs, in the additive model for example, reveals that the genetic relative risk enters into the formula in the form of a squared discrepancy from one (i.e., $(R_1 - 1)^2$) times the variance of the allele. In the case-random design, on the other hand, the square root of that term is involved. If the product of the genetic relative risk and the variance of the QTL is less than one, as is usually the case, then taking its square root will increase its value. (On the other hand, the correlation $r$ is usually much smaller than the correlation that arises from recombination during a few generations within a family, so that markers in an association study must be much closer to the trait locus than in a linkage study.)

Although they are relatively simple to assemble and have some important statistical advantages, concerns have arisen with respect to the reliability of population-based association studies. These concerns question the validity of the basic assumptions of a population-based mapping approach. In this section we will try to investigate the effects on the statistical properties of the test statistic of departures from these assumptions. We will start by looking at the relatively simple case where the population is composed of two sub-populations with different disease prevalence and different allele frequencies. One concern is that failure to take into account the subject's genetic background (i.e., the sub-population to which the subject belongs) may produce artificial associations between SNPs and the disease status. The second issue that will be discussed is the effect of dependencies between the sampled chromosomes due to relatedness among the participating subjects. The concern here is that even distant relatedness may lead to dependence that increases the variability of the test statistic and leads to elevated frequencies of false positives.

### 12.2.1 A Population Substructure

In order to simplify the discussion let us assume that the target population is composed of two sub-populations. Geographical, social, or racial barriers may subdivide a population and reduce the homogenizing effect of random mating. Consequently, differences in the original genetic structure or random drift may create genetic and phenotypic variability between the sub-populations. We assume that the sub-populations are each randomly mating within, but they may differ genetically and phenotypically from each other. In particular, they may have different frequencies for a given allele of a given bi-allelic marker. Denote the allele frequencies at the marker for the two sub-populations by $f_1$ and $f_2$, respectively. Assume that the given marker is not in linkage disequilibrium with any trait locus, so one should expect that the noncentrality parameter of the associated test statistic equals zero.

A randomly selected gamete may originate from the first or from the second sub-population. Denote by $\pi_r$ the probability that a gamete in the random control originates from the first population and by $1 - \pi_r$ the probability it originated from the second. (Presumably $\pi_r$ is the proportion of the first sub-population within the general population.) If the prevalence of the disease among members of the first sub-population is different from that among members of the second, we may find that the frequency among affecteds of gametes originating from the first sub-population, denoted here by $\pi_a$, differs from the same frequency among random controls. As we now demonstrate, the result may be a cryptic association between the marker and the disease allele which is reflected in a noncentrality parameter differing from zero for the test statistic.

The frequency of the marker allele among the affected cases is given by

$$f_a = f_1\pi_a + f_2(1 - \pi_a) .$$

Likewise, the frequency of the allele among the random controls is

$$f_r = f_1\pi_r + f_2(1 - \pi_r) .$$

By subtracting the second expression from the first we obtain that the difference in the expected allele frequency between cases and controls is:

$$f_a - f_r = f_1(\pi_a - \pi_r) + f_2((1 - \pi_a) - (1 - \pi_r)) = (f_1 - f_2)(\pi_a - \pi_r) .$$

If both terms in the product on the right-hand side are non-zero, the result is a non-zero difference in marker allele frequencies, which leads to a noncentrality parameter differing from zero.

From these calculations we see that statistical analysis that fails to take into account the composition of the population may lead to erroneous conclusions. This should come as no surprise if we recognize the fact that case-random studies are a form of observational studies. It is well known that failing to take into account important covariates in the analysis of observational studies may bias the interpretation of the results. We have found that population substructures are potentially important covariates.

Let us demonstrate the issue of population substructure using a simulated example. We generate a sample of cases and a sample of controls from a population composed of two sub-populations. The probability of originating from the first sub-population among cases is $\pi_a = 0.6$ and the parallel probability among controls is $\pi_r = 0.4$:

```
> n <- 500
> pheno <- c(rep(1,n),rep(0,n))
> pop <- c(rbinom(n,1,0.6),rbinom(n,1,0.4))+1
```

The next step is to generate the pair of gametes for each of the subjects. The frequency of the allele is 0.3 if the subject originates from the first sub-population and it is 0.7 if he or she originated from the second. The gametes are stored in a matrix with two columns:

```
> gam <- matrix(nrow=length(pheno),ncol=2)
> gam[pop==1,] <- rbinom(2*sum(pop==1),1,0.3)
> gam[pop==2,] <- rbinom(2*sum(pop==2),1,0.7)
```

In the previous section we applied the chi-square test of independence to a frequency table. An alternative is to apply the test to two vectors of factors, $y$ and $x$, giving the levels for the two variables for each observation:

```
> y <- rep(pheno,2)
> x <- as.vector(gam)
> z <- rep(pop,2)
> chisq.test(y,x)
```

```
Pearson's Chi-squared test with Yates' continuity correction

data:  y and x
X-squared = 17.6814, df = 1, p-value = 2.612e-05
```

Note that an observation is associated with a gamete, thus generating two observations per subject. This is implemented by turning the matrix into a vector (of length 2,000) and duplicating the vector of group status. Observe that the chi-square test produced a significant association. The association is created since both $y$ and $x$ are associated with the sub-population variable $z$, which is not accounted for.

In some cases we may be able to identify the factors that make the population heterogeneous. When this is so, one can use more sophisticated statistical techniques to correct for the problems encountered in the preceding analysis. Below we describe very briefly some methods that correct for the effect of covariates and some R functions that can be used in order to implement these methods.

When applied to the formulation discussed above, which there led to the allele test, other approaches to test the hypothesis $p_a = p$ formulate the problem in terms of the odds, $\text{ODDS}_a = p_a/(1 - p_a)$ and $\text{ODDS} = p/(1 - p)$ favoring the $D$ allele over the $d$ allele in the affected and control populations, respectively. The *odds ratio*, defined as

$$\text{OR} = \frac{\text{ODDS}_a}{\text{ODDS}} = \frac{p_a(1 - p)}{p(1 - p_a)} \ ,$$

is a very useful way of measuring the relation of $p_a$ and $p$. In terms of this parameter, the null hypothesis $p_a = p$ becomes $\text{OR} = 1$. Recall that in the multiplicative model with Hardy-Weinberg equilibrium, $p_a = R_1p/(R_1p+1-p)$, which some algebra shows is equivalent to $\text{OR} = R_1$.

In a structured population as discussed above, where we assume that different sub-populations mate randomly among themselves, but much less frequently among different sub-populations, we expect the allele frequencies $p_a$ and $p$ to vary among sub-populations. However, $R_1$, which reflects the increase in risk of the disease when a $D$ allele replaces a $d$ allele, can reasonably be assumed not to vary among sub-populations. The Mantel-Haenszel test deals with this situation: there are a number of two by two tables (i.e., cases and controls divided into sub-populations), with $p_a$ and $p$ varying from table to table but with the odds ratio the same for all tables. It allows us to test the hypothesis OR = 1 that there is no statistical difference between cases and controls. This test is implemented within R by the function "mantelhaen.test", where "z" denotes a factor that defines the different two by two tables.

```
> mantelhaen.test(y,x,z)

    Mantel-Haenszel chi-squared test with continuity correction

data:   y and x and z
Mantel-Haenszel X-squared = 0.0585, df = 1, p-value = 0.8089
alternative hypothesis: common odds ratio is not equal to 1
95 percent confidence interval:
 0.8455211 1.2543541
sample estimates:
common odds ratio
         1.029846
```

Note that the output gives an estimator of the common odds ratio and a 95% confidence interval, which in this case includes the value of 1. Hence when the sub-population is accounted for, the spurious association gets eliminated and the p-value (= 0.8089) is no longer significant.

The Mantel-Haenszel chi-square test can in some cases help us solve the problem of population substructures. However, covariates that are associated both with the phenotype and with the distribution of the alleles need not come in the form of discrete factors. For example, the onset of many diseases are strongly affected by age. Although one may consider grouping individuals according to age, an alternative is to introduce age as a continuous covariate and account for it. It would be convenient to have at our disposal a statistical tool that can handle both discrete factors and continuous covariates.

Logistic regression is used in order to find a link between the distribution of a dichotomous response and covariates. The tool may be applied both in the case where subjects are selected based on the values of their covariates or, in the case of retrospective trials, where subjects are selected based on their response. In the latter case, there are technical complications involved in the interpretation of the intercept ($\beta_0$ in the following discussion) but tests for the effects of regression coefficients ($\beta_1, \beta_2, \ldots$) are completely valid. The case-random design is an example of a retrospective trial.

In logistic regression the odds are assumed to follow a multiplicative model of the form

$$\text{ODDS} = \exp(\beta_0 + \beta_1 x + \beta_2 z + \cdots) ,$$

where the $\cdots$ indicate additional covariates that may be included in the model. In our applications $x$ will be 0 or 1, indicating whether an individual is a control or a case, respectively, and $z$ will denote the sub-population. The odds ratio relevant to the case-control comparison is the ODDS for $x = 1$ divided by the ODDS for $x = 0$ with all other covariates held constant, or $\exp(\beta_1)$. A logistic regression can be fit in R with the aid of the function "glm" (meaning *generalized linear models*). The primary arguments to the function are a model, which identifies the response and covariates, and a family argument that identifies the distribution of the response. If "family=binomial()" is selected, then logistic regression is applied. The output of the function is an object that contains the details of the fitted model. We apply the function to our simulated trial, where at first we ignore the population structure described by z.

```
> glm(y~x,family=binomial())

Call:  glm(formula = y ~ x, family = binomial())

Coefficients:
(Intercept)              x
    -0.1863         0.3814

Degrees of Freedom: 1999 Total (i.e. Null);  1998 Residual
Null Deviance:       2773
Residual Deviance: 2755           AIC: 2759
```

Observe that the model is introduced via: "**response ~ covariates**". Several covariates, as well as interactions, may be introduced on the right hand-side of the model equation.

The generic function "**summary**", when applied to the output of the "**glm**" function, produces summaries of the model. Among these summaries are tests for the significance of the coefficients. The output of the function is a list. Under the name "**coefficients**" in that list is a matrix with the output of the test for each coefficient:

```
> summary(glm(y~x,family=binomial()))$coef
              Estimate Std. Error    z value      Pr(>|z|)
(Intercept) -0.1862649 0.06280168 -2.965923 3.017764e-03
x            0.3813542 0.08987357  4.243230 2.203255e-05
```

As we can see, the coefficient x, which represents in this example the genetic marker, is highly significant. The p-value is very close to the p-value we obtained when we used Pearson's chi-square test. This follows from the fact that the test statistics can be shown to be asymptotically equivalent. The estimate of $\beta_1$ (coefficient of x) is 0.381, so the odds ratio would be estimated to be $\exp(0.381) = 1.46$.

Using logistic regression modeling we can also introduce non-genetic co-variates, such as the sub-population identifier z:

```
> summary(glm(y~x+z,family=binomial()))$coef
                Estimate Std. Error     z value      Pr(>|z|)
(Intercept) -1.29590830  0.1449098 -8.9428622 3.792186e-19
x            0.02941846  0.1006359  0.2923259 7.700375e-01
z            0.85730836  0.1006142  8.5207463 1.585284e-17
```

Observe, that just as in the case of the Mantel-Haenszel chi-square test, the x covariate is no longer sigificant (p-value = 0.770) when the population substructure is accounted for. The estimator of the odds ratio is $\exp(0.0294) = 1.0298$, which is consistent with the Mantel-Haenszel procedure. Note also that the coefficient of $z$ is highly significant, as a reflection of the population structure.

*Remark 12.1.* In the preceding discussion, we were concerned about the effect of admixture on the false positive rate and on techniques to mitigate this problem. Unfortunately none of these techniques is helpful if we fail to recognize the factors underlying population heterogeneity. A safeguard, although not a certain cure for the problem, is to use instead of a random control group, a control group that is matched to the cases. For example, if the cases are obtained from hospital records, one could look in the same records to find controls that are not affected by the disease of interest, but who resembles the cases in a list of demographic characteristics: age, sex, home address, etc. One still runs the risk of omitting unrecognized confounding factors.

*Remark 12.2.* One problem that can be difficult to recognize and can adversely affect the *power* of a test to detect association is heterogeneity in the trait locus itself. Suppose there are two different functional polymorphisms at two different places in the trait locus. They have alleles $D_i$ and $d_i$ for $i = 1, 2$. If $D_1$ is associated with marker allele $B$ and $D_2$ with marker allele $b$, their contributions to the noncentrality parameter of (for example) the allele test would have opposite signs that could even cancel completely. One can try to guard against this by using several nearby markers, in the expectation that they would not all be subject to the same difficulty. This problem does not arise in family-based linkage studies if the frequencies of the disease alleles are sufficiently small, that it is unlikely both would arise independently in the same family. Allelic heterogeneity would have little effect on the noncentrality parameter for linkage, since identity by descent for either $B$ or $b$ would contribute positively to the statistic.

## 12.2.2 Inbreeding

The statistics given above all involve assumptions of independence. In particular, the allele statistic is made up of an estimator of $p_a - p$ in the numerator, standardized in the denominator by an estimator of the standard deviation of the numerator. This estimate of the standard deviation is based on an assumption that the different gametes entering into the numerator are independent, both within and between individuals. However, if subjects are distantly related to one another, their genotypes may be positively correlated, which may result in an increase in the probability of false positive errors.

In order to quantify the effect of relatedness it is convenient to reintroduce Wright's coefficients of inbreeding and of kinship, which were defined in Chap. 3. The coefficient of inbreeding is defined as the probability that the two homologous copies of a locus within an individual are identical by descent (IBD), while the coefficient of inbreeding of two individuals is the probability that at a given locus two randomly selected alleles, one from each individual, are IBD. Obviously the coefficient of inbreeding of an individual equals the coefficient of relatedness of that person's parents. For reasons of cultural tradition or merely because the finiteness of a population limits the number of possible mates, there usually is some low average level of kinship/inbreeding within populations, especially small populations. We denote the coefficient of kinship of a randomly selected couple or equivalently the inbreeding coefficient of a randomly selected individual by $F$.

Consider a pair of gametes in the random control sample and consider the possible combination of $D$ and $d$ by conditioning whether the two gametes are identical by descent or not. When the gametes are identical by descent, the alleles must be identical. When they are not, the alleles are independent. Consequently:

$$\Pr(DD) = (1 - F)p^2 + Fp \ ,$$
$$\Pr(Dd) = (1 - F)2p(1 - p) \ ,$$
$$\Pr(dd) = (1 - F)(1 - p)^2 + F(1 - p) \ .$$

Let $x_i$ be the indicator that the allele of gamete $i$ is $D$. We see that $E(x_i x_j) = (1 - F)p^2 + Fp$. Observe that the marginal distributions of $x_i$ and $x_j$ are Bernoulli, with a probability $p$ of "success". It follows that $E(x_i) = E(x_j) = p$ and $\mathrm{var}(x_i) = p(1 - p)$. From the description of the joint distribution we see that:

$$\mathrm{cov}(x_i, x_j) = (1 - F)p^2 + Fp - p^2 = F(p - p^2) = Fp(1 - p) \ ,$$

for $i \neq j$.

The fraction of $D$ in the random sample, $\hat{p}_r$, is an estimate of the frequency, $p$, of $D$ in the population. The variability of the sample fraction is given by:

$$\mathrm{var}(\hat{p}_r) = \frac{1}{(2n_r)^2} \sum_{i=1}^{2n_r} \sum_{j=1}^{2n_r} \mathrm{cov}(x_i, x_j) = p(1 - p)\big[F + (1 - F)/(2n_r)\big].$$

Observe that if the population is completely outbred, so $F = 0$, we obtain the variance of an independent sample. If $F > 0$, then the variance is larger, and as a function of the sample size, it converges not to zero but to $p(1-p)F$ as $n_r \to \infty$.

In performing an association study, we implicitly assume that many of the affecteds are somehow related to each other, and since they are a subset of the larger population, the inbreeding coefficient among the affected cases is expected to be at least as large and perhaps larger than in the general population. Under the null assumption of no association between the locus under investigation and the phenotype we get that the variance of the sample proportion of $D$ among affected is

$$\mathrm{var}(\hat{p}_a) = p(1-p)\big(F_a + (1 - F_a)/(2n_a)\big) \,,$$

where $F_a \geq F$ is the paramcter of inbreeding among affected.

Consider the variability of the difference $\hat{p}_a - \hat{p}_r$ under the null hypothesis that $p_a = p$, and assume for simplicity that $F = 0$, while $F_a > 0$. Then we find that

$$\mathrm{var}(\hat{p}_a - \hat{p}_r) = p(1-p)\left(F_a \frac{2n_a - 1}{2n_a} + \frac{1}{2n_a} + \frac{1}{2n_r}\right).$$

The variability of $\hat{p}_a - \hat{p}_r$ is larger than the term $p(1-p)(1/(2n_a)+1/(2n_r))$, which is used for standardization. Miss-specification of the true variability results in a statistic, which under the null distribution has a normal distribution with a variance equal to the ratio between the true variance and the assumed variance. To explore the effect of a non-zero coefficient of inbreeding on the probability of false positives, we assume a trial with 500 affected individuals and 500 controls who are genotyped at 10,000 markers. A Bonferroni correction is applied:

```
> na <- nr <- 500
> n.mark <- 10^5
> x <- qchisq(1-0.05/n.mark,df=1)
> F <- seq(0,4*10^(-4),length=100)
> v0 <- 1/(2*na) + 1/(2*nr)
> v1 <- F*(2*na-1)/(2*na) + v0
> sig <- 1-exp(-(1-pchisq(x*v0/v1,df=1))*n.mark)
> plot(range(F),range(c(0,sig)),type="n",xlab="F",
+      ylab="Sig. level")
> lines(F,sig,lty=2)
> abline(h=0.05)
> legend(0,max(sig),legend=c("Assumed significance",
+      "Actual significance"),lty=1:2)
```

The results are presented in Fig. 12.2. Observe that even small values of the coefficient of inbreeding may substantially increase the probability of false detection of a QTL. The effect is noticeable once the coefficient of inbreeding is above one in 10,000.

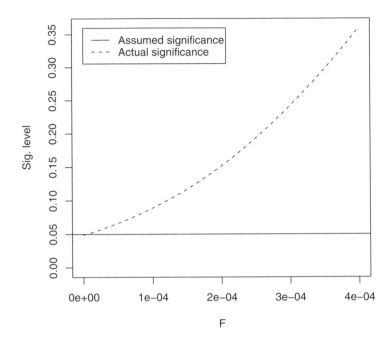

**Fig. 12.2.** The actual significance level of the allele test as a function of the coefficient of inbreeding.

The conclusion must be that the issue of cryptic relatedness in population-based association trials cannot be taken lightly. Naturally, the problem is more severe if the target population is likely to be inbred, for example if it is an isolated population which was formed by a relatively small number of founders. In a larger and more outbred population the problem is likely to be less severe. For example, assume the extreme situation where 100 pairs in our sample of 500 affected individuals are related. If these pairs were all first cousins (with a coefficient of relatedness of $2^{-4}$), we will get that value of $F$ in our sample is equal to:

```
> (1/2^4)*100/choose(500,2)
[1] 5.01002e-05
```

and if all were second cousins (with a coefficient of relatedness of $2^{-6}$), we will get that value of $F$ is:

```
> (1/2^6)*100/choose(500,2)
[1] 1.252505e-05
```

In both cases the effect on the significance level appears not to be a serious problem.

## 12.3 The Transmission Disequilibrium Test

Concerns regarding the sensitivity of population-based association studies to population assumptions has motivated the development of more robust study designs. One such design is the transmission disequilibrium test (TDT). The basic sampling units in this design are an affected offspring with both parents. Only parents that are heterozygous at the marker are considered. The number of copies (0, 1, or 2) of a given allele, which were transmitted to the affected offspring, is the basic building block of the test statistic. The null distribution of this building block, conditional on the heterozygosity status of the parents, is determined by the transmission probabilities only. It is binomial with a probability of one-half for "success", regardless of the distribution of the allele in the general population. Therefore, the hope is that the null distribution of the test statistic will be more robust in terms of population assumptions.

In order to define the test formally, consider trios combined of pairs of parents and an affected offspring indexed by $i$, $1 \leq i \leq n$. Consider a bi-allelic marker and identify one of the alleles. For each trio, let $F_i$ be the indicator that the father is heterozygous at the marker and let $M_i$ be the indicator that the mother is heterozygous. Let $x_{\mathrm{F}i}$ be the indicator of the event that the identified allele is transmitted to the offspring from the father and let $x_{\mathrm{M}i}$ be the parallel indicator of transmission from the mother. For each trio we compute the total number of copies of the allele that were transmitted to the offspring from heterozygous parents: $F_i x_{\mathrm{F}i} + M_i x_{\mathrm{M}i}$. The test statistic accumulates the number of copies across all trios and standardizes it according to its conditional null distribution, given the status of the $F$ and $M$ indicators:

$$Z = \frac{\{\sum_{i=1}^{n}[F_i(x_{\mathrm{F}i} - 1/2) + M_i(x_{\mathrm{M}i} - 1/2)]\}}{[\sum_{i=1}^{n}(F_i + M_i)/4]^{1/2}} .$$

Indeed, observe that a term is added to the sum in the numerator only when a parent is heterozygous. When added, the term is a centered Bernoulli random variable. The total number of centered Bernoulli random variables is $\sum_{i=1}^{n}(F_i + M_i)$, with a variance of 1/4 each. The asymptotic null distribution of the test statistic, when $n$ is large, is standard normal. Since we do not know which marker allele is associated with the disease-predisposing allele, the test should be based on $|Z|$, or equivalently $Z^2$.

Next, let us investigate the noncentrality parameter associated with this statistic. Start with the case where the marker coincides with the QTL, which we assume also to be bi-allelic. Denote by $A$ the event that the offspring is affected and consider the conditional expectation

$$\mathrm{E}\big[F(x_{\mathrm{F}} - 1/2) + M(x_{\mathrm{M}} - 1/2)\,\big|\,A\big] = 2 \cdot \mathrm{E}\big[F(x_{\mathrm{F}} - 1/2)\,\big|\,A\big],$$

where the equality follows by symmetry. The indicator $F$ corresponds to the event that the father is $Dd$-heterozygous. The event associated with $x_F$ corresponds to the transmission of the allele $D$ from the father to the offspring. Compute the probability:

$$\Pr(F = 1, x_F = 1, A) = \Pr(A \mid F = 1, x_F = 1)\Pr(F = 1, x_F = 1)$$
$$= [g_2 p + g_1(1 - p)]p(1 - p) .$$

The term in the square brackets is the conditional probability that the child is affected given that the father contributes one $D$ allele to the offspring. With probability $p$ the allele segregated from the mother is also $D$; otherwise it is $d$, which occurs in probability $1 - p$. Similar arguments lead to the relation:

$$\Pr(F = 1, x_F = 0, A) = [g_1 p + g_0(1 - p)]p(1 - p) .$$

Finally, under Hardy-Weinberg equilibrium the probability of being affected is given by
$$\Pr(A) = [g_2 p^2 + g_1 2p(1 - p) + g_0(1 - p)^2] .$$

The combination of these results and some algebra produce:

$$\mathrm{E}\big[F(x_F - 1/2) + M(x_M - 1/2) \mid A\big] = p(1 - p)\frac{g_2 p + g_1(1 - p) - g_1 p - g_0(1 - p)}{g_2 p^2 + g_1 2p(1 - p) + g_0(1 - p)^2} .$$

In the multiplicative model the expression simplifies to:

$$\mathrm{E}\big[F(x_F - 1/2) + M(x_M - 1/2) \mid A\big] = p(1 - p)\frac{R_1 - 1}{p(R_1 - 1) + 1} .$$

The denominator of $Z$ may be approximated under the null hypothesis and local alternatives by the square root of $n\mathrm{E}(F + M)/4 \approx np(1 - p)$. Therefore, for the multiplicative model,

$$\mathrm{E}(Z) \approx n^{1/2}\frac{(R_1 - 1)[p(1 - p)]^{1/2}}{p(R_1 - 1) + 1} . \tag{12.5}$$

Observe the similarity between the expression for the noncentrality of the TDT statistic and the parallel expression for the case-random design given in (12.2). In the first order of approximation the two expressions differ only in the way sample sizes are accounted for. In the TDT we use $n^{1/2}$, where $n$ is the number of trios, and in the case-random statistic we use $(2n_a n_r/n)^{1/2}$, where $n$ is the total number of cases and random controls. While it is easier to recruit a large sample of cases and unrelated controls, the TDT does not require the same level of concern about spurious associations, as we now show.

We proceed with the assessment of the statistical properties of the TDT test by considering the noncentrality parameter at a marker, which is correlated with, but not identical to, the functional polymorphism. Let $B$ and $b$ be the alleles of such a bi-allelic marker and concentrate on the expectation of the term associated with the contribution of the father. With some abuse of notation denote by $x_t$ the indicator of the transmission of the allele $B$ to the affected offspring and by $F_t$ the indicator of heterozygosity of the father at the marker. We compute the expectation in question by conditioning on the genotype at the trait locus and on the allele of the trait locus that was transmitted to the offspring. From considerations of symmetry it follows that when the father's genotype at the trait locus is either $DD$ or $dd$, then the conditional distribution of the statistic computed at the marker has a zero mean. We consider, therefore, the case where the father is heterozygous at the trait locus, denoted here by $F_\tau$. Identify the allele $D$ and indicate with $x_\tau$ if it has been the one transmitted to the offspring.

Consider the equation $\mathrm{E}\big[F_t(x_t-1/2)\,\big|\,A\big] = \mathrm{E}\big[\mathrm{E}\{F_t(x_t-1/2)\,\big|\,F_\tau, x_\tau\}\,\big|\,A\big]$. The critical point here is the fact that the internal expectation on the right-hand side is computed under the null distribution. The two non-zero values of the random variable are $1/2$ and $-1/2$. Consequently, readapting the notation from Table 12.4, we have

$$2\mathrm{E}\big\{F_t(x_t - 1/2)\,\big|\,F_\tau = 1, x_\tau = 1\big\} =$$
$$\Pr(F_t = 1, x_t = 1 \,|\, F_\tau = 1, x_\tau = 1) - \Pr(F_t = 1, x_t = 0 \,|\, F_\tau = 1, x_\tau = 1)$$
$$= (1-\theta)\frac{p_{\mathrm{BD}}}{p}\frac{p_{\mathrm{bd}}}{1-p} + \theta\frac{p_{\mathrm{bD}}}{p}\frac{p_{\mathrm{Bd}}}{1-p} - (1-\theta)\frac{p_{\mathrm{bD}}}{p}\frac{p_{\mathrm{Bd}}}{1-p} - \theta\frac{p_{\mathrm{BD}}}{p}\frac{p_{\mathrm{bd}}}{1-p}$$
$$= (1-2\theta)\frac{p_{\mathrm{BD}}p_{\mathrm{bd}} - p_{\mathrm{bD}}p_{\mathrm{Bd}}}{p(1-p)} = (1-2\theta)\,r\left[\frac{f(1-f)}{p(1-p)}\right]^{1/2},$$

where $r$ is the correlation coefficient given in (12.3) and $\theta$ is the recombination fraction between the marker and the trait locus. In a similar fashion:

$$2\mathrm{E}\big\{F_t(x_t - 1/2)\,\big|\,F_\tau = 1, x_\tau = 0\big\} = -(1-2\theta)\,r\left[\frac{f(1-f)}{p(1-p)}\right]^{1/2}.$$

Four other terms, which arise from conditioning on $F_\tau = 0$ and $x_\tau =$ either 0 or 1, all cancel out. In summary:

$$\mathrm{E}\big\{F_t(x_t - 1/2)\,\big|\,F_\tau, x_\tau\big\} = (1-2\theta)\,r\left[\frac{f(1-f)}{p(1-p)}\right]^{1/2}F_\tau(x_\tau - 1/2)\,.$$

Taking the expectation of this expression shows that the noncentrality parameter under the multiplicative model of the test statistic at a marker, computed over the entire sample of trios, is:

$$\mathrm{E}(Z_t) = (1 - 2\theta) \times r \times \mathrm{E}(Z_\tau) = (1 - 2\theta) \times r \times \xi\,, \qquad (12.6)$$

where $1 - 2\theta$ is the correlation coefficient associated with linkage and $r$ is the correlation coefficient associated with linkage-disequilibrium.

Consequently, in order to have power to detect a trait locus, both correlations should be non-zero. However, since linkage disequilibrium acts at much smaller distances compared to linkage, in actuality the TDT will have no power unless $\theta \approx 0$, which gives us back the relation for noncentrality at a bi-allelic marker in the case-random design. Note, however, that since there must be linkage ($\theta < 1/2$) in order for there to be a non-zero noncentrality, we do not have the concern that we had for a case-random design that there might be completely spurious association of markers unlinked to the trait locus. Put another way, even if there is spurious association in the population, it will not contribute to the noncentrality parameter unless there is linkage; and then it could actually help.

## 12.4 Discussion

Like the preceding chapter, this one is only an introduction to a complex subject. A comparison of the noncentrality parameters for association studies and for linkage studies and the decay in those parameters for markers some distance away from the genes themselves makes it clear why association studies have become increasingly popular as it has become feasible to genotype larger and larger numbers of markers. However, like admixture studies, the properties of association studies depend on population history. While family-based association studies avoid some difficulties associated with population history, producing them is roughly as complicated as organizing linkage studies.

## 12.5 Bibliographical Comments

The case-control method is standard in epidemiological studies. An elegant discussion of the multiplicative model and its usefulness in understanding the case-control method in genetic epidemiology is given by Risch and Merikangas [67].

The transmission disequilibrium test (TDT) was introduced by Spielman, McGinnis, and Ewens [79]. For additional developments see Spielman and Ewens [80], Rabinowitz [61], Fulkerson et al. [32] and Abecasis et al. [1].

## Problems

**12.1.** Derive the conditional probabilities of the three possible genotypes among affected for the additive, dominant, and recessive models (assuming Hardy-Weinberg equilibrium in the population).

**12.2.** Investigate and compare via simulations the statistical power of the test based on the frequencies of alleles and the test based on genotypes for the additive, dominant, and recessive models.

**12.3.** Define a random variable $X$ to be 0 or 1 according as the allele at one locus is $b$ or $B$ and a second random variable $Y$ to be 0 or 1 according as the allele at a second locus is $d$ or $D$. Show that the correlation coefficient $[E(XY) - E(X)E(Y)]/[\text{var}(X)\text{var}(Y)]^{1/2}$ is given by $r$ defined in (12.3).

**12.4.** Prove the relation:

$$\text{var}(\hat{p}_r) = (1 + F)p_r(1 - p_r)/(2n_r)$$

in the case where different individuals within the sample are independent, but the coefficient of inbreeding of each individual is $F$.

**12.5.** Investigate the effect of relatedness of different affected individuals on the two-degrees-of-freedom test of equality of the genotype frequency between cases and random controls.

**12.6.** Develop a formula for the noncentrality parameter of the TDT that is applicable when the families come from different (randomly mating) substrata of the population, so the allele frequencies in the $i$th family depend on $i$, say $f_i$ at the marker and $p_i$ at the trait locus. The algebraic expression you obtain will be somewhat simpler if you assume that $p_i(R_1 - 1) \ll 1$ for all $i$, so can be neglected in the denominator of (12.5). Specialize your formula to the case where there are only two substrata, with proportions $\pi$ and $1 - \pi$ of affecteds.

**12.7.** Develop a version of the TDT for QTL detection under the conditions of the text (known parental genotypes). What is the noncentrality parameter at a marker linked to a QTL? Adapt this statistic for use with sib pairs. (Recall that the phenotypic values of sibs are correlated because of identity by descent and perhaps also a common environment.)

**12.8.** Show that under the conditions of the text the numerator of the TDT can be expressed alternatively as $\sum_{i=1}^{n}[x_{\text{F}i} - \bar{x}_{\text{F}i} + x_{\text{M}i} - \bar{x}_{\text{M}i}]$, where $\bar{x}_{\text{F}i}$ and $\bar{x}_{\text{M}i}$ are the average number of distinguished alleles in the father and mother, respectively. Note that $F_i/4 = \bar{x}_{\text{F}i}(1 - \bar{x}_{\text{F}i})$, so the denominator can also be expressed in this alternative notation. (The alternative notation can be helpful in cases where parents are unavailable for genotyping, and one uses, for example, siblings instead.)

# 13

# Inferring Haplotypes from Genotypes and Testing Association

In the previous chapter we discussed the association between phenotype and genotype for tests based on a single SNP marker. In this chapter we would like to consider tests of association that combine information from several markers.

The basic notion in the joint analysis of several loci is that of an haplotype. Given a sequence of loci on the same chromosome, an haplotype is a sequence of the alleles of a given DNA molecule. In such a case, we also say that the alleles are in the same *phase*. Unfortunately, in most cases haplotypes cannot be measured directly. Indeed, standard techniques of genotyping involve the determination of the genotype of each SNP. This genotype consists of the two alleles from the person's pair of homologous DNA molecules. The *phase* of two heterozygous loci may be ambiguous. This ambiguity can be partly resolved by the use of statistical tools.

A major subject in this chapter is the description of such a tool, namely the expectation-maximization (EM) algorithm that was introduced in a different context in Chap. 10. In the current setting, where partial information on the haplotype is observed but some of the phases are ambiguous, the EM algorithm can be used in conjunction with an assumption of Hardy-Weinberg equilibrium to compute the maximum likelihood estimates of haplotype frequencies in the population, based on the genotypic data of a sample.

In principle, for a given collection of $k$ bi-allelic markers the total number of different haplotypes may include $2^k$ distinct elements, a number that grows very fast with the number of loci considered. This restricts the applicability of the EM algorithm to dealing simultaneously with no more than about 10 loci. In reality, however, fewer haplotypes are actually present in the population in substantial numbers. Consequently, one may hope that appropriate parametrization of the distribution of haplotypes may allow the joint analysis of more than a score of markers. Indeed, modeling the distribution in Chap. 10 reduced the number of parameters that need to be estimated to two, regardless of the number of markers, and allowed for analysis on a genome-wide basis. Unfortunately, the distribution of haplotypes in a particular population may be a result of a complex evolutionary history and is difficult to model. We do use some data-reduction techniques, which have a marginal contribution, in the application of the EM algorithm. But still, as far as we know, providing a representative and manageable parametric model for the distribution of haplotypes is an open problem.

In the first section we describe the determination of the phase between two loci. We apply this in order to compute the correlation coefficient between the loci. The next section develops the EM algorithm for the determination of haplotype distributions over a larger collection of markers.

Once haplotypes are obtained, they can serve for testing association with a trait. In the last section we suggest test statistics based on haplotypes for the data presented in Sect. 12.1.6.

## 13.1 Determining the Phase Between Two Loci

In this section we assume that we are provided with a sample of unrelated individuals from some target population. A pair of bi-allelic markers are genotyped for each of the individuals in the sample with the goal of determining the distribution of the two-locus haplotypes. In order to clarify the issues involved and to make the presentation more targeted we consider a numerical example as we walk through the details of the discussion. Return to the artificial data presented in the previous chapter, which was saved as an R dataframe under the name "CR". This dataframe contains genotype information from 3260 individuals collected for three SNPs. In the chapter we tested the association between these SNPs and the disease status and found strong association between snp1 and the disease and between snp3 and the disease. In light of these results one may raise a question regarding the relationship between these two markers: are both markers so strongly correlated with each other that each should be considered equivalent to a single marker, or is each marker providing independent information with respect to the association with the disease?

In order to address this question we would like to assess the correlation coefficient $r$ between snp1 and snp2. Here we are motivated by the presumption that a correlation coefficient close to one (in absolute value) is an indication that the first possibility is correct and a correlation coefficient closer to zero supports the second possibility. Recalling the computation of the correlation coefficient between a bi-allelic QTL and a bi-allelic marker, which was conducted with the aid of Table 12.4 of haplotype distributions, we are tempted to address the issue of estimating the correlation coefficient from genotypic data by first computing a parallel table for the distribution of the haplotypes that can be formed from the two SNPs, and then using the estimated distribution in order to compute the correlation coefficient.

Let us initiate the process by taking a second look at the numerical example:

```
> CR <- read.table("CaseRandom.csv",header=TRUE,sep=",")
> table(CR$snp1,CR$snp3)

      C/C T/C T/T
  A/A 172 265 119
  T/A 228 880 492
  T/T  74 440 590
```

The table we formed represents the joint distribution of genotypes in our sample. Each person in the sample appears in one of the table entries. Each person, however, is represented by a pair of haplotypes associated with the pair of copies of the given autosome. In order to make the point, we rewrite the same table in a different format in Table 13.1

**Table 13.1.** Joint distribution of genotypes for a pair of markers

| snp1 | snp2 | frequency | haplotype 1 | haplotype 2 |
|------|------|-----------|-------------|-------------|
| A/A  | C/C  | 172       | A-C         | A-C         |
| A/A  | T/C  | 265       | A-T         | A-C         |
| A/A  | T/T  | 119       | A-T         | A-T         |
| A/T  | C/C  | 228       | A-C         | T-C         |
| A/T  | T/C  | 880       | A-T and T-C or A-C and T-T? | |
| A/T  | T/T  | 492       | A-T         | T-T         |
| T/T  | C/C  | 74        | T-C         | T-C         |
| T/T  | T/C  | 440       | T-T         | T-C         |
| T/T  | T/T  | 590       | T-T         | T-T         |

Observe that the actual frequency in the sample of haplotypes can be partially inferred from the genotypes. For example, we can infer that each of the 172 subjects with a genotype A/A in **snp2** and a genotype C/C in **snp1** must carry a pair of A-C haplotypes. It is also the case that each of the 265 subjects who are heterozygote at **snp1** but A/A-homozygote at snp2 must carry a single copy of the A-C haplotype (and a single copy of the A-T haplotype). Similarly, it can be inferred that each of the 228 subjects of the 4th row of Table 13.1 carries a single copy of the A-C haplotype (and a single copy of the T-C haplotype).

However, for the 880 double-heterozygote at the 5th row of the table one cannot determine the haplotype composition, since both the pair (A-T,T-C) and the pair (A-C,T-T) are consistent with the genotype. The other subjects in the sample do not carry the A-C haplotype.

One may conclude, thus, that the frequency of the A-C haplotype may be any number between $2 \times 172 + 265 + 228 = 837$ and $837 + 880 = 1717$, out of a total of $2 \times 3260 = 6520$ haplotypes in the sample.

Denote by $0 \leq \tilde{\vartheta} \leq 1$ the proportion in the sample of double-heterozygote individuals which have the combination (A-C,T-T) of haplotypes. Given the value of $\tilde{\vartheta}$ we can conclude that the frequency of the haplotype A-C in the sample is $837 + \tilde{\vartheta} \times 880$. Likewise, for a given value of $\tilde{\vartheta}$, the frequencies of the other 3 haplotypes in the sample are given by the entries in Table 13.2:

**Table 13.2.** The frequency in the sample of the four haplotypes, given the proportion of (A-C,T-T) double-heterozygotes

| Haplotype | Frequency |
|-----------|-----------|
| A-C | $837 + \tilde{\vartheta} \times 880$ |
| A-T | $995 + (1 - \tilde{\vartheta}) \times 880$ |
| T-C | $816 + (1 - \tilde{\vartheta}) \times 880$ |
| T-T | $2112 + \tilde{\vartheta} \times 880$ |

Denote the probabilities of the four haplotypes by $p_i$, $1 \leq i \leq 4$, according to the four rows of Table 13.2. Natural estimates of these probabilities as functions of $\tilde{\vartheta}$ are the corresponding relative frequencies, i.e., the entries of the table divided by the total number of 6520. Since we assume that haplotypes are inherited as a unit (i.e., without recombination), each one can be regarded as a single allele. Assuming they are in Hardy-Weinberg equilibrium, one can express the unknown $\tilde{\vartheta}$ in terms of the allele frequencies. The probability of obtaining a double-heterozygote is $p_1 p_4 + p_2 p_3$. The probability of a person have the pair of haplotypes (A-C,T-T), given that he or she is double-heterozygote, is:

$$\vartheta = \frac{p_1 p_4}{p_1 p_4 + p_2 p_3} .$$
(13.1)

Equating $\vartheta$ with $\tilde{\vartheta}$ we get from Table 13.2 the relation:

$$\vartheta = \frac{(837 + \vartheta\,880)(2112 + \vartheta\,880)}{(837 + \vartheta\,880)(2112 + \vartheta\,880) + (995 + (1 - \vartheta)\,880)(816 + (1 - \vartheta)\,880)} ,$$
(13.2)

which can be solved in order to obtain a numerical value for $\vartheta$, and thereby numerical values for the haplotype frequencies.

Before providing a more general justification of the proposed procedure let us implement it in the numerical example

```
> N <- c(837,995,816,2112)
> H <- 880
> hetero <- function(th,N,H)
+ {
+      f <- N + c(th,1-th,1-th,th)*H
+      t <- f[1]*f[4]/(f[1]*f[4] + f[2]*f[3])
+      return(th - t)
+ }
> th <- uniroot(hetero,c(0,1),N=N,H=H)$root
> th
```

[1] 0.7806138

The function "uniroot" finds in this case the value of "th" which solves the given equation. This value can be used to determine the distribution of haplotypes and the value of $r$:

```
> f <- N + c(th,1-th,1-th,th)*H
> p <- f/sum(f)
> c <- p[1]*p[4]-p[2]*p[3]
> p1 <- p[1]+p[2]
> p2 <- p[1]+p[3]
> r <- c/sqrt(p1*(1-p1)*p2*(1-p2))
> r
[1] 0.3002771
```

Observe that although the correlation coefficient is not zero, it is also definitely not equal to one. Hence, it is likely that each of the markers is only partially associated with a trait locus. Alternatively, it may be that each is associated with a different trait locus.

Solving (13.2) provides us with a heuristic approach for the estimation of the haplotype frequencies. We can justify this heuristic approach by appealing to the statistical principle of maximum likelihood estimation. Indeed, our data are composed of the nine frequencies of genotypes in Table 13.1. These data can be thought of as emerging from a multinomial distribution. Under the assumption of Hardy-Weinberg equilibrium of haplotypes, the cell probabilities may be parametrized with the aid of the distribution of haplotypes in the population to produce the likelihood:

$$p_1^{2\times172}(p_1p_2)^{265}p_2^{2\times119}(p_1p_3)^{228}(p_1p_4+p_2p_3)^{880}(p_2p_4)^{492}p_3^{2\times74}(p_3p_4)^{440}p_4^{2\times590},$$

which simplifies to

$$L = e^{\ell(p_1,\ldots,p_4)} \propto p_1^{837}p_2^{995}p_3^{816}p_4^{2112}(p_1p_4+p_2p_3)^{880}.$$

(Observe that in the computation of the likelihood we ignore the multinomial coefficient and powers of 2, which are irrelevant for the determination of the maximum-likelihood estimates of $p_i$, $1 \le i \le 4$.)

One could maximize the likelihood (or its log) as a function of $\vartheta$ defined in (13.1). However, it is a better preparation for the more general EM algorithm if we proceed indirectly. First observe that the haplotype probabilities must sum to one, which gives rise to a maximization problem with constraints and suggests the use of Lagrange multipliers.

The application of Lagrange multipliers in the current setting involves several steps: First we differentiate the Lagrangian, which is equal to $\ell(p_1,\ldots,p_4)-\lambda(p_1+\cdots+p_4-1)$, with respect to $\lambda$ and $p_i$, for $i = 1,2,3,4$. The solution of the constrained maximization problem is then obtained by equating these partial derivatives to zero and solving the system of equations. In order to have a more convenient representation of the system one can multiply by $p_i$ the equation associated with the partial derivative with respect to that $p_i$ and obtain, upon using the definition (13.1), the set of equations:

$$837 + 880\,\vartheta = \lambda p_1$$
$$995 + 880\,(1 - \vartheta) = \lambda p_2 \qquad (13.3)$$
$$816 + 880\,(1 - \vartheta) = \lambda p_3$$
$$2112 + 880\,\vartheta = \lambda p_4\ .$$

Since $p_1 + \cdots + p_4 = 1$, $\lambda$ is equal to the sum of the terms on the left-hand sides of these equations. By substituting the resulting expressions for $p_1, \ldots, p_4$ in (13.1), we obtain (13.2). Therefore, solving (13.2) is the key to obtaining the maximum-likelihood estimate of the distribution of haplotypes based on genotypic data.

In the numerical example we applied the function "uniroot" in order to solve (13.1). This function uses an efficient algorithm for the computation of a root of a general function. An alternative approach for finding a root, which may be applied in our case, is to set an initial value for $\vartheta$ and then iterate (13.1). Let us try this approach:

```
> th <- 0.5
> for (i in 1:100)
+ {
+       f <- N + c(th,1-th,1-th,th)*H
+       th <- f[1]*f[4]/(f[1]*f[4] + f[2]*f[3])
+ }
> th
[1] 0.7805995
```

Indeed, we get an answer which is very close to the solution obtained by the application of "uniroot".

In the next section we describe the application of a general iterative algorithm, the EM algorithm, which can be used in order to compute maximum-likelihood estimates of multi-loci haplotype frequencies using genotypic information.

## 13.2 The Expectation-Maximization Algorithm

Generally stated, the problem of estimating haplotype frequencies based on genotypic data can be thought of as a problem of missing data. Indeed, had we the means to establish the haplotype of each individual in the sample, and thus the frequency of each haplotype in our sample, then a natural estimate of the haplotype distribution in the population would have been their distribution in the sample. Unfortunately, in many cases we cannot resolve the pair of haplotypes from the genotype of a person since more than one pair is consistent with the observed genotype.

Inference from partial information requires a statistical model for the full information. The EM algorithm, which we met before in Chap. 10, was developed in order to obtain maximum-likelihood estimates of the parameters of the statistical model in a setting of partial information.

For our current problem, we assume Hardy-Weinberg equilibrium, so the distribution of haplotypes in a random individual is multinomial with a sample size of two. This model is parametrized by the population frequency of the haplotypes. An iterative procedure is applied in order to obtain the maximum-likelihood estimate of these population frequencies, given the genotypic information in the sample.

Consider the set $\{h\}$ of the different haplotypes that may be present in the population and denote by $p$ the vector of population probabilities of these haplotypes. Let $G_i$ be the observed genotypes of the $i$th person in the sample and let $H_i$ be the pair of haplotypes of that person. The log likelihood in the case of full information is $\log \Pr_p(H_i)$. Since $H_i$ is not observable, we use as a surrogate the *expected* log likelihood, conditional on what is observed (namely, $G_i$). But, in order to compute this conditional expectation, we must know a value for the parameter $p$, which we do not. To get out of this circular reasoning, we proceed iteratively. Given a current estimated value $p^*$ of the population haplotype frequencies the EM algorithm computes an updated estimated value in two steps, called the *expectation* and *maximization* steps. In the expectation step the function:

$$Q(p, p^*) = \sum_{i=1}^{n} \mathrm{E}_{p^*}\big[\log \Pr_p(H_i \mid G_i)\big]$$

is obtained. In the maximization step this function is maximized with respect to $p$. The current $p^*$ is replaced by the maximizing $p$. Starting from an initial value for $p^*$, the algorithm is iterated until it converges to a fixed point and the values of $p^*$ stop changing. It can be shown that the EM algorithm works in the sense that the sequence of values of $p^*$ created by the algorithm increases the likelihood of the observed data at each iteration, although it is possible that the process converges before $p^*$ reaches the true maximum likelihood estimator.

Denote by $I_i(h_1, h_2)$ the indicator of the (ordered) pair of haplotypes in an individual. Observe that the expectation of the indicator is $p_{h_1} p_{h_2}$ and the log of the likelihood of the (unobserved) full information is

$$\log \Pr_p(H_i) = \sum_{h_1} \sum_{h_2} I_i(h_1, h_2)[\log(p_{h_1}) + \log(p_{h_2})] \ .$$

Upon summing over $i = 1, \ldots, n$ and taking the conditional expectation, we obtain:

$$Q(p, p^*) = \sum_{i=1}^{n} \sum_{h_1} \sum_{h_2} \Pr_{p^*}\left[H_i = (h_1, h_2) \,\middle|\, G_i\right]\left[\log\left(p_{h_1}\right) + \log\left(p_{h_2}\right)\right].$$

The maximization with respect to the vector $p$ may be obtained again with the aid of a Lagrange multiplier, which now involves the solution of the following equations:

$$2\sum_{i=1}^{n} \sum_{h_1} \Pr_{p^*}\left(H_i = (h_1, h) \,\middle|\, G_i\right) = \lambda p_h, \quad h \in \{h\},$$

and $\sum_h p_h = 1$, which implies that $\lambda = 2n$. Finally, from the fact that

$$\Pr_{p^*}\left(H_i = (h_1, h_2) \,\middle|\, G_i\right) = \frac{p^*_{h_1} p^*_{h_2}}{\sum_{(h_1, h_2) \in G_i} p^*_{h_1} p^*_{h_2}}, \quad \text{for } (h_1, h_2) \in G_i$$

we obtain the EM updated value:

$$p_h = \frac{1}{n} \sum_{i=1}^{n} \sum_{h_1} \frac{p^*_{h_1} p^*_h I_{\{(h_1, h) \in G_i\}}}{\sum_{(h_1, h_2) \in g} p^*_{h_1} p^*_{h_2}} = \sum_{g} \sum_{\{h_1 : (h_1, h) \in g\}} \frac{p^*_{h_1} p^*_h (n_g/n)}{\sum_{(h_1, h_2) \in g} p^*_{h_1} p^*_{h_2}},$$
(13.4)

where $n_g$ corresponds to the number of subjects with genotype $g$, and the sum on $g$ extends over all genotypes that occur in the sample.

Returning to the numerical example, we see that the application of the EM algorithm produces the iterative equations

$$\frac{837}{6520} + \frac{880}{6520} \times \frac{p^*_1 p^*_4}{p^*_1 p^*_4 + p^*_2 p^*_3} = p_1$$

$$\frac{995}{6520} + \frac{880}{6520} \times \frac{p^*_2 p^*_3}{p^*_1 p^*_4 + p^*_2 p^*_3} = p_2$$

$$\frac{816}{6520} + \frac{880}{6520} \times \frac{p^*_2 p^*_3}{p^*_1 p^*_4 + p^*_2 p^*_3} = p_3$$

$$\frac{2112}{6520} + \frac{880}{6520} \times \frac{p^*_1 p^*_4}{p^*_1 p^*_4 + p^*_2 p^*_3} = p_4,$$

which are equivalent to the iterative relation used for determining $\vartheta$.

Let us program the EM algorithm for a general number of loci. Again, we use for illustration the numerical example. This time we consider the haplotype of all three SNPs. First we determine what the different genotypes in our sample are.

```
> genotype <- paste(CR$snp1,CR$snp2,CR$snp3,sep=":")
> geno.freq <- table(genotype)
> geno.freq
```

```
genotype
A/A:A/A:T/T A/A:G/A:C/C A/A:G/A:T/C A/A:G/A:T/T A/A:G/G:C/C
          1          16          39          20         156
A/A:G/G:T/C A/A:G/G:T/T T/A:A/A:C/C T/A:A/A:T/C T/A:A/A:T/T
        226          98           4          30          52
T/A:G/A:C/C T/A:G/A:T/C T/A:G/A:T/T T/A:G/G:C/C T/A:G/G:T/C
        117         673         391         107         177
T/A:G/G:T/T T/T:A/A:C/C T/T:A/A:T/C T/T:A/A:T/T T/T:G/A:C/C
         49          18         154         442          31
T/T:G/A:T/C T/T:G/A:T/T T/T:G/G:C/C T/T:G/G:T/C T/T:G/G:T/T
        250         137          25          36          11
```

Note that there are 25 distinct genotypes, where now the term *genotype* refers to combinations of the genoytpes in the three loci.

The next task is to identify, for each of the genotypes, the set of pairs of haplotypes that will produce it. For that purpose we create a dataframe "hap.pair" with the required information. Observe that for a given locus the phase is uncertain if the genotype at the locus is heterozygous. We build the dataframe genotype by genotype. For each genotype we move from one SNP to the next, peeling off the genotype with the function "substr", which reads off substrings of a character string; and gradually we build the haplotype with the aid of the function "paste". At each heterozygous locus the number of possible haplotypes is doubled:

```
> n.mark <- 3
> geno.name <- names(geno.freq)
> hap.pair <- data.frame(geno.index=NULL,hap1=NULL,hap2=NULL)
> for(geno.index in 1:length(geno.name))
+ {
+     gn <- geno.name[geno.index]
+     hap1 <- hap2 <- NULL
+     for(i in 1:n.mark)
+     {
+         t <- (i-1)*4
+         g1 <- substr(gn,1+t,1+t)
+         g2 <- substr(gn,3+t,3+t)
+         if(g1==g2)
+         {
+             hap1 <- paste(hap1,g1,sep="")
+             hap2 <- paste(hap2,g2,sep="")
+         }
+         else
+         {
+             h1.1 <- paste(hap1,g1,sep="")
+             h2.1 <- paste(hap2,g2,sep="")
+             h1.2 <- paste(hap1,g2,sep="")
```

```
+                  h2.2 <- paste(hap2,g1,sep="")
+                  hap1 <- c(h1.1,h1.2)
+                  hap2 <- c(h2.1,h2.2)
+             }
+          }
+      hap.pair <- rbind(hap.pair,
+          cbind(geno.name[geno.index],hap1,hap2))
+ }
> colnames(hap.pair) <- c("genotype","hap1","hap2")
> rownames(hap.pair) <- 1:nrow(hap.pair)
> nrow(hap.pair)
[1] 61
```

Observe that the total number of distinct haplotype pairs is 61. Let us present the first 7 rows of "hap.pair":

```
> hap.pair[1:7,]
      genotype hap1 hap2
1 A/A:A/A:T/T  AAT  AAT
2 A/A:G/A:C/C  AGC  AAC
3 A/A:G/A:C/C  AAC  AGC
4 A/A:G/A:T/C  AGT  AAC
5 A/A:G/A:T/C  AAT  AGC
6 A/A:G/A:T/C  AGC  AAT
7 A/A:G/A:T/C  AAC  AGT
```

In the first genotype, which is "A/A:A/A:T/T", all loci are homozygous. It produces, therefore, a single row in the "hap.pair" data frame. The second genotype, "A/A:G/A:C/C", has one heterozygous locus and it produces two rows. The third genotype "A/A:G/A:T/C" produces four rows since two of the loci are heterozygous. The principle applies to the rest of the rows of "hap.pair".

The EM algorithm is initiated by setting a starting value for the vector of parameters. The values in the vector are updated in each iteration until convergence is obtained. The parameters in this problem are haplotype probabilities. We use as an initial value the uniform distribution over haplotypes:

```
> haplo <- levels(hap.pair$hap1)
> hp <- rep(1/length(haplo),length(haplo))
> names(hp) <- haplo
> hp
   AAT   AAC   AGC   AGT   TAC   TAT   TGC   TGT
 0.125 0.125 0.125 0.125 0.125 0.125 0.125 0.125
```

Observe that the vector "hp" is indexed by haplotypes, represented here as character strings.

The function "EM.haplo" applies the EM algorithm in order to estimate the distribution of the haplotypes. The first argument to the function is the

initial value of the parameter, which is indexed by haplotype identifiers. The
second argument is a vector of sampled genotypes. Recall that the term "genotype" refers here to the combination of observed alleles over the set of markers
considered. Each entry in this vector represents a single individual. The next
two entries, "hap.pair" and "geno.name", connect the coding of the haplotype and the coding of the genotype. The vector "geno.name" gives the list of
genotype codes and the dataframe "hap.pair" identifies the pairs of haplotype codes associated with each genotype code. Finally, "max.iter" indicates
the maximal number iterations of the EM algorithm and "tol" the tolerance
limit for convergence:

```
> EM.haplo <- function(hp,genotype,hap.pair,geno.name,
+                 max.iter=10^3,tol=10^(-3))
+ {
+     geno.freq <- table(factor(genotype,levels=geno.name))
+     geno.freq <- as.vector(geno.freq/sum(geno.freq))
+     iter <- 1
+     d <- Inf
+     while(d > tol & iter < max.iter)
+     {
+         hp.old <- hp
+         hap.pair.prob <- hp[hap.pair$hap1]*hp[hap.pair$hap2]
+         hap.pair.norm <- tapply(hap.pair.prob,
+             hap.pair$genotype,sum)
+         hap.pair.norm <- hap.pair.norm/geno.freq
+         hap.pair.prob <- hap.pair.prob/
+             hap.pair.norm[hap.pair$genotype]
+         hp <- tapply(hap.pair.prob,hap.pair$hap1,sum)
+         iter <- iter+1
+         d <- sum((hp-hp.old)^2)/sum(hp.old^2)
+     }
+     return(list(haplo.prob=hp,iter=iter-1))
+ }
```

Let us walk through the steps of the algorithm. The first step is to compute
the relative frequencies of the different genotypes in the sample. Observe that
as a precaution, since some genotypes may be absent in our sample, we identify
the genotypes as factors with the levels given in "geno.names".

The EM algorithm is applied with a "while" loop, after we have set initial
values for "iter", the count of the number of iterations, and "d", the measure
of divergence. In the first line of the algorithm the current distribution of
haplotypes is stored. In the second line two vectors are formed and multiplied.
The length of the first vector is equal to the number of rows in the dataframe
"hap.pair" and it contains the probabilities of the first haplotype in the
pair of haplotypes. The second is the probability of the second haplotype and
the product is the probability of the pair. In the following line the function

"`tapply`" is used. This function takes as arguments a vector, a factor of the same length, and a function. It applies the function to subsets of the vector, which are determined by the levels of the factor. It returns a list or a vector of length equal to the number of levels. In this application the result is the vector that has as coordinates the sum of haplotype-pair probabilities for each genotype. In the following line the by-genotype-sums are divided by the genotype relative frequencies. The conditional probabilities are computed in line 5. The probability of pairs is divided in each genotype by the sum of the probabilities of pairs. The updated vector of the haplotype distribution is computed via another application of the function "`tapply`". This time the partial sums are with respect to the levels of the first haplotype. Finally, the iteration counter is incremented and the measure of divergence is calculated.

After completing the "`while`" loop the output of the function is produced. The output is in the form of a list, which contains the final vector of haplotype probabilities and the number of iterations applied by the EM algorithm to produce that vector.

Consider the application of the EM algorithm to our data set:

```
> out <- EM.haplo(hp,genotype,hap.pair,geno.name)
> signif(out$haplo.prob,3)
    AAT     AAC     AGC     AGT     TAC     TAT     TGC     TGT
 0.0211 0.00977 0.2230  0.1620  0.0718  0.3690  0.0841  0.0591
> out$iter
[1] 4
```

Convergence was obtained in this case after four iterations. Observe that there are three common haplotypes – "TAT", "AGC", "AGT" – which cover among themselves more than 75% of the distribution.

## 13.3 Testing Association with Haplotypes

In the previous sections we dealt with the issue of estimating population haplotype frequencies. In this section we will consider briefly tests for association that apply to the estimated haplotype frequencies. For illustration we will continue with the analysis of the dataframe "`CR`", which contains phenotype and genotype information for a set of 3260 individuals.

In Chap. 12 we considered chi-square tests for association between the disease status and each of the three SNPs for which measurements were taken. The highest association was found with `snp1`, which produced a p-value of $8.1 \times 10^{-5}$ for the allele test. A straightforward extension of the same approach to haplotypes is to replace counts of alleles by expected counts of haplotypes and again apply the chi-square test:

```
> c.geno <- genotype[CR$group=="CASE"]
> r.geno <- genotype[CR$group=="RAND"]
```

```
> c.hp <- EM.haplo(hp,c.geno,hap.pair,geno.name)$haplo.prob
> r.hp <- EM.haplo(hp,r.geno,hap.pair,geno.name)$haplo.prob
> hap.count <- 2*rbind(c.hp*length(c.geno),
+                      r.hp*length(r.geno))
> round(hap.count,1)
        AAT   AAC    AGC   AGT   TAC    TAT   TGC   TGT
[1,]   32.3   9.0  351.2 197.4  97.4  423.3  89.4  68.0
[2,]  100.4  52.7 1109.4 859.5 360.6 2000.3 463.3 305.7
> chisq.test(hap.count)


        Pearson's Chi-squared test

data:  hap.count
X-squared = 34.2847, df = 7, p-value = 1.523e-05
```

Observe that we applied the function "EM.haplo" in order to estimate the relative frequency of the different haplotypes among the cases and then among the random controls. To account for the fact that each subject contributes a pair of haplotypes, the resulting frequencies are multiplied by twice the appropriate sample sizes to produce expected counts of haplotypes. The chi-square test is carried out by the application of the function "chisq.test" to the table of counts. The resulting p-value is $1.5 \times 10^{-5}$, which is slightly smaller than the p-value produced by the test that is based on a single SNP.

One should remember that Hardy-Weinberg equilibrium was assumed in the estimation of haplotype frequencies. Therefore, the given test is analogous to the allele test, which was derived under the same assumption. The conditions discussed in Chap. 12 that might lead to a violation of Hardy-Weinberg would be a problem here as well.

The chi-square test takes into account the sampling variability of haplotypes, but it ignores the uncertainty that results from the reconstruction of haplotypes based on the partial information provided by the genotypes. More sophisticated analyses can be used in order to take this extra source of variability into account, but we do not pursue this point.

The test we considered measured the overall association between disease status and the entire set of haplotypes. Next we turn to the investigation of the association with each of the haplotypes separately, by computing for each of the haplotypes an appropriate $2 \times 2$ table:

```
> p <- rep(1,8); names(p) <- colnames(hap.count)
> for(h in 1:8)
+ {
+     t <- cbind(hap.count[,h],apply(hap.count[,-h],1,sum))
+     p[h] <- chisq.test(t)$p.value
+ }
> signif(p,1)
   AAT   AAC   AGC   AGT   TAC   TAT   TGC   TGT
```

```
2e-01 4e-01 6e-07 5e-01 3e-01 2e-03 5e-02 6e-01
```

The haplotype "AGC" stands out as being the most significantly associated with the trait. Its significance is still apparent even if we make a Bonferroni correction for multiple testing by multiplying by 8:

```
> signif(min(p)*8,1)
[1] 5e-06
```

The difference between the p-values of the overall chi-square statistic and of the Bonferroni-corrected haplotype specific statistic is not large for these data. There is a potential for a substantially larger difference when the number of haplotypes is larger, and in general there is no good reason for preferring one statistic to the other in all situations. Indeed, consider the similar, but simpler, problem of testing whether a large number of normal populations with a common variance, say equal to one, all have mean value zero, against the alternative that at least one has a mean value different from zero. If only one population has a mean value different from zero, this situation can be detected more efficiently by looking at individual test statistics and making a Bonferroni correction; but if a substantial number of the populations have means differing slightly from zero, a chi-square statistic that combines all the data is more powerful.

In the next example we would like to consider the application of haplotype frequencies in a regression analysis. Recall that the data frame "CR" included the sex of the subjects. One may be interested, for example, in the investigation of the effect of both the sex and the "AGC" haplotype on the disease status.

It can be shown that the score test for a linear regression model with an unobserved predictor would replace that predictor by its expected value given the observed information, where the expected value is computed under the null distribution. (To deal with missing information in Chaps. 7, 9, and 10, we have implicitly used this fact without mentioning the statistical principle.) Hence we will produce a regression analysis with the count of the number of "AGC" haplotypes for each subject replaced by the conditional expected count, given the person's genotype.

Since the computation of the expected count is carried out under the null assumption of no trait effect, we may pool both samples together when computing the distribution of haplotypes:

```
> hp <- EM.haplo(hp,genotype,hap.pair,geno.name)$haplo.prob
```

The expected count of the "AGC" haplotype, given each possible genotype, is conducted by multiplying the probability of a pair of haplotypes by the count of "AGC" in the pair – 0, 1, or 2. For each genotype the product is summed over all the pairs that produce that genotype and the result divided by the sum of probabilities of those pairs:

```
> hap1 <- hap.pair$hap1
```

```
> hap2 <- hap.pair$hap2
> c.AGC <- hp[hap1]*hp[hap2]*((hap1=="AGC")+(hap2=="AGC"))
> c.AGC <- tapply(c.AGC,hap.pair$genotype,sum)
> n.AGC <- hp[hap1]*hp[hap2]
> n.AGC <- tapply(n.AGC,hap.pair$genotype,sum)
> con.expec.AGC <- c.AGC/n.AGC
> con.expec.AGC
A/A:A/A:T/T A/A:G/A:C/C A/A:G/A:T/C A/A:G/A:T/T A/A:G/G:C/C
  0.0000000   1.0000000   0.7516409   0.0000000   2.0000000
A/A:G/G:T/C A/A:G/G:T/T T/A:A/A:C/C T/A:A/A:T/C T/A:A/A:T/T
  1.0000000   0.0000000   0.0000000   0.0000000   0.0000000
T/A:G/A:C/C T/A:G/A:T/C T/A:G/A:T/T T/A:G/G:C/C T/A:G/G:T/C
  0.9522479   0.8601234   0.0000000   1.0000000   0.4837625
T/A:G/G:T/T T/T:A/A:C/C T/T:A/A:T/C T/T:A/A:T/T T/T:G/A:C/C
  0.0000000   0.0000000   0.0000000   0.0000000   0.0000000
T/T:G/A:T/C T/T:G/A:T/T T/T:G/G:C/C T/T:G/G:T/C T/T:G/G:T/T
  0.0000000   0.0000000   0.0000000   0.0000000   0.0000000
```

The vector "con.expec.AGC" contains the expected count given the genotypes. It is indexed by the levels of the genotypes.

It is straightforward now to produce for each subject its expected count of the given allele and to carry out the regression analysis:

```
> h <- con.expec.AGC[genotype]
> fit1 <- glm(CR$group ~ CR$sex + h,family=binomial())
> summary(fit1)$coef
               Estimate Std. Error   z value     Pr(>|z|)
(Intercept)  1.2044101 0.08652173 13.920319 4.767702e-44
CR$sexM      0.5780918 0.09519701  6.072583 1.258690e-09
h           -0.3927976 0.07731632 -5.080397 3.766461e-07
```

Correction for the effect of sex did not substantially alter the effect of the haplotype. The significant p-value for sex itself is an indication that the sampling was not balanced with respect to the subject's sex.

*Remark 13.1.* The case-random is a retrospective design. Logistic regression is applicable to such designs since hypothesis testing for the effect of predictors (other than the intercept) are valid even when the response is predetermined by the sampling design.

*Remark 13.2.* Unlike the QTL regression models of previous chapters where the tests were constructed with respect to the variability of the response, the situation here is such that variability is given in terms of random genotypes. Still, it can be shown that the score test for assessing genetic and environmental effects is obtained by substituting haplotype counts by their conditional count, given the genotypes. The counts are centered about the average expected count. Moreover, when considering genome scans and when considering robust statistics, we replaced variability in the phenotype by variability in the genotypes. This observation shows that the procedures that were proposed in Chaps. 7, 9, and 10 when facing the issue of missing genotypic information, are also score statistics.

## 13.4 Bibliographical Comments

The application of the EM algorithm for the determination of the frequency of haplotypes from genotypes that was presented here is based on the work of Excoffier and Slatkin [26]. The application of generalized linear models for haplotype analysis was proposed by Schaid et al. [69].

## Problems

**13.1.** Investigate using simulation the variability of the estimator of the correlation coefficient between a pair of loci. Compare the reduction in accuracy between the case where haplotypes are known to the case where they must be inferred from genotypes.

**13.2.** Assume you are given a sample of $n$ sib-pairs taken from some population. The goal is to estimate the frequencies of haplotypes in that population based on the observed genotypes of the siblings. We assume Hardy-Weinberg equilibrium in the parents and that markers are so close so we can ignored parental recombinations.

(a) Compute the likelihood of the genotypes at a pair of loci as a function of $\vartheta$ and propose an algorithm for the maximization of the likelihood. Investigate the properties of the proposed estimator via simulations.

(b) Extend the proposed algorithm in order to deal with more than two loci.

**13.3.** The analysis given in 13.3 was motivated by the multiplicative model. Develop parallel test statistics for the detection of haplotypes that act according to a recessive model in the sense that a pair of identical haplotypes are needed to produce an effect.

# References

1. Abecasis, G.R., Cardon, L. R., Cookson, W.O.C.: A general test of association for quantitative traits in nuclear families. Am. J. Hum. Genet. **66**, 279–292 (2000)
2. Abecasis, G.R., Cherny, S.S., Cookson, W.O., Cardon,L.R.: Merlin – rapid analysis of dense genetic maps using sparse gene flow trees. Nat. Genet. **30**, 97–101 (2002)
3. Allison, D.B., Thiel, B., St. Jean, P., Elston, R. C., Infante, M. C., Schork, N. J.: Multiple Phenotype Modeling in Gene-Mapping Studies of Quantitative Traits: Power Advantages. Am. J. Hum. Genet. **63**, 1190–1201 (1998).
4. Almasy, L., Blangero, J.: Multipoint quantitative-trait linkage analysis in general pedigrees. Am. J. Hum. Genet. **62**, 1198–1211 (1998)
5. de Andrade, M., Amos, C.I.: Ascertainment issues in variance components models. Genet Epidemiol. **19**, 333–344 2000
6. Beaty, T.H., Liang, K.Y.: Robust inference for variance components models in families ascertained through probands: I. conditioning on the proband's phenotype. Genet. Epidemiol. **4**, 203–210 (1987)
7. Blangero, J., Almasy, L.: Multipoint oligogenic linkage analysis of quantitative traits. Genet. Epidemiol. **14**, 959–964 (1997)
8. Bogdan, M., Doerge, R., Ghosh, J.K.: Modifying the Schwarz Bayesian information criterion to locate multiple interacting quantitative trait loci. Genetics **167**, 989–999 (2004)
9. Broman, K.W., Speed, T.P.: A model selection approach for the identification of quantitative trait loci in experimental crosses (with discussion). J. Roy. Stat. Soc. B **64**, 641–656, 731–775 (2002)
10. Chen, W-M., Broman, K., Liang, K.Y.: Quantitative trait linkage analysis by generalized estimating equations: unification of variance components and Haseman-Elston regression. Genet. Epidem. **26**, 265–272 (2004)
11. Churchill, G.A., Doerge, R.W.: Empirical threshold values for quantitative trait mapping. Genetics **138**, 963–971 (1994)
12. Cordell, H.J., Todd, J.A., Bennett, S.T., Kawaguchi, Y., Farrall, M.: Two-locus maximum LOD score analysis of a multifactorial trait: joint consideration of IDDM2 and IDDM4 with IDDM1 in type 1 diabetes. Am. J. Hum. Genet. **57**, 920–934 (1995)

13. Cordell, H.J., Wedig, G.C., Jacobs, K.B., Elston, R.: Multilocus linkage tests based on affected relative pairs. Am. J. Hum. Genet. **66**, 1273–1286 (2000)

14. Cox, N.J., Frigge, M., Nicolae, D.L., Concannon, P., Hanis, C.L., Bell, G.I., Kong, A.: Loci on chromosomes 2 (NIDDM1) and 15 interact to increase susceptibility to diabetes in Mexican Americans. Nature Genetics **21**, 213–215 (1999)

15. Cox, D.R., Hinkley, D.V.: Theoretical Statistics. Chapman and Hall, London (1974)

16. Crow, J.F., Kimura, M.: An Introduction to Population Genetics Theory. Harper and Row, New York (1970)

17. Darvasi, A., Weinreb, A., Minke, V., Weller, J.I., Soller, M.: Detecting marker-QTL linkage and estimating QTL gene effect and map location using a saturated genetic map. Genetics **134**, 943–951 (1993)

18. Darvasi, A. and Soller, M.: Advanced intercross lines, an experimental population for fine genetic mapping. Genetics **141**, 1198–1207 (1995)

19. Doerge, R. W., Zeng, Z-B., Weir, B.: Statistical issues in the search for genes affecting quantitative traits in experimental populations. Statistical Science **12**, 195–219 (1997).

20. Diao, G., Lin, D. Y.: A powerful and robust method for qapping quantitative trait loci in general pedigrees, Am. J. Hum. Genet. **77**, 97–111 (2005).

21. Dupuis, J., Siegmund, D.: Boundary crossing probabilities in linkage analysis. In: Bruss, F.T., Le Cam, L. (ed) Game Theory, Optimal Stopping, Probability and Statistics. Institute of Mathematical Statistics, Hayward, California (2000)

22. Dupuis, J. and Siegmund, D.: Statistical methods for mapping quantitative trait loci from a dense set of markers. Genetics **151**, 373–386 (1999)

23. Dupuis, J., Brown, P., Siegmund, D.: Statistical methods for linkage analysis of complex traits from high resolution maps of identity by descent. Genetics **140**, 843–856 (1995)

24. Elston, R.C., Stewart J.: A general model for the genetic analysis of pedigree data, Human Heredity, **21**, 523–542 (1971)

25. Ewens, W.J.: Mathematical Population Genetics I. Theoretical Introduction. (2nd Edition) Springer-Verlag, New York (2004)

26. Excoffier, L., Slatkin, M.: Maximum-likelihood estimation of mulecular haplotype frequencies in a diploid population. Mol. Biol. Evol. **12**, 921–927 (1995)

27. Falconer, D.S., Mackay, T.F.C.: Introduction to Quantitative Genetics (4th Edition). Longman, Harlow, UK (1996)

28. Feingold, E., Brown, P.O., Siegmund, D.: Gaussian models for genetic linkage analysis using complete high resolution maps of identity-by-descent. Am. J. Hum. Genetics **53**, 234–251 (1993)

29. Feller, W.: An Introduction to Probability Theory and Its Applications, Vol. I, John Wiley and Sons, NY (1968)

30. Fisher, R.A.: The correlation between relatives on the supposition of Mendelian inheritance. Trans. of the Roy. Soc. Edinburgh **52**, 399–433 (1918)

31. Forrest, B., Feingold, E.: Composite statistics for QTL mapping with moderately discordant sibling pairs. Am. J. Hum. Genet. **66**, 1642–1660 (2000)

32. Fulker, D.W., Cherny, S.S., Sham, P.C., Hewitt, J.K.: Combined linkage and association analysis for quantitative traits. Am. J. Hum. Genet. **64**, 259–267 (1999)

33. Gudbjartsson, D.F., Jonasson, K., Frigge, M.L., Kong, C.A.: Allegro, a new computer program for multipoint linkage analysis. Nat. Genet. **25** 12–13 (2000)

34. Gudbjartsson, D.F., Thorvaldsson, T., Kong, C.A., Gunnarsson, G., Ingolfsdottir, A.: Allegro version 2. Nat. Genet. **37** 1015–1016 (2005)

35. Haldane, J.B.S., Waddington, C.H.: Inbreeding and linkage. Genetics. **16**, 357–374 (1931)

36. Haley, C.S., Knott, S.A.: A simple regression method for mapping quantitative trait loci in line crosses using flanking markers. Heredity **69**, 315–324 (1992)

37. Haseman, J.K., Elston, R.C.: The investigation of linkage between a quantitative trait and a marker locus. Behavior Genetics **2** 3–19 (1972)

38. Holmans, P.: Asymptotic properties of affected sib-pair linkage analysis. Amer. J. Hum. Genet. **52**, 362–374 (1993)

39. Hopper, J.L., Mathews, J.D.: Extensions to multivariate normal models for pedigree analysis. Ann. Hum. Genet. **46** 373–383 (1982)

40. Jackson, A.U., Galecki, A.T., Burke, D.T., Miller, R.A.: Mouse loci associated with life span exhibit sex-specific and epistatic effects. J. Gerontol. Biological Sciences **57A** B9–B15 (2002)

41. Jackson, A.U., Fornes, A, Galecki, A.T., Miller, R.A., Burke, D.T.: Multiple trait QTL analysis using a large mouse sibship. Genetics **151**, 785–795 (1999)

42. James, J.W.: Frequency in relatives for an all-or-none trait. Ann. Hum. Genet. **35**, 47–48 (1971)

43. John, S., Shephard, N., Liu, G., Zeggini, E., Cao, M., Chen, W., Vasavda, N., Mills, T., Barton, A., Hinks, A., Eyre, S., Jones, K.W., Ollier, W., Silman, S., Gibson, N., Worthington, J., Kennedy, G. C.: Whole-genome scan, in a complex disease, using 11,245 single-nucleotide polymorphisms: comparison with microsatellites, Am. J. Hum. Genet. **75**, 54–64 (2004)

44. Kempthorne, O.: An Introduction to Genetic Statistics. John Wiley and Sons, New York (1957)

45. Kimura, M.: A probability method for treating inbreeding systems, especially with linked genes. Biometrics **19**, 1–17 (1963)

46. Kruglyak L., Daly M.J., Reeve-Daly M.P., Lander E.S.: Parametric and nonparametric linkage analysis: a unified multipoint approach. Amer. J. Hum. Genet. **58**, 1347–1363 (1996)

47. Lander, E.S., Botstein, D.: Mapping Mendelian factors underlying quantitative traits using RFLP linkage maps. Genetics **121**, 185–199 (1989)

48. Lander, E.S., Kruglyak L.: Genetic dissection of complex traits: guidelines for interpreting and reporting linkage results. Nature Genetics **11**, 241–247 (1995)

49. Lange, K., Mathematical and Statistical Methods for Genetic Analysis (2nd Edition). Springer-Verlag, NY (2001)

50. Long, J.C.: The genetic structure of admixed populations. Genetics **127**, 417–428 (1991)

51. Lynch, M., Walsh, B.: Genetics and Analysis of Quantitative Traits, Sinauer Associates, Sunderland, MA (1998)

52. Mather, K., Jinks, J.L.: Biometrical Genetics (3rd Edition). Chapman and Hall, London (1982)

53. McKeigue, P.M.: Mapping genes that underlie ethnic differences in disease risk by linkage disequilibrium in recently admixed populations. Am. J. Hum. Genet. **60**, 188–196 (1997)

54. McKeigue, P.M.: Mapping genes that underlie ethnic differences in disease risk: methods for detecting linkage in admixed populations, by conditioning on parental admixture. Am. J. Hum. Genet. **63** 241–251 (1998)

55. McPeek, M.S.: Optimal allele-sharing statistics for genetic mapping using affected relatives. Genet. Epidem. **16**, 225–249 (1999)

56. Morton, N.: Sequential tests for the detection of linkage, Am. J. Hum. Genet. **7**, 277–318 (1955)

57. Ott, J.: Analysis of human genetic linkage (3rd Edition). Johns Hopkins Press, Baltimore (1999)

58. Penrose, L. S.: The detection of autosomal linkage in data which consist of pairs of brothers and sisters of unspecified parentage. Ann. Eugen. **8**, 133–138 (1935)

59. Peng, J., Siegmund, D.: Mapping quantitative traits with ascertained sibships. PNAS **101**, 7845–7850 (2004)

60. Peng, J., Tang, H.K., Siegmund, D.: Genome scans when there is gene-covariate interaction, Genetic Epidemiol. **25**, 173–184 (2005)

61. Rabinowitz, D.: A transmission disequilibrium test for quantitative trait loci. Hum. Hered. **47**, 342–350 (1997)

62. Rebai, A., Goffinet, B., Mangin, B.: Approximate thresholds of interval mapping test for QTL detection. Genetics **138**, 235–240 (1994).

63. Rebai, A., Goffinet, B., Mangin, B.: Comparing power of different methods for QTL detection. Biometrics **51** 87–99 (1995).

64. Rice, J.A.: Mathematical Statistics and Data Analysis (2nd Edition). Duxbury Press, Belmont, CA (1995)

65. Risch, N.: Linkage strategies for genetically complex traits I, II, III. The power of affected relative pairs, Am. J. Hum. Genet. **46**, 222–228, 229–241, 242–253 (1990)

66. Risch, N.: Corrections to "Linkage Strategies for Genetically Complex Traits. III. The Effect of Marker Polymorphism on Analysis of Affected Relative Pairs". Am. J. Hum. Genet. **51**, 673–675 (1992)

67. Risch, N., Merikangas, K.: The future of genetic studies of complex human diseases. Science **273**, 1516–1517 (1996)

68. Ross, S.M.: A First Course in Probability (7th Edition). John Wiley and Sons, NY (2005)

69. Schaid, D.J., Rowland, C.M., Tines, D.E., Jackobson, R.M, Poland, G.A.: Score tests for association between traits and haplotypes when linkage phase is ambiguous. Am. J. Hum. Genet. **70**, 425–434 (2002)

70. Sen, S., Churchill, G.: A statistical framework for quantitative trait mapping, Genetics, **159**, 371–87 (2001)

71. Sham, P.C., Purcell, S., Cherny, S.S., Abecasis, G.R.: Powerful regression-based quantitative-trait linkage analysis of general pedigrees. Am. J. Hum. Genet. **71**, 238–253 (2000)

72. Shi, J.: Two problems in quantitative trait mapping, Ph. D. thesis, Stanford University (2006).

73. Siegmund, D.: Genome wide significance. In: Elston, R., Olson, J., Palmer, L. (ed) Biostatistical Genetics and Genetic Epidemiology. John Wiley and Sons, Chichester (2002)

74. Siegmund, D.: Upward bias in estimation of genetic effects. Am. J. Hum. Genet. **71** 1183–1188 (2003)

75. Siegmund, D.: Model selection in irregular problems: applications to mapping QTLs. Biometrika **91**, 785–800 (2004)

76. Siegmund, D., Yakir, B.: Significance level in interval mapping. In: Zhang, H., Huang, J. (ed) Development of modern statistics and related topics in celebration of Yaoting Zhang's 70th birthday. World Scientific, Singapore (2003)

77. Silver, L.M.: Mouse Genetics: Concepts and Applications, Oxford University Press, Oxford (1995)

78. Skatkiewicz, J.P., T.Cuenco, K., Feingold, E.: Recent advances in human quantitative-trait-locus mapping: comparison of methods for discordant sibling pairs. Am. J. Hum. Genet. **73**, 874–885 (2003)

79. Spielman, R.S., McGinnis, R.E., Ewens, W.J.: Transmission test for linkage disequilibrium: the insulin gene region and insulin-dependent diabetes mellitus (IDDM). Am. J. Hum. Genet. **52**, 506–516 (1993)

80. Spielman, R.S., Ewens, W.J. The TDT and other family-based tests for linkage disequilibrium and association. Am J Hum Genet **59**, 983–989 (1996)

81. Schwarz, G.: Estimating the dimension of a model. Ann. Statist. **6**, 461-464. (1978)

82. Tang, H.K., Siegmund, D.: Mapping quantitative trait loci in oligogenic models. Biostatistics **2**, 147–162 (2001)

83. Tang, H.K., Siegmund, D.: Mapping multiple genes for quantitative and complex traits. Genetic Epidemiology **22**, 313–327 (2002)

84. T.Cuenco, K., Szatkiewicz, J.P., Feingold, E.: Recent advances in human quantitative-trait-locus mapping: comparison of methods for selected sibling pairs. Am. J. Hum. Genet. **73**, 863–873 (2003)

85. Teng, J., Siegmund, D.: Combining information within and between pedigrees for mapping complex traits. Am. J. Hum. Genet. **60**, 979–992 (1997)

86. Teng, J. and Siegmund, D.: Multipoint linkage analysis using affected relative pairs and partially informative markers (with discussion). Biometrics **54** 379–411 (1998)

87. Venables, W.N., Ripley, B.D.: Modern Applied Statistics with S-plus (3rd Edition). Springer, NY (2001)

88. Wang, K., Huang, J.: A score-statistic approach for the mapping of quantitative-trait loci with sibships of arbitrary size, Am. J. Hum. Genet., **70**, 412–424 (2002)

89. Whittemore, A., Halpern, J.: A class of tests for linkage using affected pedigree members. Biometrics **50**, 118–127 (1994)

90. Wright, S.: Inbreeding and recombination. PNAS **19**, 420–433.

91. Yang, R., Tian, Q., Xu, S.: Mapping quantitative trait loci for longitudinal traits in line crosses, Genetics **173**, 2339–2356 (2006)

92. Yi, N., Xu, S., Allison, D. B.: Bayesian model choice and search strategies for mapping interacting quantitative trait loci. Genetics **165**, 867–883 (2003).

93. Zhu, X., Cooper, R.S., Elston, R.C.: Linkage analysis of a complex disease through the use of admixed populations. Am. J. Hum. Genet. **74**, 1136–1153 (2004)

# Subject Index

# Index of R Functions

## Multiple Testing Procedures and Applications to Genomics

**Sandrine Dudoit and Mark J. van der Laan**
This book establishes the theoretical foundations of a general methodology for multiple hypothesis testing and discusses its software implementation in R and SAS. The methods are applied to a range of testing problems in biomedical and genomic research, including the identification of differentially expressed and co-expressed genes in high-throughput gene expression experiments, such as microarray experiments; tests of association between gene expression measures and biological annotation metadata (e.g., Gene Ontology); sequence analysis; and the genetic mapping of complex traits using single nucleotide polymorphisms.

2007. 584 p. (Springer Series in Statistics) Hardcover ISBN 978-0-387-49316-9

## Statistical Methods in Bioinformatics: An Introduction
## Second Edition

**W. Ewens and G. Grant**
Advances in computers and biotechnology have had a profound impact on biomedical research, and as a result complex data sets can now be generated to address extremely complex biological questions. This book provides an introduction to some of these new methods. The second edition features new chapters on microarray analysis and on statistical inference, including a discussion of ANOVA, and discussions of the statistical theory of motifs and methods based on the hypergeometric distribution. Much material has been clarified and reorganized.

2004. 588 p. (Statistics for Biology and Health) Hardcover
ISBN 978-0-387-40082-2

## Computational Genome Analysis: An Introduction

**R. Deonier, S. Tavaré, and M. Waterman**
Computational Genome Analysis: An Introduction presents the foundations of key problems in computational molecular biology and bioinformatics. It focuses on computational and statistical principles applied to genomes, and introduces the mathematics and statistics that are crucial for understanding these applications. The book is appropriate for a one-semester course for advanced undergraduate or beginning graduate students, and it can also introduce computational biology to computer scientists, mathematicians, or biologists who are extending their interests into this exciting field.

2005. 512 p. (Statistics for Biology and Health) Hardcover ISBN 978-0-387-98785-9

**Easy Ways to Order►** Call: Toll-Free 1-800-SPRINGER • E-mail: orders-ny@springer.com • Write: Springer, Dept. S8113, PO Box 2485, Secaucus, NJ 07096-2485 • Visit: Your local scientific bookstore or urge your librarian to order.